Advances in Behavior Genetics

Series Editor
Yong-Kyu Kim

For further volumes:
http://www.springer.com/series/10458

Deborah Finkel · Chandra A. Reynolds
Editors

Behavior Genetics of Cognition Across the Lifespan

Springer

Editors
Deborah Finkel
Department of Psychology
School of Social Sciences
Indiana University
New Albany
USA

Chandra A. Reynolds
Department of Psychology
University of California
Riverside
USA

ISBN 978-1-4614-7446-3 ISBN 978-1-4614-7447-0 (eBook)
DOI 10.1007/978-1-4614-7447-0
Springer New York Dordrecht Heidelberg London

Library of Congress Control Number: 2013943279

Printed on acid-free paper

Springer is part of Springer Science+Business Media (www.springer.com)

Preface

Along with psychopathology, cognition has been one of the primary phenotypic focal points of the field of behavior genetics since its inception (Plomin et al. 2012). Darwin (1871) discussed commonalities between the mental powers of humans and other animals, implying a genetic basis to cognitive function. Francis Galton's 1869 examination of eminent families in Britain was one of the earliest attempts to investigate whether cognitive achievements run in families. As definitions and assessments of intelligence were developed and refined over the following decades (for a review see Chap. 1), behavior genetic investigation of intelligence experienced parallel increases in sophistication. In 1963, Erlenmeyer-Kimling and Jarvik summarized results from the first 50 years of modern behavioral genetic research on intelligence and concluded that the pattern of correlations among pairs of varying types of genetic relatedness "closely approaches the theoretical value predicted on the basis of genetic relationship alone" (p. 1477). They were careful to conclude that although genetic factors may play a large role in potential intellectual achievement, environmental factors will contribute to ultimate cognitive performance. Reactions to strong consistent evidence for genetic influences on intellectual ability prompted many behavioral geneticists to painstakingly explain the concept of heritability. Edited volumes on heredity, environment, and intelligence from that era focused on the perceived incongruity between behavioral genetic and socialization theories of intelligence and the correct interpretation of heritability (Ljungman 1975; Vandenburg 1968; Vernon 1979). In the meantime, behavioral genetic studies continued to accumulate data and refine their approaches to the issue.

Only 20 years after Erlenmeyer-Kimling and Jarvik (1963), a meta-analysis of familial studies of intelligence included twice as many studies (111 vs. 52) and four times as many correlational pairings (113,942 vs. more than 30,000; Bouchard and McGue 1981). Similar to the earlier review, the authors concluded that the pattern of correlations was remarkably consistent with polygenic theory, but did not discount the importance of environmental factors. Ten years later the results were verified using structural equation modeling, allowing the direct estimation of significant nonadditive as well as additive genetic variance (Chipuer et al. 1990). The development of molecular genetic methodologies over the last 20 years have allowed the field to move beyond anonymous genetic variance to the attempt to identify specific genes or gene loci that contribute to intellectual functioning. A recent genome-wide association study agreed with previous conclusions that genetic variation makes a significant contribution to intellectual performance (Davies et al. 2011). The results are consistent with the accumulated impact of many small genes having small effects on cognitive function.

Parallel with these advances, behavioral geneticists were still arguing that "developmental psychologists should find room for behavioral genetics" as recently as 1994 (McGue 1994). Two volumes edited by Grigorenko and Sternberg in the late 1990s represent the accumulated state of knowledge at that time (Grigorenko and Sternberg 1997; Sternberg and Grigorenko 2001). The volumes can be considered a matched set, with one focusing on the impact of family environment on intelligence while acknowledge the role of genetics (Grigorenko and Sternberg 2001) and the other attempting to establish that the field has moved beyond the overly simplistic nature vs. nurture controversy with regard to intellectual functioning (Sternberg and Grigorenko 1997). Scarr (1997) wrote of reconciliation between behavioral genetic and socialization theories; but, in the same volume Bidell and Fischer (1997) argued that the basic techniques of behavioral genetics are flawed. Hunt (1997) declared that he did not want to hear the phrase "nature vs. nurture controversy" ever again, while proposing that the argument was more properly political rather than scientific.

Science moves swiftly and we believe that the 15 years intervening since the Sternberg and Grigorenko (1997) volume has brought many changes, both in the field of behavioral genetics of cognition and in its acceptance by scientists generally. In his chapter discussing unresolved questions in the behavioral genetics of intelligence at that time, Waldman (1997) listed: (a) developmental behavioral genetics, (b) gene by environment interaction and correlation, (c) behavior genetics as a tool for examining the construct validity of intelligence, and (d) behavior genetics as a tool for examining causal influences on intelligence. Significant advances in the last decade on all four fronts merit collection in a new volume dedicated to summarizing the current state of the art in behavioral genetic investigations of cognition across the lifespan. Two of the most significant advances in the field guide the structure of the current volume. First, developmental behavioral genetics forms the basic structure of the book, which is divided into sections on childhood and adolescence, middle adulthood, and aging. Recent advances in both collection and statistical modeling of twin data, particularly longitudinal twin data, make this an especially advantageous moment to produce a work that presents a collection of the groundbreaking research on cognitive abilities across the lifespan. Second, two chapters focus specifically on investigations of gene by environment interplay in childhood (Chap. 2) and adulthood (Chap. 6). Increasing sophistication in statistical modeling and molecular genetic methods combine to allow for nuanced investigations of gene by environment correlation and interaction.

The current volume is presented as a survey of the current research in the field of behavior genetics of cognition. This volume presents an overview of the current state of quantitative and molecular genetic investigations into the many facets of cognitive performance and functioning across the lifespan. In the past, it may have been easier to identify distinct fields of study or approaches within behavioral genetics. Now, borders are more fluid and researchers are working together across boundaries. To divide the topic of behavior genetics of cognition across the lifespan into individual chapters is fundamentally arbitrary and the chapters collected here will overlap to some extent. As these chapters were prepared, it became clear that even defining the end of childhood and the beginning of adulthood—or the boundary between middle adulthood and aging—was not straightforward. Similarly, dividing discussions of aging into normative, nonnormative, gene by environment interplay, and even brain morphology does not accurately capture the cross-pollination that exists in these fields today. Our goal is to ensure that each chapter can both stand alone and work with the other chapters to present the field as the integrated whole it is becoming.

Part I focuses on childhood and Chap. 1 by Wadsworth, Corley, and DeFries provides a summary of the history of conceptions of intelligence, an introduction to behavioral genetic methods of investigation, and a discussion of behavior genetic and molecular genetic investigations of normative intellectual development in childhood. Chapter 2, by Turkheimer and Horn, reviews the evidence that heritability of intelligence in childhood is moderated by parental socioeconomic status. In Chap. 3, Carlier and Roubertoux summarize the current state of the field in understanding genetic influences on atypical intellectual development in childhood.

In Part II, the emphasis is on adulthood, including both middle-adulthood and aging. Midlife has long been perceived as a period of "latency" in which functioning is maintained before the transition to aging, per se. As Kremen, Moore, Franz, Pannizon, and Lyons demonstrate in Chap. 4, researchers are beginning to focus their attention on midlife as a potential source of subsequent changes in intellectual functioning and changes in genetic and environmental influences on functioning. With the aging of the baby-boomer generation, genetic and environmental influences on cognitive aging have drawn increasing interest over the last two decades and multiple recent reviews exist (Finkel and Reynolds 2009, 2010; McGue and Johnson 2008). In Chap. 5, Johnson, McGue, and Deary focus on overarching trends in behavior genetics of cognitive aging and recent molecular genetic advances. The focus of Chap. 6 by Reynolds, Finkel, and Zavala is gene by environment interplay in normative cognitive aging. Chapter 7 by Gatz, Jang, Karlsson, and Pedersen summarizes the current state of the art in behavioral and molecular genetic investigations of dementia.

Part III focuses on the contributions made by advances in biological and neurobiological approaches to our understanding of the nature of cognition and genetic and environmental influences on cognitive functioning. In Chap. 8, Chavarría-Siles, Fernández, and Posthuma review the impact that recently developed neuroimaging techniques has on our knowledge of brain morphology and function, and subsequent advances in our understanding of genetic and environmental influences on biological underpinnings of cognitive function. The longitudinal perspective of this volume is evident in the advances in animal models of cognition and cognitive aging reviewed by Galsworthy, Arden, and Chabris in Chap. 9.

Finally, in Chap. 10 we try to build on and respond to Waldman's (1997) stellar discussion of the unresolved issues and future directions in behavior genetic studies of cognition across the lifespan. We identify the issues raised by Waldman that are still in need of attention or resolution and identify new directions that we feel the field is prepared to explore.

The editors would like to thank the following for serving as reviewers of chapters in this volume: Kaare Christensen, Gene Fisch, Boo Johansson, Wendy Johnson, William Kremen, Matthew Panizzon, Stephen Petrill, Sally Wadsworth, and Irwin Waldman.

References

Bidell, T. R., & Fischer, K. W. (1997). Between nature and nurture: The role of human agency in the epigenesist of intelligence. In R. J. Sternberg & E. Grigorenko (Eds.), *Intelligence, heredity, and environment* (pp. 193–242). Cambridge: Cambridge University Press.

Bouchard, T. J., & McGue, M. (1981). Familial studies of intelligence: A review. *Science, 212,* 1055–1059.

Chipuer, H. M., Rovine, M. J., & Plomin, R. (1990). LISREL modeling: Genetic and environmental influences on IQ revisited. *Intelligence, 14*(1), 11–29.

Darwin, C. (1871). *The descent of man, and selection in relation to sex* (1st ed.). London: John Murray.

Davies, G., Tenesa, A., Payton, A., Yang, J., Harris, S. E., et al. (2011). Genome-wide association studies establish that human intelligence in highly heritable and polygenic. *Molecular Psychiatry, 16*(10), 1–10.

Erlenmeyer-Kimling, L., & Jarvik, L. F. (1963). Genetics and intelligence: A review. *Science, 142,* 1477–1479.

Finkel, D., & Reynolds, C. A. (2009). Behavioral genetic investigations of cognitive aging. In Y.-K. Kim (Ed.), *Handbook of behavior genetics* (pp. 101–112). New York: Springer.

Finkel, D., & Reynolds, C. A. (2010). Cognitive and physical aging pathways: Contributions from behavioral genetics. In J. C. Cavanaugh & C. K. Cavanaugh (Eds.), *Aging in America: Volume 2: Physical and mental health* (pp. 26–56). Denver: Praeger Publishing.

Galton, F. (1869). *Hereditary genius: An inquiry into its laws and consequences*. London: Macmillan.

Grigorenko, E., & Sternberg, R. J. (Eds.). (2001) *Family environment and intellectual functioning: A life-span perspective*. Mahwah: Lawrence Erlbaum Associates.

Hunt, E. (1997). Nature vs. nurture: The feeling of *vujà dé*. In R. J. Sternberg & E. Grigorenko (Eds.), *Intelligence, heredity, and environment* (pp. 531–551). Cambridge: Cambridge University Press.

Ljungman, C. G. (1975). *What is IQ? Intelligence, heredity, and environment*. London: Gordon Cremonesi.

McGue, M. (1994). Why developmental psychology should find room for behavioral genetics. In C. E. Nelson (Ed.), *Threats to optimal development: integrating biological, psychological, and social risk factors* (pp. 105–149). Hillsdale: Lawrence Erlbaum Associates.

McGue, M., & Johnson, W. (2008). Genetics of cognitive aging. In F. I. M. Craik & T. A. Salthouse (Eds.), *The handbook of aging and cognition* (3rd ed., pp. 55–96). New York: Psychology Press.

Plomin, R., DeFries, J. C., Knopik, V., & Neiderheiser, J. (2012). *Behavioral genetics* (6th ed.). New York: Worth.

Scarr, S. (1997). Behavior-genetic and socialization theories of intelligence: Truce and reconciliation. In R. J. Sternberg & E. Grigorenko (Eds.), *Intelligence, heredity, and environment* (pp. 3–41). Cambridge: Cambridge University Press.

Sternberg, R. J., & Grigorenko, E. (Eds.) (1997). *Intelligence, heredity, and environment*. Cambridge: Cambridge University Press.

Vandenberg, S. G. (1968). The nature and nurture of intelligence. In D. C. Glass (Ed.), *Biology and behavior: Genetics* (pp. 3–58). New York: Sage Foundation.

Vernon, P. E. (1979). *Intelligence: Heredity and environment*. New York: Freeman.

Waldman, I. D. (1997). Unresolved questions and future directions in behavior-genetic studies of intelligence. In R. J. Sternberg & E. Grigorenko (Eds.), *Intelligence, heredity, and environment* (pp. 552–570). Cambridge: Cambridge University Press.

Contents

Contributors

Rosalind Arden, Ph.D. Institute of Psychiatry, King's College London, London, UK
e-mail: r.arden@kcl.ac.uk

Michèle Carlier, Ph.D. Aix-Marseille Université, CNRS UMR 7290 Psychologie Cognitive, Fédération de Recherche 3C - Comportement Cerveau Cognition, case D, Bât 9, 3 Place Victor Hugo, 13331 Marseille Cedex 3, Marseille, France
e-mail: michele.carlier@univ-amu.fr

Christopher F. Chabris, Ph.D. Department of Psychology, Union College, Schenectady, NY USA
e-mail: chabrisc@union.edu

Iván Chavarría-Siles, Ph.D. Department of Psychiatry, Mount Sinai School of Medicine, One Gustave L. Levy Place, Box 1230, New York, NY 10029, USA
e-mail: ivan.chavarria-siles@mountsinai.org

Robin P. Corley, Ph.D. Institute for Behavioral Genetics, University of Colorado at Boulder, 447 UCB, Boulder, CO 80309-0447, USA
e-mail: corley@colorado.edu

Ian J. Deary, Ph.D. Department of Psychology, University of Edinburgh, 7 George Square, Edinburgh, EH8 9JZ, UK
e-mail: i.deary@ed.ac.uk

John C. DeFries, Ph.D. Institute for Behavioral Genetics, University of Colorado at Boulder, 447 UCB, Boulder, CO 80309-0447, USA
e-mail: defries@colorado.edu

Guillén Fernández, Ph.D. Donders Institute, for Brainm, Cognition, and Behavior, Radboud University Nijmegen Medical Center, P.O. Box 9101 6500 HB, Nijmegen, The Netherlands
e-mail: g.fernandez@donders.ru.nl

Deborah Finkel, Ph.D. Department of Psychology, School of Social Sciences, Indiana University Southeast, 4201 Grant Line Road, New Albany, IN 47150, USA
e-mail: dfinkel@ius.edu

Carol E. Franz, Ph.D. Department of Psychiatry, University of California, San Diego, 9500 Gilman Drive (MC 0738), La Jolla, CA 92093-0738 USA
e-mail: cfranz@ucsd.edu

Michael J. Galsworthy, Ph.D. Department of Applied Health Research, University College London, London, UK
e-mail: m.galsworthy@ucl.ac.uk

Margaret Gatz, Ph.D. Department of Psychology, University of Southern California, Los Angeles, CA 90089, USA
e-mail: gatz@usc.edu

Erin E. Horn, M.A. Department of Psychology, University of Virginia, PO Box 400400, Charlottesville, VA 22904-4400, USA
e-mail: horn@virginia.edu

Jung Yun Jang, M.A. Department of Psychology, University of Southern California, Los Angeles, CA 90089, USA
e-mail: jungyjan@usc.edu

Wendy Johnson, Ph.D. Department of Psychology, University of Edinburgh, 7 George Square, Edinburgh, EH8 9JZ, UK
e-mail: Wendy.johnson@ed.ac.uk

Ida K. Karlsson, M.Sc. Department of Medical Epidemiology and Biostatistics, Karolinska Institutet, Stockholm 17177, Sweden
e-mail: ida.karlsson@ki.se

William S. Kremen, Ph.D. Department of Psychiatry, University of California, San Diego, 9500 Gilman Drive (MC 0738), La Jolla, CA 92093-0738 USA
e-mail: wkremen@ucsd.edu

Center of Excellence for Stress and Mental Health, VA San Diego Healthcare System La Jolla, 92093-0738 CA, USA

Michael J. Lyons, Ph.D. Department of Psychology, Boston University, 64 Cummington Mall, Room 149, Boston, USA
e-mail: mlyons@bu.edu

Matt McGue, Ph.D. Department of Psychology, University of Minnesota, 75 E. River Road, Minneapolis, MN 55455, USA
e-mail: mcgue001@umn.edu

Caitlin S. Moore, M.A. Department of Psychology, Boston University, 64 Cummington St., Boston, MA 02215-2407 USA
e-mail: csmoore@bu.edu

Matthew S. Panizzon, Ph.D. Department of Psychiatry, University of California, San Diego, 9500 Gilman Drive (MC 0738), La Jolla, CA 92093-0738 USA
e-mail: mspanizzon@ucsd.edu

Center of Excellence for Stress and Mental Health, VA San Diego Healthcare System La Jolla, 92093-0738 CA, USA

Nancy L. Pedersen, Ph.D. Department of Psychology, University of Southern California, Los Angeles, CA 90089, USA
e-mail: nancy.pedersen@ki.se

Department of Medical Epidemiology and Biostatistics, Karolinska Institutet, Stockholm 17177, Sweden

Danielle Posthuma, Ph.D. Center for Neurogenomics and Cognitive Research (CNCR) & Department of Medical Genomics, Vrij University Amsterdam VUMC W&N building, room-439 De Boelelaan 1085, 1081 HV Amsterdam, The Netherlands
e-mail: Danielle.posthuma@cncr.vu.nl

Chandra A. Reynolds, Ph.D. Department of Psychology, University of California, Riverside, CA 92521, USA
e-mail: chandra.reynolds@ucr.edu

Pierre L. Roubertoux, Ph.D. Aix-Marseille Université, Génétique et Neurosciences, INSERM U 910, Génétique Médicale, Génomique Fonctionnelle, 27 bvd Jean Moulin, 13385 Marseille Cedex 5, Marseille, France
e-mail: pierre.roubertoux@univ-amu.fr

Eric Turkheimer, Ph.D. Department of Psychology, University of Virginia, PO Box 400400, Charlottesville, VA 22904-4400, USA
e-mail: turkheimer@virginia.edu

Sally J. Wadsworth, Ph.D. Institute for Behavioral Genetics, University of Colorado at Boulder, 447 UCB, Boulder, CO 80309-0447, USA
e-mail: wadswort@colorado.edu

Catalina Zavala, M.A. Department of Psychology, University of California-Riverside, Riverside, CA 92521, USA
e-mail: catalina.zavala@email.ucr.edu

Part I
Childhood

Chapter 1
Cognitive Abilities in Childhood and Adolescence

Sally J. Wadsworth, Robin P. Corley and John C. DeFries

The primary objective of this introductory chapter is to provide an overview of the genetics of cognitive abilities in childhood and adolescence. We begin with a brief introduction to the constructs of general and specific cognitive abilities, and then discuss theories of cognitive development, both historical and current. This is followed by descriptions of state-of-the-art methods in developmental behavioral genetics, and their application to salient issues in child and adolescent cognitive development and academic achievement. We then conclude with a brief discussion of possible future directions for both basic and applied research in this rapidly developing interdisciplinary field.

1.1 General and Specific Cognitive Abilities

General cognitive ability (intelligence) is associated with important life outcomes such as educational and occupational attainment, social mobility, and even health (Gottfredson 1997; Gottfredson and Deary 2004). Although the phrase "general cognitive ability" and the word "intelligence" are synonymous, the latter has been defined in many different ways. For example, Binet and Simon (1905), developers of the first test of global intelligence, suggested that "to judge well, to comprehend well, to reason well, these are the essential activities of intelligence" (reprinted in Binet and Simon 1916, p. 43). Later, Spearman (1923) stated that "everything intellectual can be reduced to some special case or other of deducing either relations or correlates"

S. J. Wadsworth (✉) · R. P. Corley · J. C. DeFries
Institute for Behavioral Genetics, University of Colorado at Boulder,
447 UCB, Boulder, CO 80309-0447, USA
e-mail: wadswort@colorado.edu

R. P. Corley
e-mail: corley@colorado.edu

J. C. DeFries
e-mail: defries@colorado.edu

D. Finkel, C. A. Reynolds (eds.), *Behavior Genetics of Cognition Across the Lifespan*,
Advances in Behavior Genetics 1, DOI 10.1007/978-1-4614-7447-0_1,

(p. 300). In 1921, when the editors of the *Journal of Educational Psychology* asked 17 leading investigators in the field to provide definitions of intelligence, they received 11 different definitions (Wasserman and Tulsky 2005). More recently, Wechsler (1958) suggested that intelligence is "the aggregate or global capacity of the individual to act purposefully, to think rationally, and to deal effectively with his environment" (p. 7). Similarly, Sternberg (2012) recently offered the following: "Intelligence is one's ability to learn from experience and to adapt to, shape, and select environments" (p. 19).

As varied as conceptualizations of cognitive ability may be, considerable agreement about the most important elements of any definition was demonstrated by the results of a survey of more than a thousand psychologists, educators, sociologists, and geneticists, who were asked to rate 13 possible characteristics of intelligence (Snyderman and Rothman 1987). The same three elements were rated as important by 96% of the participants: (1) abstract thinking or reasoning, (2) the capacity to acquire knowledge, and (3) problem-solving ability. Other qualities about which there was substantial agreement (60–80%) were adaptation to one's environment, creativity, general knowledge, linguistic competence, mathematical competence, memory, and mental speed. Thus, there are key elements, or skills, upon which most can agree. However, fewer respondents *expressed* confidence in how well those characteristics are measured. Nearly half felt that the capacity to acquire knowledge was not adequately measured. Although the measurement of problem-solving ability and abstract reasoning were viewed as less problematic, the majority of respondents reported that other qualities, such as adaptation to one's environment and creativity, were inadequately measured.

The most commonly used tests of cognitive abilities are individually administered standardized tests such as the Stanford-Binet (Terman 1973), the Wechsler Intelligence Scale for Children (Wechsler 1974, 1991, 2003), and the Woodcock-Johnson Tests of Achievement (Woodcock et al. 2001), assessing verbal and nonverbal skills including general knowledge, vocabulary, quantitative reasoning, working memory, spatial ability, and processing speed. The Woodcock-Johnson III also includes assessment of executive function skills (i.e., planning, decision making, problem solving, working memory, and inhibition). Critics often suggest that intelligence tests simply measure test-taking skills, making them little more than achievement tests. However, it is important to note that intelligence tests were originally developed as screening tools to be used in the prediction of scholastic success. In this capacity, they have performed quite well. The skill sets measured by these tests may be considered individually as specific cognitive abilities, or collectively as a measure of general cognitive ability. Although the latter is often referred to as "intelligence" or "aptitude," we shall use the less culturally dependent phrase "general cognitive ability" ("g") in this chapter.

1.2 Theories of Cognitive Development

Among the many theories of cognitive development (see Sternberg and Kaufman 2011 for reviews), there are several "traditions", theories, or schools of thought that have been highly influential. These include the psychometric (factor-analytic) tradition, with its hierarchical structure of general and specific abilities; the information processing tradition, focusing on the development and differentiation of specific cognitive components; the epistemological tradition with its hallmark stages of development (e.g., Piagetian theory); and the contextualist tradition, with an emphasis on the contexts of learning and assessment. More recent, and moving to the forefront of theories of cognitive ability, are systems theories such as Sternberg's triarchic theory (2012; 1985), with its emphasis on creative, analytical, practical, and wisdom-based skills; Gardner's theory of multiple intelligences (Gardner 2006), which argues that there is no ***g***, but rather eight different intelligences (linguistic, math, spatial, musical, kinesthetic, naturalist, interpersonal, and intrapersonal); dynamic skill theory, which analyzes variability in patterns of behavior, both systematic change over time and moment-by-moment fluctuations (Rose and Fischer 2011); planning, attention, simultaneous, successive, and a model of executive functions, which fits loosely in the factor-analytic tradition (Willis et al. 2011); and finally, biologically based theories that seek to explain intelligence in terms of brain-based mechanisms (Haier 2011).

Most salient to our discussion of the behavioral genetics of cognition is the factor-analytic tradition. Many of the most widely accepted theories fall into this category. It is most relevant to behavioral genetics because factor-analytic theories emphasize individual differences. These theories are roughly grouped into two basic subtheories: (1) that of a general factor, ***g***, which underlies all mental abilities, but assumes there are factors specific to each individual ability and (2) that of multiple factors, often overlapping, but not necessarily due to a common factor.

1.2.1 Historical Context

Although Galton was the first to suggest a general mental ability underlying specific abilities and tasks (Galton 1869; Sattler 1992), it was Spearman who first introduced methods by which this and other factor-analytic theories could be tested. Spearman's (1927) theory included one general factor underlying all the ability dimensions, with specific factors for each dimension. In contrast, Thorndike (1927) proposed a multifactor theory wherein individuals possess a number of interrelated, but distinct intellectual abilities. He suggested that some of these abilities have elements in common, and thereby cluster together, but each retains its own unique elements. More divergent from Spearman's general factor theory was Thurstone's (1938) multidimensional theory, in which cognitive ability, or intelligence, is composed of as many as seven independent factors, known as primary mental abilities. However, he later found these factors to be correlated, leading to the hypothesis of a second-order factor related to ***g*** (Thurstone and Thurstone 1941).

1.2.2 Current Theories

More recent theorists posit variations on the hierarchical model. For example, Carroll (1993) suggested a three-stratum model, beginning at the item level. Factor

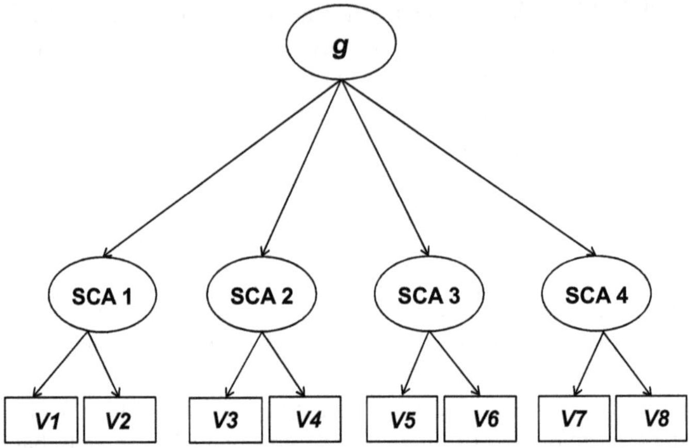

Fig. 1.1 Model of the hierarchical relationship between general cognitive ability (g) and specific cognitive abilities (SCA) and their indicators (V)

analysis of the many items would result in many narrow first-order (stratum I) factors. A number of broader second-order (stratum II) factors would result from factor analysis of the narrow first-order factors, followed by a few general factors, or ***g***, (stratum III) at the top of the hierarchy.

In contrast, Cattell (1963) characterized the structure of cognitive functioning based upon a two-factor model of "fluid" (G_f) and "crystallized" (G_c) intelligence. "Fluid intelligence" refers to aptitude and capacity for new learning, which is considered to be relatively free of cultural influences. Alternatively, "crystallized intelligence" refers to acquired knowledge or skill, related to achievement, and is more heavily dependent on cultural exposure. G_f-G_c theory was eventually expanded to include eight or nine primary abilities, with remarkable similarity to Carroll's broad stratum II factors. This similarity led to the synthesis of these theories of intellectual development into what is known today as Cattell–Horn–Carroll (CHC) theory (McGrew 2005).

Much current research in behavioral genetics employs a hierarchical representation with specific measures or abilities forming the base for broad correlated factors that combine to represent ***g*** or general cognitive ability (Fig. 1.1). From a statistical perspective, there is little controversy about the existence of ***g***. However, there is still debate regarding its practical application and theoretical usefulness (Sternberg 2012). So current is the topic, that it even has its own Facebook page!

For proponents of the view that cognitive ability is multifaceted, research tends to focus on the developmental courses of the specific cognitive processes, such as verbal ability, spatial ability, memory, perceptual speed, or the extensions of G_f and G_c. The primary focus of this research is on how these processes change with age and the factors that influence stability or change, especially genetic and environmental influences, including those factors that may modify estimates of genetic and environmental influences.

In the sections that follow, we provide illustrations of several current topics of developmental behavior genetic research on cognitive abilities across childhood and adolescence, including both quantitative and molecular approaches. We also discuss the relationship between cognitive abilities and academic achievement, one of the primary occupations of childhood and adolescence.

1.3 Behavioral Genetic Analyses of Cognitive Ability: Quantitative Approaches

From the time of Galton's classic study of the relatives of eminent men, *Hereditary Genius* (Galton 1869), the etiology of individual differences in general cognitive ability has been an important focus of behavioral science. Although it was at the center of heated controversy for decades (Bereiter 1969; Jensen 1969a; 1969b; 1972; Kagan 1969), a survey of over 1,000 psychologists and educational specialists later indicated overwhelming support for the heritable nature of cognitive ability (Snyderman and Rothman 1987).

A number of family studies have been conducted demonstrating significant and substantial familial resemblance for general cognitive ability (e.g., DeFries et al. 1979). However, because members of intact families share both genetic and family environmental influences, results of family studies alone are not sufficient for assessing the extent to which familial resemblance is the result of genetic or environmental influences. In contrast, analyzing data from twins and adoptive families facilitates estimates of the magnitudes of genetic and environmental influences. This section provides a brief overview of how twin, adoption, and family data are used in behavioral genetic research.

1.3.1 The Twin Model

The basis of the twin design lies in the comparison of the resemblance between scores of identical (monozygotic, MZ) twins, who are nearly identical genetically, and those of fraternal (dizygotic, DZ) twins who share, on average, half of their segregating genes. Assuming that shared family environments are no more similar for MZ than for DZ twins, and that mating is random (i.e., parent scores on the trait of interest are uncorrelated), the correlation between scores for MZ twins is a function of the heritability of the trait (i.e., an index of the extent to which individual differences in the trait are due to genetic influences, and represents that proportion of total variance due to genetic variance) plus shared family environmental influences, whereas that for DZ twins is a function of one-half the heritability of the trait plus shared family environmental influences. Thus, heritability may be estimated by doubling the difference between the MZ and DZ correlations, and shared

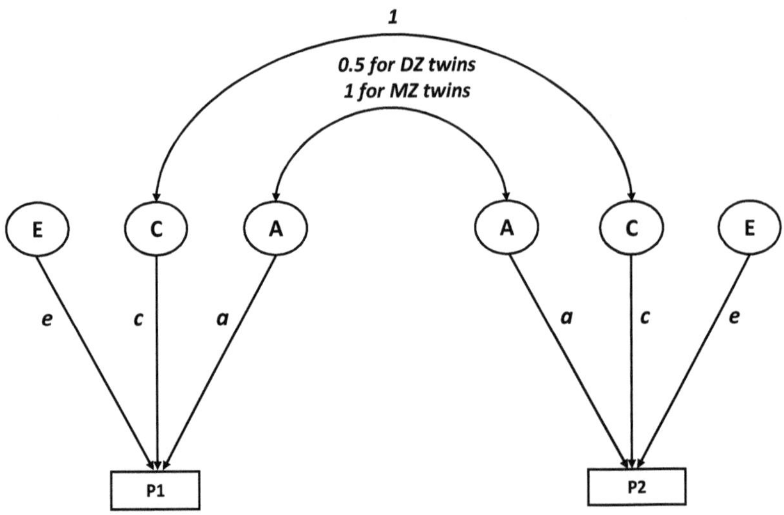

Fig. 1.2 Model of twin resemblance

family environmental influences may be estimated as the MZ correlation minus the heritability. Figure 1.2 illustrates the application of structural equation modeling (SEM) to twin data. This simplified univariate model depicts the effects of additive genetic (A), shared environmental (C), and nonshared environmental (E) influences on the phenotype (i.e., the observed trait or phenotype, P) of each twin. Heritability is estimated as the square of the genetic path (*a*), while shared and nonshared environmentalities are estimated as the square of the shared environmental (*c*) and nonshared environmental (*e*) paths, respectively. The genetic correlation between DZ twins for the same trait is 0.5 (because they share half of their segregating genes), whereas that for MZ twins is 1.0 (because they are nearly genetically identical). The shared environmental correlation is 1.0 for both MZ and DZ twins. Dominance, a type of nonadditive genetic variance involving the interaction of alleles at a locus, may also contribute to estimates of heritability, and is indicated if the DZ correlation is less than half the MZ correlation. See Plomin et al. (2013) for a detailed discussion of nonadditive genetic variance.

Twin analyses can also include data from families of twins. For example, data from siblings of twins have been used to test for "special twin environments" (Astrom et al. 2011; Koeppen-Schomerus et al. 2003; Medland et al. 2003; Wadsworth et al. 2011); to the extent that DZ twins are more similar than the twins with their cosibs, special twin environments are indicated. Less common is the use of data from children of identical twins. Although they are cousins, children of identical twins are as genetically related as half-siblings (Iacono 1999).

1.3.2 The Adoption Design

Comparing resemblance of related and unrelated family members' adoption designs provides the most direct evidence of both shared family environmental influences and heritability. Adoption studies typically use either the sibling model, the parent-offspring model, or a model that combines both relationships.

1.3.2.1 The Sibling Model

The basis of the adoptive sibling design lies in the comparison of the correlations of unrelated siblings to those of related siblings. Whereas adoptive siblings are genetically unrelated, related siblings share, on average, half of their segregating genes. Therefore, the phenotypic correlation between related siblings is a function of one-half the heritability of the trait, plus shared environmental influences, which are assumed to be similar for related and unrelated siblings. In contrast, the phenotypic correlation between genetically unrelated adoptive siblings arises only from shared environmental influences (assuming no selective placement, in which case, the adoptive and birth parents' scores should be uncorrelated). By analyzing these correlations, the contributions of genetic, shared environmental, and nonshared environmental influences (i.e., those environmental influences, which affect members of a sib pair randomly) can be estimated (Plomin et al. 1988). The genetic correlation between related siblings for the same trait is 0.5 (because they share half of their segregating genes), whereas for adoptive siblings it is 0.0 (because they are not genetically related). The shared environmental correlation is 1.0 for both related and unrelated siblings.

1.3.2.2 Parent-Offspring Models

The parent-offspring design includes resemblances of adoptive parents and adopted offspring, and/or biological parents and adopted-away offspring, and parents and offspring from intact control families. Adoptive parents and offspring share only home environmental influences; thus, in the absence of selective placement, the correlation between these family members for a given measure provides a direct estimate of shared family environmentality for parents and offspring. In contrast, biological parents and adopted-away offspring share, on average, half of their segregating genes. Therefore, the correlation between these family members estimates one-half the heritability of the measure. Control parents and offspring share both family environment, and half of their segregating genes, so that their correlation estimates one-half the heritability of the trait plus shared environmentality (for parents and offspring). By comparing intact control parent-offspring correlations with those from either biological parents and adopted-away offspring or adoptive parents and adopted offspring, the variance in the measure can be partitioned into portions due to genetic, shared environmental, and nonshared environmental influences.

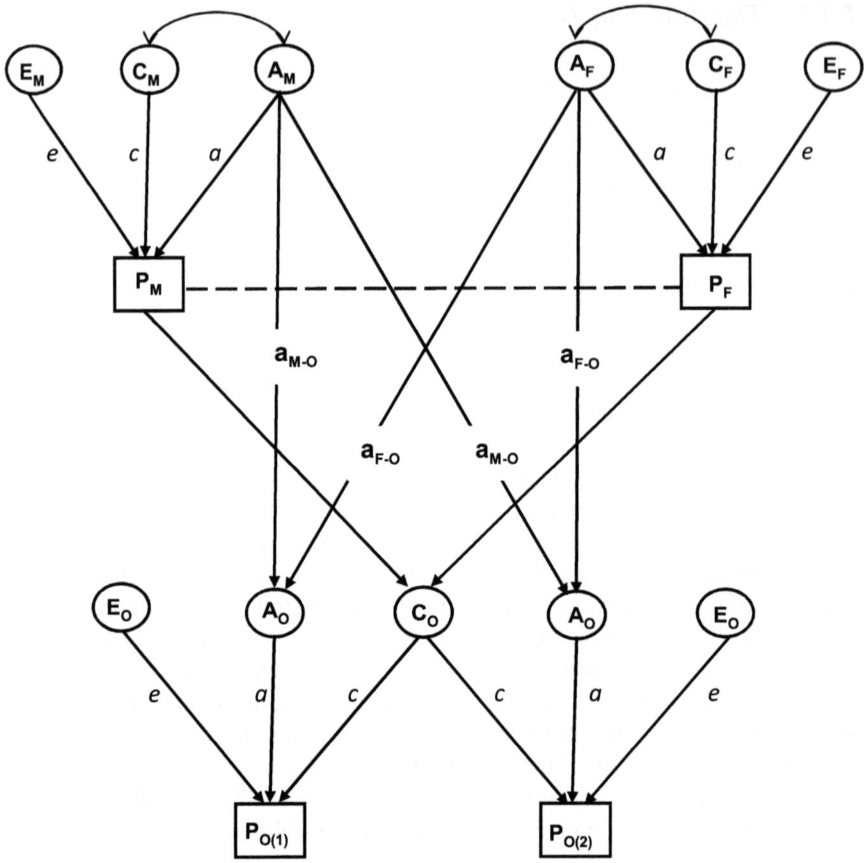

Fig. 1.3 Model of parent-offspring resemblance. $a_{M\text{-}O}$ and $a_{F\text{-}O}=0$ for adoptive parents and offspring and 0.5 for biological parents and offspring

However, analyzing data from all three family types simultaneously facilitates more powerful tests of alternative hypotheses (Fig. 1.3).

1.3.3 Combined and Extended Analyses

Twin data may be combined with adoption data for even more powerful tests of hypotheses. Moreover, the twin and adoption analyses can be extended to consider multivariate relationships among phenotypes and/or developmental processes. Multivariate extensions typically include structural equation models similar to those described above, but assess these relationships among multiple variables simultaneously using models such as the Cholesky decomposition, in which each phenotypic factor is represented by a genetic, shared and nonshared environmental factor, or by more theory-driven models, with the hypothesized relationships among the variables specified.

Developmental analyses may assess heritability and environmentality at different ages, or may involve models, such as Cholesky decomposition or Simplex models, that assess genetic and environmental contributions to stability and change over time (Neale and Cardon 1992; Neale and Maes 2002). Multiple regression models, such as DeFries–Fulker analyses (DeFries and Fulker 1985, 1988), described in Sect. 1.6.6, may be used to assess differential genetic influence as a function of age. More recently, adapted for use in behavior genetic studies is latent variable growth curve (LGC) modeling, which provides a basis for evaluating individual differences in rates of growth and patterns of change over time (McArdle 2006; McArdle et al. 1998; Neale and McArdle 2000).

1.4 Behavioral Genetic Analyses of Cognitive Ability: Molecular Approaches

An exciting use of the quantitative methods just described is the identification of phenotypes with high heritability to facilitate efforts to localize and identify genes that contribute to variability in complex behavioral traits. The past two decades have seen an explosion in the accessibility of molecular methods for the study of complex phenotypes such as cognitive ability. In this section, we briefly describe some approaches that are currently being used to localize or identify genes for cognitive ability.

1.4.1 Linkage/Quantitative Trait Loci Approaches

The traditional approach to gene localization for diseases and single-gene traits begins with linkage analysis. Linkage is the tendency of two genes or DNA sequences in close proximity on the same chromosome to be transmitted or inherited together. This cotransmission can be traced through families, particularly through multigeneration pedigrees. With a few hundred DNA markers of known location spread evenly throughout the genome, it is possible to localize a chromosomal region for the gene or genes of interest. For a recent explication, see Plomin et al. (2013).

Linkage analysis has been used to find genes influencing or causing many diseases and disorders. Of particular relevance to the study of cognition, are phenylketonuria (PKU) and fragile X syndrome, both of which result in mental retardation (for more information, see Chap. 3). Traditional multigenerational linkage analysis is appropriate for detecting genes of large effect, and not those of relatively small effect. Cognitive ability, like most human behaviors, is a quantitative trait (i.e., continuously distributed), which is likely influenced by many genes of small effect and many environmental factors (Plomin et al. 2013).Thus, other methods have been employed to assess linkage for cognitive abilities. For example, linkage-based Quantitative Trait Loci (QTL) methods analyze data from many pairs of close relatives, usually siblings. If a deoxyribonucleic acid (DNA) marker is closely linked to a gene that influences a quantitative trait, siblings who are more similar for the trait will share more alleles (variations) of the DNA marker (0, 1, or 2) inherited from their parents (identical by descent). The first linkage for a complex trait to be identified using this method was a QTL for reading disability on chromosome 6 (Cardon et al. 1994). A few studies have used sib-pair QTL analysis in the search for genes influencing cognitive ability (Doyle et al. 2008; Luciano et al. 2006; Posthuma et al. 2005), although most have examined cognitive ability in the presence of a disorder such as ADHD, autism, or reading disability.

1.4.2 Association

Although sib-pair QTL analysis can detect genes of smaller effect than traditional pedigree linkage analysis, it may still lack power to detect loci for complex traits such as cognitive abilities (Plomin et al. 2013; Risch and Merikangas 1996). The largest effect sizes for behavioral traits in children for reading, math, and general cognitive ability explain less than 0.5% of the variance (Butcher et al. 2008; Docherty et al. 2010; Meaburn et al. 2008). Fortunately, newer methods have been developed to analyze such data. For example, association analysis compares the frequency of an allele or alleles in different groups of people who may or may not be related, such as children with reading difficulties (RD) and those without, or children with high cognitive ability and those with average or below average cognitive ability. Association analysis is very powerful, typically making use of large samples and can, therefore, detect genes of very small effect (Plomin et al. 2013). Initially, it was used primarily to test associations with genes whose function was known and there existed some prior hypothesis for involvement of the locus (i.e., candidate genes). However, with the advent of microarrays, chips with millions of DNA markers facilitate genome-wide association studies (GWAS), which have become the method of choice for localizing QTL (Plomin et al. 2013).

Recently, it has been noted that GWAS account for only a portion of the heritability of a trait (Deary et al. 2012; Plomin 2012; Yang et al. 2011). If the whole genome has been scanned, why doesn't the sum of all the small effects equal the heritability of the trait? This has been dubbed the "missing heritability problem" and has led to concerns that perhaps twin and adoption studies have resulted in overestimates of heritability of complex traits (e.g., Zuk et al. 2012). A number of explanations for the missing heritabilities have been suggested, such as G × G interaction, because estimates of total heritability assume no G × G; nonfunctional genetic variants, new/de novo mutations, including copy number variations, i.e., variations in the number of repeated segments of DNA; sequence content (Eichler et al. 2010); genetic heterogeneity; and G × E interaction (Kutalik et al. 2011). Novel methods of approaching the problem have been developed. For example, Kutalik et al. (2011) have developed a maximum likelihood method that makes it possible to infer the explained variance, which in some cases, e.g., height, has been found to be as high as 10 times that estimated from single-nucleotide polymorphisms (SNP) alone. Yang et al. (2011) and Deary et al. (2012) describe genome-wide complex trait analysis (GCTA), which uses SNP data from unrelated individuals, but instead of identifying SNPs associated with a trait, it estimates the proportion of phenotypic variance explained by all the SNPs considered simultaneously (Yang et al. 2011). Results from GCTA, which do not rely on family-based analyses, suggest that the "missing" heritability is probably not due to overestimation of genetic effects from family-based designs, e.g., comparisons of MZ and DZ twin resemblance, but is more likely the result of many small genetic contributors, with effect sizes too small to estimate reliably using available sample sizes. Such analyses have accounted for a substantial proportion of the missing heritability of cognitive ability (Davies et al. 2011).

1.5 Behavioral Genetics of General Cognitive Ability (*g*) in Childhood and Adolescence: Findings of Family, Twin, and Adoption Studies

1.5.1 A Brief History

Family, twin, and adoption studies have been reviewed in detail in previous publications. The most frequently cited review is that by Erlenmeyer-Kimling and Jarvik (1963), which summarized correlations on measures of cognitive ability for 16 different family relationships (including adoptive and nonadoptive families, and twins), with subjects of different ages, spanning two generations. Analyzing data from over 30,000 correlational pairings, the combined results of these studies indicated increased resemblance for cognitive ability with increased degree of genetic relatedness. Median correlations closely approached those expected under a purely genetic model. In addition, the authors concluded that the pattern exhibited by the combined datasets was consistent with a polygenic hypothesis.

During the next two decades, results of several studies of twins and siblings, as well as adoptive and biological parents and their offspring, lent support to findings of significant and substantial heritability for general cognitive ability in children. Loehlin and Nichols (1976) reported twin correlations from 19 studies of general cognitive ability, and obtained median correlations of 0.86 for MZ twin pairs, and 0.62 for DZ pairs.

In a comparison of data published between 1976 and 1979 to those published prior to 1963, Plomin and DeFries (1980) summarized the results of over 4,600 pairings from family studies, over 2,500 pairings using the adoption design, and over 2,100 pairs of twins. Results of this review suggested that, based on the newer data, genetic influences accounted for about 50 % of the variance in cognitive ability (vs. 70 % based on the earlier data), with nonshared environmental influences accounting for most of the remaining variance. The studies included subjects ranging in age from 5 to 6 years in the Louisville Twin Study (Wilson 1983) to high-school age in the National Merit Scholarship Qualifying Tests study (Loehlin and Nichols 1976).

It was noted that the pattern of correlations for MZ and DZ twin pairs was identical for these two age groups. The authors suggested that the difference in heritability estimates obtained between the later and earlier studies was likely due to multiple factors, including sample size and methodology.

Subsequently, in a comprehensive review of the world literature on familial studies of intelligence, Bouchard and McGue (1981) summarized data from 111 studies, 59 of which were published in the 17 years after the Erlenmeyer-Kimling and Jarvik (1963) review. Included were 113,942 familial pairings comprising over 10,000 twin pairs reared together, as well as 65 pairs of MZ twins reared apart, over 26,000 sibling pairs and more than 12,000 parent-offspring pairs, as well as numerous other pairings, such as half-sibling pairs and cousins. The majority of the subjects were 20 years of age and younger (McGue et al. 1993). Similar to the findings of Plomin and DeFries (1980), the combined results of these studies suggested a heritability for general cognitive ability of about 0.50, with shared family environmentality of about 0.20.

Results of studies of twins and siblings, as well as adoptive and biological parents and their offspring, have continued to confirm findings of significant and substantial heritability for general cognitive ability. Furthermore, the combined results of these studies support the conclusions of these later reviews, suggesting that heritable influences account for about 50 % of the variance in cognitive ability in childhood and adolescence with shared environmental influences accounting for 10–20 % of the variance (Bishop et al. 2003; Cardon et al. 1992a; Cherny et al. 1997; Fulker and Cherny 1995; McGue et al. 1993; Petrill and Deater-Deckard 2004; Petrill 2004; Plomin and DeFries 1985, 1998; Plomin et al. 1988; Plomin et al. 1997).

Taken together, therefore, family, twin, and adoption studies of the etiology of individual differences in general cognitive ability in childhood and adolescence suggest moderate-to-strong genetic influences. In addition, environmental influences appear to be important, with about 20 % of the variance being due to environmental influences shared by family members.

1.5.2 Beyond Nature Versus Nurture

General cognitive ability is one of the most studied phenotypes in psychological science. However, although knowing the relative importance of genetic and environmental influences is important, it tells us little about developmental processes or its relationship to other phenotypes. While research on the heritability of ***g*** was being conducted, investigators in North America and Europe were developing new methods to assess multivariate relationships and the developmental processes underlying these effects—see Neale and Cardon (1992) and Fulker et al. (1993) for accounts of the history of the development of these methods. Evidence for the explosion in new methods for studying such relationships can be seen in the special issue of *Behavior Genetics* (1986, Vol. 16, No. 1) on Multivariate Behavioral Genetics and Development. The special issue included several papers describing path models developed to analyze multivariate and longitudinal relationships, many of which are still currently in use.

In the 1980s and 1990s, SEM (path analysis) became the gold standard for analysis of individual differences in quantitative phenotypes. Utilizing maximum likelihood estimation of parameters, its estimates are highly precise, and confidence intervals are obtained with relative ease. In addition, assumptions are explicitly expressed and tests of nested and competing models are facilitated (Neale and Cardon 1992; Neale and Maes 2002; Plomin et al. 2013). The application of SEM to cognitive abilities has facilitated the study of numerous salient issues, from the simple univariate proportion of the variance in general cognitive ability due to genetic influences, to genetic and environmental influences on continuity and change during development and the relationships among specific cognitive abilities. More recently, Latent variable growth curve (LGC) modeling has been applied to longitudinal data. LGC modeling can be used to evaluate individual differences in patterns and rates of growth in performance on tests of general and specific cognitive abilities, academic achievement, and other developing characteristics, and has been adapted for use with genetically informative data (McArdle et al. 1998).

1.5.3 Current Topics

An excellent in-depth review of the recent literature on the genetic bases of cognitive ability has been provided by Deary et al. (2009). Therefore, in the sections that follow, we discuss selected studies, which illustrate current topics and findings in the study of general cognitive ability in childhood and adolescence.

1.5.3.1 Heritability and age

Many studies included in the reviews discussed previously have established conclusively that genetic influences on ***g*** are substantial and significant; thus, current studies are addressing more complex questions (Plomin et al. 2013). For example, are the magnitudes of genetic and environmental influences similar in early and middle childhood and adolescence? Are there *qualitative* differences between genetic and environmental influences at the different ages (i.e., are different genetic influences expressed at different ages)? Are the etiological influences the same for different aspects of cognitive ability? Do specific cognitive abilities share common genetic or environmental influences? Does the genetic etiology of cognitive abilities differ for boys and girls?

A number of studies have recently addressed the topic of age effects on the magnitude of genetic and environmental influences on cognitive ability and on the etiology of age-to-age stability and change, with most finding increasing heritability and decreasing shared environmental influence, and genetic influences on stability. The first study that will be described utilizes a unique sample, including both twins and adoptive and nonadoptive siblings. As noted previously, the most direct test of shared

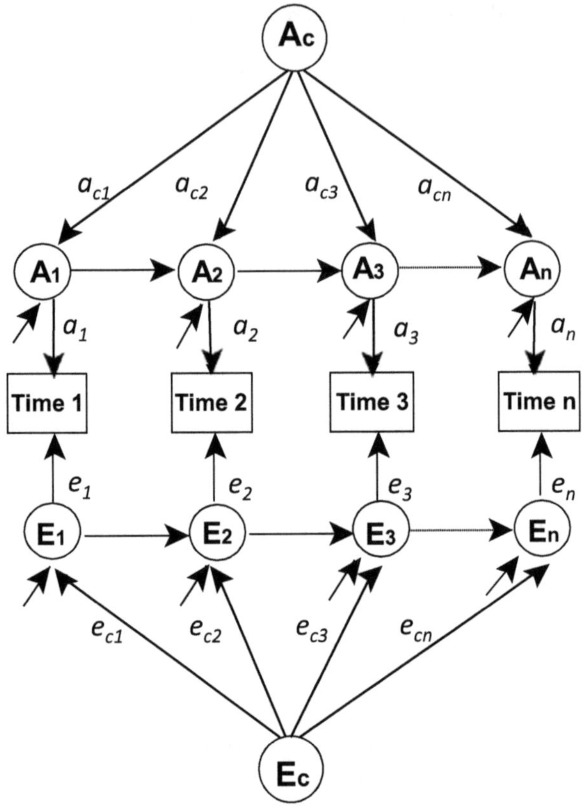

Fig. 1.4 Simplex common factor model of development

environmental influences comes from a comparison of scores of adoptive (unrelated) siblings to those of related siblings. By combining the power of the twin design with the adoptive sibling design, more powerful tests of salient hypotheses can be conducted than would be possible with either genetically informative design alone.

Extending the work of Cherny et al. (1997), Bishop et al. (2003) conducted developmental SEM analyses on measures of general cognitive ability spanning infancy to early adolescence (Bishop et al. 2003). The subjects were participants in either the Colorado Adoption Project (CAP; $N=200$–213 sibling pairs; DeFries 1994; Petrill et al. 2003; Plomin and DeFries 1985; Plomin et al. 1988) or the Colorado Longitudinal Twin Study ($N=386$–415 pairs; Emde and Hewitt 2001) Twins, adoptive siblings, and nonadoptive siblings were tested at ages 1, 2, 3, 4, 7, 9, and 10, and adoptive and nonadoptive siblings were also tested at age 12 (Bishop et al. 2003).

Using a developmental model such as the genetic simplex factor model shown in Fig. 1.4, the magnitude of genetic and environmental influences at each age, as well as age-to-age genetic and environmental transmission, age-specific genetic and environmental influences, and genetic and environmental influences common to all ages were assessed.

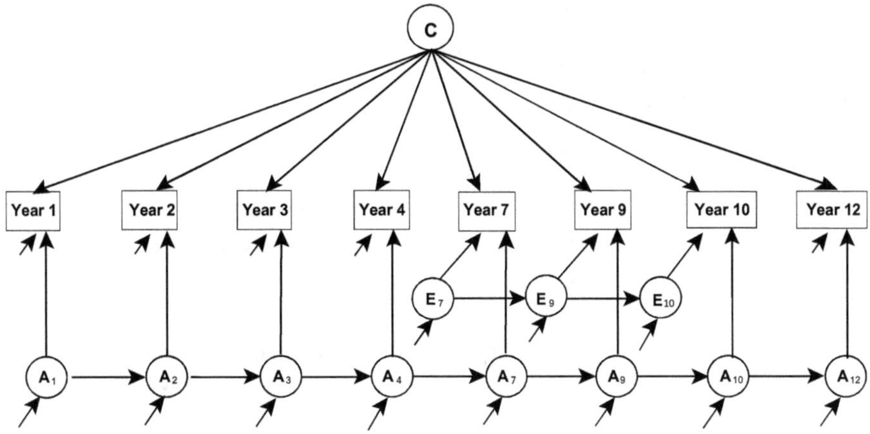

Fig. 1.5. Reduced model of cognitive development (from Bishop et al. 2003)

Estimates of heritability were somewhat variable but generally increased with age, ranging from 0.42 at age 3 to 0.74 at age 10, but dropping back to 0.49 at age 12. As heritability increased, shared environmental influences decreased. Although previous CAP analyses had shown that for general cognitive ability, there is evidence for both genetic and shared environmental continuity as well as genetic discontinuity throughout childhood (Cardon et al. 1992b; Cherny and Cardon 1994; Cherny et al. 1997; Fulker et al. 1993; Petrill et al. 1998; Petrill et al. 1997), results of this study indicated that during the transition to adolescence (after age 9), genetic factors contributed only to continuity (Fig. 1.5).

Subsequent analyses extending to age 16 with data from only sibling pairs (Petrill et al. 2004) confirmed moderate-to-strong genetic and nonshared environmental influences at each age, and no evidence of new genetic influence at age 16. Similar to the findings of Bishop et al. (2003), genetic influences were found to be responsible for continuity and nonshared environmental influences were responsible for change. However, more recently, Brant et al. (2009) analyzed cognitive data from only the same-sex twin pairs ($n=483$) assessed in person at ages 1, 2, 3, 4, 7, 12, and 16 in the Colorado Longitudinal Twin Study. When the model was fit to the data, large and stable genetic influences emerged that increased from early childhood to adolescence, accounting for 31–66% of the variance in ***g***, largely due to age-to-age transmission. Shared environmental influences were also stable, accounting for 19–46% of the variance. Nonshared environmental influence almost exclusively accounted for change. When different models were compared, the best fitting model at ages 1, 2, 3, 4, and 7 was one that included shared (C) and nonshared (E) environmental common factors, an additive genetic (A) age-to-age transmission factor, and A, C, and E innovations at each age. High age-to-age genetic transmission, as well as the large genetic innovations at each age, suggested that although there are stable genetic influences throughout childhood and adolescence, there are also new influences operating at each age. Furthermore, large estimates for the C common

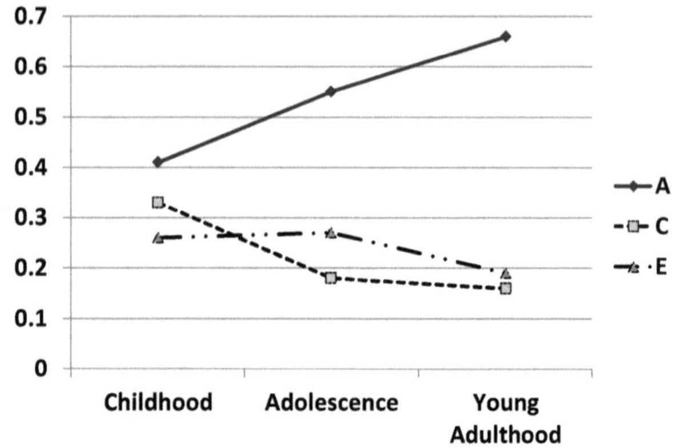

Fig. 1.6 Genetic and environmental influences on general cognitive ability in childhood, adolescence and young adulthood. (after Haworth et al. 2010)

factor suggested that stability is also conferred by shared environment, and low innovations suggest these are largely the same environmental influences throughout childhood.

Combining data from four countries to create a larger sample than all previous studies of the genetics of cognitive abilities combined, Haworth et al. (2010) recently analyzed data from 11,000 twin pairs from six samples in four countries (Australia, the Netherlands, the United Kingdom, and the United States) to test the hypothesis that the heritability of general cognitive ability increases linearly with age. The sample was split into three age groups: childhood, adolescence, and young adulthood. Subjects in the childhood group ranged in age from 4 to 10 years, with a mean of 9 years; those in the adolescence group were between 11 and 13 years of age (mean = 12 years), and those in the young adulthood group ranged in age from 14 to 34 years, with a mean age of 17 years. Both intraclass correlations and SEM indicated a linear increase in heritability and decrease in shared environment with age, such that in childhood, 41 % of the variance in general cognitive ability was due to genetic influences and 33 % to shared environmental influences. Heritability increased to 55 % in adolescence while shared environment dropped to 18 %. Finally, in young adulthood, heritable influences accounted for 66 % of the variance in general cognitive ability whereas shared environment accounted for 16 %.

The differences in the magnitude of heritability estimates at each age proved to be significant. Furthermore, although the influence of shared environment diminished across age groups, it also was significant at each age. Figure 1.6 depicts the relative changes in etiological influence from childhood to young adulthood. Although the samples, ages, and models differed among the four countries, their results were remarkably similar: increasing genetic influence and decreasing shared environmental influence from childhood into adolescence, genetic stability, and small but significant stability of shared environmental influences, with nonshared environmental influence limited to change.

In brief, the results of each of these studies are consistent with those of previous studies, which found increasing heritability with age for measures of cognitive ability as well as genetically mediated stability (e.g., Bishop et al. 2003; Deary et al.

2009; Fulker et al. 1993; Fulker et al. 1988; Haworth et al. 2010; McGue et al. 1993; Plomin et al. 1997).

1.5.3.2 High Cognitive Ability

Heritabilities and environmentalities represent averages across distributions. In addition to these full population parameters, researchers are interested in the etiologies of the extremes of a distribution and whether they are differentially influenced by subsets of genes or environments that would result in different etiological profiles than found in unselected populations.

The etiology of high cognitive ability is a particularly exciting topic, in part because of its relevance to finding genes for cognitive ability. It has been hypothesized that the genes involved in the development of exceptionally high cognitive ability are the same genes influencing development of normal range cognitive ability. This has been termed the "continuity hypothesis" (Plomin and Haworth 2009). If the continuity hypothesis is correct, individuals with high cognitive ability should have more increasing alleles that tend to enhance cognitive ability; thus, power to detect associations with genes of small effect could be increased by analyzing data from large numbers of individuals with high cognitive ability.

In the study by Haworth et al. (2009a) with longitudinal data from 11,000 twin pairs from four countries, liability threshold models were used to estimate the genetic and environmental etiologies of general cognitive ability in the upper 15 % of the distribution. Results indicated that the heritability of high cognitive ability was substantial, explaining 50 % of the variance, whereas shared environmental influences explained 28 % of the variance. The authors concluded that genetic variation contributes substantially to high ***g*** in the four countries from which the sample was drawn (i.e., Australia, the Netherlands, the United Kingdom, and the United States).

In the study by Brant et al. (2009a) in which longitudinal cognitive data were analyzed from 483 same-sex twin pairs participating in the Colorado Longitudinal Twin Study, an analysis of data from only those scoring in the top 15 % also used a threshold model to determine if the structure and relative importance of genetic and environmental factors was similar to those obtained for the full range of scores. The results were variable due to the small sample size at this threshold, but the patterns did not differ significantly. A model, which included additive genetic transmission (but no A common factor), shared and nonshared environmental common factors, and A, C, and E innovations resulted in a nonsignificant reduction in fit. Thus, there was no significant difference in developmental architecture between high cognitive ability and cognitive ability within the normal range. This suggests that high cognitive ability may be a useful phenotype for gene-finding efforts, since the etiologies of high and normal range cognitive abilities may be similar.

The two preceding studies used liability threshold models to estimate heritabilities of cognitive abilities in the upper 15 % of the distribution of scores. DeFries and Fulker (1985, 1988) suggested another approach, specifically for use with selected samples, such as children with high or low performance on cognitive tasks.

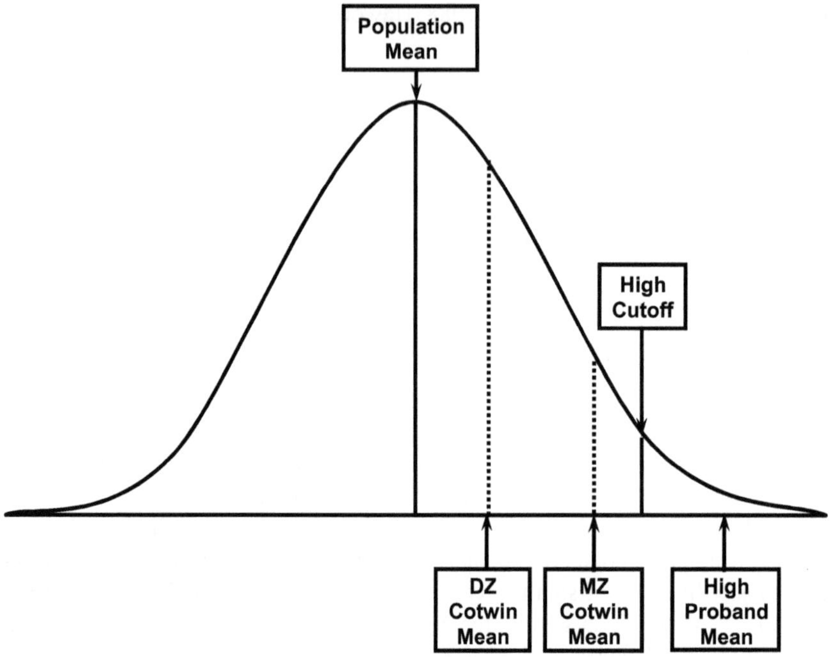

Fig. 1.7 Hypothetical distribution for general cognitive ability of an unselected sample of twins, and of the MZ and DZ cotwins of probands with high cognitive ability

Although for diseases or disorders a comparison of MZ and DZ concordance rates can provide an objective test of genetic etiology, cognitive ability is based on a continuous measure, with somewhat arbitrary cutoffs for "high" and "low" groupings. Transformation of a continuous measure into a categorical variable (high cognitive ability vs. cognitive ability within the normal range) results in a loss of information about the continuum of variation.

Therefore, DeFries and Fulker (1985) proposed that when probands are selected because of extreme scores on a continuous measure, such as cognitive ability, the scores of both MZ and DZ cotwins are expected to regress toward the mean of the unselected population. To the extent that extreme scores are heritable, we would expect to see a differential regression between scores of MZ and DZ cotwins (Fig. 1.7). If the MZ and DZ proband means are approximately equal, a *t*-test of the difference between the means of the MZ and DZ cotwins could provide an adequate test of genetic etiology.

However, the multiple regression analysis of twin data provides a more general/flexible and statistically more powerful test. Thus, DeFries and Fulker (1985, 1988) proposed two regression models. First, the basic model predicts a cotwin's score (C) from the proband's score (P) and the coefficient of relationship (R; 1.0 for MZ twins and 0.5 for DZ twins) as follows:

$$C = B_1P + B_2R + A, \tag{1.1}$$

where B_1 is the partial regression of cotwin's score on proband's score. B_2 is the partial regression of cotwin's score on the coefficient of relationship and estimates twice the difference between the means of the MZ and DZ cotwins after covariance adjustment for any difference between MZ and DZ proband means, providing a test for genetic etiology. A is the regression constant. When each individual's score is appropriately transformed (i.e., expressed as a deviation from the mean of the unselected population and divided by the difference between the mean of the probands and that of the unselected population), $B_2 = h_g^2$, an estimate of the extent to which the extreme scores of the probands are due to heritable influences.

DeFries and Fulker (1985, 1988) also proposed a second, augmented model

$$C = B_3 P + B_4 R + B_5 PR + A, \tag{1.2}$$

where PR is the product of the proband's score and the coefficient of relationship. B_5 estimates h^2, a measure of the extent to which individual differences within the selected group are due to genetic influences. When this augmented model is fit to the data, B_3 estimates c^2 and B_4 equals the difference between h_g^2 and h^2 when the twin data are transformed prior to multiple regression analysis, thereby providing a test for the difference between the etiologies of extreme scores versus individual differences. The multiple regression analysis (often called "DeFries-Fulker" or "DF" analysis) is also highly flexible; both the basic and augmented models can be extended to include covariates such as sex, SES, or high/low performance in order to test for differential h_g^2 and h^2 as a function of group membership (e.g., Friend et al. 2008; Hawke et al. 2007; Wadsworth et al. 2010). Furthermore, a bivariate extension can be employed to assess contemporaneous bivariate relationships or longitudinal relationships, estimating "bivariate heritability", an index of the extent to which the proband score on the selected variable is due to genetic factors, which also influence the correlated variable, and genetic correlation, an estimate of the extent to which the same genetic influences contribute to both variables (e.g., Astrom et al. 2007; Bishop 2001; Light and DeFries 1995). Most recently, the model has been extended to facilitate the inclusion of data from nontwin siblings of twin pairs (Astrom et al. 2011; Wadsworth et al. 2012).

Bivariate DF analysis can be used to address a central question in research on general cognitive ability: whether the magnitudes of genetic influences at the extremes of a distribution are the same as those for individual differences within the normal range. This has relevance for "generalist genes".

That genes influencing one aspect of learning also influence other aspects of learning (Plomin and Kovas 2005), has been demonstrated for behaviors within the normal range. However, less is known about whether such generalist or pleiotropic genes operate within the extremes of the distribution. In order to test the hypothesis that generalist genes affect high ability and achievement, Haworth et al. (2009a) analyzed data from 4,000, 12-year-old twin pairs participating in the Twins Early Development Study (TEDS; Kovas et al. 2010), to investigate the genetic and environmental overlap among Internet-based tests of cognitive ability, reading, math, and language in the upper 15 % of the distribution. The data were subjected to bivariate DF extremes analysis and genetic correlations were estimated for pairs of measures. The genetic correlations among the measures ranged from 0.52 to 0.63, indicating substantial genetic overlap among the measures. Thus, the authors suggested that generalist genes are just as evident at the extremes of a population as they are throughout the full range of performance.

Analyzing data from a combined sample of adolescent twins, adoptive siblings, and biological siblings, Kirkpatrick et al. (2009) estimated the shared environmentality for both high cognitive ability and cognitive ability within the full range of ability. The goal of this study was to answer three questions: first, is there shared environmental influence on high cognitive ability in adolescence? Second, if so, do shared environmental influences differ for high cognitive ability and ability within the full range? Finally, are there specific measurable family-level variables that contribute to high cognitive ability in adolescence? To answer the first two questions, three separate analyses were conducted: biometric model-fitting, DF (1985, 1988) multiple regression using data from the high-ability subsample, and DF multiple regression using data from the full-range sample. Estimates of shared environmental influences were moderate, ranging from 0.17 to 0.24 from the three separate analyses, indicating that shared environmental influences on high cognitive ability are similar to those on the full range of ability. To answer the third question, data from the adoptive sibling sample were analyzed. The family-level variables included parental occupation, parental education, and disruptive life events. Results indicated that only parental education showed a clear association with children's general cognitive ability. The authors suggested that future studies of environmental effects on general cognitive ability will require large samples, genetically sensitive designs, sophisticated methods, and a variety of environmental measures.

1.5.3.3 *g* and Genes

The past two decades have seen a burgeoning interest in finding genes for cognitive abilities. As noted previously, the advent of GWAS has facilitated a number of studies with the goal of finding genes of small effect related to cognitive abilities (e.g., Davies et al. 2011; Davis et al. 2010; Loo et al. 2012; Luciano et al. 2006; Plomin et al. 2001; Posthuma et al. 2005). Unfortunately, while a number of associations have been reported, few have been replicated (for reviews, see Deary et al. 2009; Mandelman and Grigorenko 2011). There are several possible reasons for this, including differences in measures and phenotype definitions, differences in methods and even false positives (Chabris et al. in press). However, although the range of findings and lack of replication warrant caution in interpretation of the results of these studies, they also make a very good case for continued study of the molecular genetics of cognitive abilities.

1.5.3.4 *g*, Genes, and Age

We now know that individual differences in cognitive ability are highly stable, and that genetic influences increase with age in childhood and adolescence (see Chaps. 4 and 5 for patterns of genetic influence in middle adulthood and aging). We also know that, as discussed in Sect. 1.4.2, attempts to identify specific genetic causes have been relatively unsuccessful. However, to estimate the genetic and environmental contributions to cognitive aging, Deary et al. (2012) used genome-wide SNP data from almost 2,000 unrelated individuals whose cognitive ability was assessed in childhood (age 11) and again in late adulthood (age 65, 70, or 79). They then used GCTA (Sect. 1.4.2) to estimate the proportion of variance at each assessment, as well as the genetic covariance, explained by causal genetic variants. Results indicated that 24% of the variation in cognitive change from childhood to late adulthood could be accounted for by genetic variants in linkage disequilibrium with common SNPs. This was considered to be a lower bound estimate of additive genetic influence on lifetime cognitive aging. Using bivariate analysis, they estimated the genetic correlation between childhood and late adulthood cognitive ability to be 0.62. The authors note that this is the first time genetic contributions to stability and change across most of the lifespan have been quantified, and suggest that these findings support the continued search for genetic causes of stability and change in cognition.

1.5.3.5 *g*, Genes, and the Extremes

In order to increase the probability of identifying genes of small effect on general cognitive ability, Davis et al. (2010) conducted a three-stage GWAS using over 350,000 SNPs in the quantitative extremes of a population sample of 7,900, 7-year-old children from the twins early development study (TEDS). Data from approximately 1,000 children from each of the extremes of the population were included in two pooling stages along with about 3,000 children from the normal range included in a third pooling stage. This approach identified 28 SNPs associated with general cognitive ability that could be used for individual genotyping in an unselected population. However, replication using larger samples in denser arrays will be necessary to identify genes for general cognitive ability.

1.6 Beyond *g*: Specific Cognitive Abilities and Academic Achievement

School achievement is the primary occupation of childhood and adolescence. It impacts occupational attainment, socioeconomic status, standing in the community, and health. While it is clear that academic achievement must necessarily be influenced by cognitive abilities, it should be equally clear that differences in academic achievement may influence cognitive abilities. Furthermore, individual differences in cognitive abilities and scholastic ability or achievement may be due to many of the same influences. Both are influenced by a range of factors, some shared and some specific. Thus, it would be reasonable to expect that SCA and scholastic abilities would share both genetic and environmental etiologies, but would also demonstrate independent influences. In this section, we explore the evidence for shared and independent etiologies of SCA and academic achievement, as well as age effects and high and low achievement.

1.6.1 Genetic and Environmental Influences on Academic Achievement

Although it may seem that genetic influences might be less important as a cause of individual differences in academic achievement than in cognitive ability (Arnold et al. 1997; Jensen, 1967; 1969a), twin and adoption studies have shown substantial genetic influence on various measures of academic achievement, with heritabilities as high or higher than those obtained for measures of general cognitive ability (Cardon et al. 1990; Gill et al. 1985; Harlaar et al. 2007; Martin 1975; Stevenson et al. 1987; Thompson et al. 1991; Wadsworth et al. 2002, 2006, 1995).

Historically, one of the largest and best known twin studies of academic achievement was that reported by Loehlin and Nichols (1976). Data from over 2,100 twin pairs who had taken the National Merit Scholarship Qualifying Test were analyzed. Genetic influences were found to be significant, with heritabilities averaging about 0.40 across the different subject areas. In addition, genetic correlations among the tests were substantial, suggesting the possibility of a general genetic factor influencing all the achievement domains. In a reanalysis of these data, Plomin and DeFries (1979) used multivariate behavioral genetic methods to assess the etiology of covariation among the five scholastic abilities. The average bivariate heritability was about 0.45. Furthermore, their findings also suggested that one general genetic factor and one general environmental factor could account for the covariance among the measures.

Support for these findings has been provided by several other twin and adoption studies of school-aged children. Heritability estimates for reading achievement, by far the most studied scholastic ability, have ranged from 0.20 to 0.70, with most estimates averaging about 0.40–0.50 (Harlaar et al. 2007; Cardon et al. 1991; Stevenson et al. 1987;Wadsworth et al. 2002; 2006). Similarly, for mathematics achievement, estimates have ranged from 0.12 to 0.69 (Thompson et al. 1991; Wadsworth et al. 1995a). Moreover, the largest twin study of learning abilities and disabilities, academic achievement, and behavior problems, the TEDS, obtained heritability estimates across domains of about 0.60.

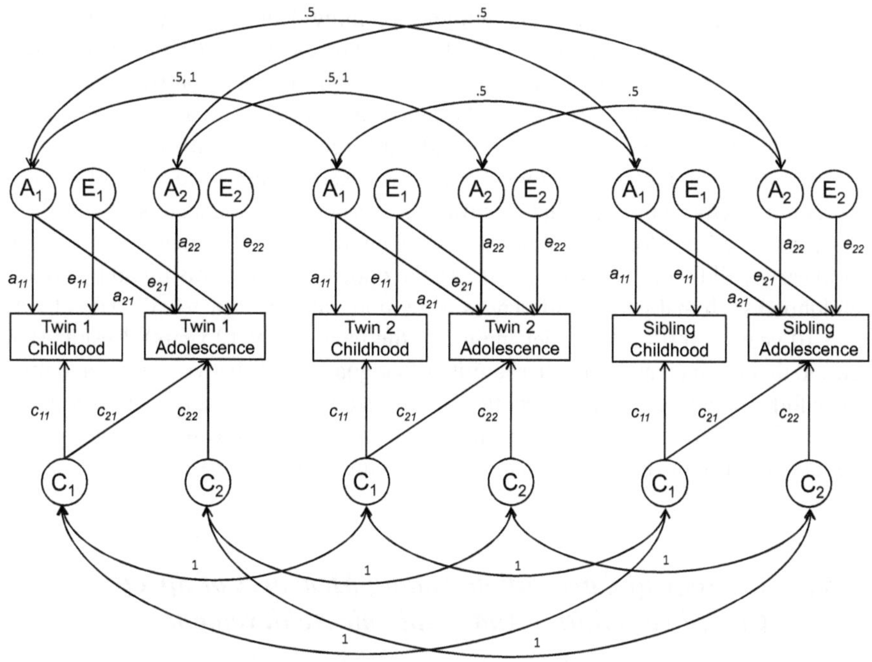

Fig. 1.8 Longitudinal twin-sibling model (from van Soelen et al. 2011)

1.6.2 Specific Cognitive Abilities, h^2, and Age

In order to assess the effects of age on the genetic and environmental structures of verbal intelligence quotient (VIQ) and performance intelligence quotient (PIQ), van Soelen et al. (2011) analyzed data from a longitudinal study in which twins and their nontwin siblings were administered the Wechsler Intelligence Scale for Children-Third Edition (WISC-III; Wechsler, 2002) at ages 9–11 and 12–14. The data for each measure were fitted separately to a longitudinal bivariate Cholesky that included both twins and their nontwin siblings. The extended twin and sibling model afforded powerful tests of differences in the sources of variance influencing Full-Scale IQ (FSIQ), VIQ, and PIQ at different ages as well as an examination of the causes of stability and change in VIQ and PIQ.

When this model (see Fig. 1.8) was fitted separately to the FSIQ, VIQ, and PIQ data at the two time points, heritabilities were 0.34 for FSIQ, 0.37 for VIQ, and 0.64 for PIQ in childhood, and 0.65, 0.51, and 0.72, respectively, in adolescence. As heritability increased, shared environmental influence decreased for FSIQ and VIQ; however, for PIQ, shared environment was not a significant factor at either age. Stability in FSIQ and VIQ was found to be due to genetic and shared-environmental influences, as evidenced by the genetic and shared environmental correlations both of which approached unity. For PIQ, stability was lower, and explained only by genetic influences.

In the first examination of the etiology of individual differences in language from early childhood to adolescence, Hayiou-Thomas et al. (2012), fitted a latent factor model to longitudinal data from nearly 7,900 twin pairs participating in the TEDS. The investigators compared the magnitudes of genetic and environmental influences on language skills in early childhood (ages 2–4), middle childhood (ages 7–10), and early adolescence (age 12) and assessed the extent to which the same genetic and environmental factors influence variation in language skills at these three stages of development. Whereas shared environmental influences were greater in early childhood ($c^2=0.74$, $a^2=0.24$), the opposite pattern was found by middle childhood such that genetic influences were more important ($a^2=0.57$–0.63 at 7, 9, and 10 years and 0.47–0.57, depending on the measure, at 12 years). Results suggested that the increase in the heritability of language skills between early and middle childhood was due to new genetic influences operating at the transition between early and middle childhood. In contrast, genetic factors were stable from middle childhood through early adolescence, and accounted for the phenotypic stability.

1.6.3 Etiology of Covariation Among Measures of Specific Cognitive Abilities And Academic Achievement

Although specific cognitive abilities and various measures of academic achievement have been shown to be correlated, little is known regarding the etiology of this relationship. Wadsworth et al. (1995a) suggested that both genetic and environmental influences may contribute to the observed relationship between specific cognitive abilities and measures of academic achievement. Because educational systems attempt to foster both ability and achievement, associations between specific cognitive abilities and achievement measures may be mediated environmentally. Indeed, preschool and compensatory educational programs have been designed based on this expectation. The success of such programs would provide support for the theory of an environmental etiology of covariation among measures of cognitive abilities and academic achievement. In addition, the home environment (including early childhood exposure to reading and arithmetic games, parental encouragement of academic achievement, and educational television) may also influence both ability and achievement. Alternatively, it is equally plausible that genetic factors, which influence scores on tests of general and specific cognitive abilities, may also affect performance on achievement tests.

Examining data from a sample of 119 adoptive and 120 nonadoptive families participating in the CAP, Cardon et al. (1990) investigated the etiology of the relationship between IQ and Reading Recognition employing a multivariate parent-offspring adoption model examining the relationship between WISC-R or WAIS-R, VIQ, PIQ (Wechsler 1974, 1981), and PIAT Reading Recognition (REC; Dunn and Markwardt 1970). As expected, the phenotypic correlation between VIQ and REC was greater than that between REC and PIQ, 0.46 and 0.27, respectively. Multivariate behavior genetic analysis revealed heritability estimates of 0.38, 0.36, and 0.41 for REC, VIQ, and PIQ in this study. The genetic correlation between REC and VIQ was 0.96 (accounting for 78 % of the phenotypic relationship between these measures), whereas that between REC and PIQ was only 0.45, explaining about 67 % of the observed correlation. These results suggest that the relationships between reading achievement and general and specific cognitive abilities are largely due to common genetic influences, with environmental influences being specific to each measure, and contributing little to the IQ-achievement relationship.

Thompson et al. (1991) examined scores of 146 MZ and 132 same-sex DZ twin pairs on measures of specific cognitive abilities (verbal, spatial, perceptual speed, and memory) as well as on the Metropolitan Achievement Test (MAT), including measures of reading, mathematics, and language skills. Although all the measures of specific cognitive abilities evidenced moderate-to-high heritabilities (ranging from 0.37 for memory to 0.74 for spatial ability), the achievement measures were less heritable, with heritabilities of 0.27 for reading, 0.17 for math, and 0.19 for language. However, genetic correlations among the measures of achievement and specific cognitive ability were high, about 0.85 between verbal ability and all three measures of scholastic achievement, accounting for more than 80 % of the phenotypic correlation with each measure. Genetic correlations between spatial ability and the three achievement measures were about 0.80, accounting for more than 80 % of the observed covariance of spatial ability with language and more than 90 % of that with reading and math. The average shared environmental correlation was zero. These findings suggest that the relationships among the measures of specific cognitive abilities and achievement used in this study are almost entirely genetically mediated.

Wadsworth (1994) analyzed cognitive ability and achievement data from 60 adoptive and 86 control sibling pairs participating in the CAP at age 7, as well as138 MZ and 95 same-sex DZ twin pairs participating in the Colorado Learning Disabilities Research Center (CLDRC) between the ages of 8 and 20 years. The cognitive ability and achievement measures included the VIQ and PIQ scales of the WISC-R or WAIS-R (Wechsler 1974, 1981), the Reading Recognition (REC) subtest of the PIAT (Dunn and Markwardt 1970), the Mathematics subtest of the PIAT (for the twin sample), and (for the sibling sample) the KeyMath Diagnostic Arithmetic Test (MATH; Connolly et al. 1976). Phenotypic correlations among the measures were substantial, ranging from 0.36 for the correlation between PIQ and MATH to 0.49 for that between VIQ and MATH. Although 65 % of the correlation between REC and MATH was mediated by VIQ, there was significant phenotypic covariation among the achievement measures independent of cognitive ability. Heritability estimates for REC and MATH were approximately equal (0.45 and 0.41, respectively) whereas those for VIQ and PIQ were somewhat higher (0.61 and 0.54, respectively). There was substantial genetic variation in both reading and mathematics performance independent of the IQ measures. Shared environmental influences were somewhat weaker, but nontrivial, with estimates ranging from 0.12 for Reading and Performance IQ, to 0.21 for Math. Estimates of genetic and environmental correlations suggested strong genetic influences on the covariation among the measures, with genetic influences accounting for 75 % of the observed correlation between VIQ and REC, and 65 % of that between VIQ and MATH. However, the genetic correlation between REC and MATH accounted for only 32 % of the

Table 1.1 Genetic and environmental influences on high reading performance. (Friend et al. 2009)

Sample	Nprs	a^2	c^2	e^2
Colorado				
Kindergarten	114	0.72	0.00	0.28
First grade	112	0.72	0.04	0.24
Second grade	105	0.52	0.17	0.31
TEDS				
First grade	1,155	0.50	0.17	0.34
Sixth grade	0 830	0.59	0.05	0.36

observed covariance. The results of this study, therefore, suggest that much of the covariance between cognitive ability and academic achievement in this sample was due to shared genetic influences. Furthermore, significant covariation was found between reading and mathematics achievement, independent of general cognitive ability, due to both shared genetic and environmental influences.

Luo et al. (2003) expanded earlier analyses by subjecting cognitive and achievement data from 277 pairs of same-sex twins to phenotypic and behavioral genetic structural equation modeling. The cognitive measures included learning, probe memory, self-paced probe learning, reaction time, discrimination time, and inspection time, from the Cognitive Assessment Tasks and the language, reading, and mathematics subtests from the MAT. The data were examined to assess the causal relationships among the cognitive and scholastic measures, as well as the extent to which these relationships vary across different domains of academic knowledge. In addition, genetic and environmental influences on the cognitive and scholastic measures, and on their association, were also assessed. The results indicated that as much as 30 % of the phenotypic variance of scholastic performance was accounted for by a cognitive general factor, which was likely related to processing speed. Covariation among the measures was primarily genetic and stable. Considering previous findings that processing speed has a substantial genetic correlation with ***g***, and that the relationship between ***g*** and scholastic achievement is primarily genetic, these findings suggest that processing speed is causally related to both ***g*** and scholastic performance, and that the etiology of this causal relationship is primarily genetic.

1.6.4 High Academic Achievement

If little is known about the etiologies of high cognitive abilities, even less is known about the etiologies of high academic achievement. However, one recent study highlighted high reading performance. In order to assess moderation of genetic influence on high reading performance by parental education level, Friend et al. (2009) estimated genetic and environmental influences on high reading performance in two different samples, the Colorado Twin Registry and TEDS, at several different grade levels. As shown in Table 1.1, the estimates are very similar to what is typically obtained for performance within the normal range, suggesting that genetic and environmental influences are continuous across the full range of ability/performance.

1.6.5 Low Academic Achievement

Unlike high academic achievement, low academic achievement has been studied for decades, primarily low reading achievement. As many as 10 % of school-aged children have difficulties learning to read, accounting for approximately 80 % of all learning disabilities. The possibility that RD may have a genetic basis was raised over a century ago when Thomas (1905) noted the familial nature of "congenital word blindness". Since that time, family studies suggested, and twin studies have obtained strong evidence for a genetic etiology of RD (e.g., DeFries 1985; DeFries et al. 1997; DeFries et al. 1991; Hallgren 1950; Harlaar et al. 2005; Stevenson et al. 1987). Initially, concordance rates were used to provide evidence of genetic influences. More recently, DF multiple regression analysis has been used to estimate the heritability of the proband deficit in reading performance (e.g., DeFries et al. 1999). With large samples, DF analysis can also be used to test the hypothesis that the etiology of extreme scores differs from that of individual differences (DeFries and Fulker 1988), or test for differential genetic etiology as a function of cognitive ability (Wadsworth et al. 2010) or gender (Hawke et al. 2006). Recent unpublished analyses of data from selected twin pairs participating in the Colorado Learning Disabilities Research Center (CLDRC) suggest that although the heritability of reading deficits is somewhat higher than that for individual differences ($B_2 = 0.63 \pm 0.07$; $B_5 = 0.47 \pm 0.22$), this difference ($B_4 = 0.16 \pm 0.23$) is not significant ($p = 0.47$). Thus, there is little evidence that the etiology of extreme scores differs from that for individual differences.

1.6.6 High and Low Academic Achievement and Genes

Both high and low math achievement, as well as achievement within the full distribution of scores, were recently explored in the first GWAS of mathematics ability and disability by Docherty et al. (2010). In over 5,000 10-year-old twin individuals SNP sets were identified. To optimize power of detecting small effects, data from high- and low-performing subjects were screened to nominate SNPs for individual genotyping. The first group consisted of 300 subjects (one from each twin pair) from the upper 16 % and 300 from the lower 16 %, representing the high and low extremes of performance. A second group comprised 300 each from the upper and lower 20 %, a nonindependent sample including 73 MZ and 83 DZ twins from the first stage. Ten independent DNA pools were created from each group, with each pool containing the DNA of 30 individuals. Extending the analysis to the nonindependent normally distributed population sample, 43 of the top-ranked 3,000 SNPs were individually genotyped. Ten of these evidenced significant linear associations with mathematics performance in a third sample spanning the full range of ability, supporting the QTL hypothesis that many genes of small effect operate across the full distribution of ability. In a subsequent analysis of data from this third sample genotyped on the 10-SNP set, Docherty et al. (2011) investigated interactions with environmental measures in the home and school. They found two significant interactions: The 10-SNP set was more strongly associated with mathematical ability in chaotic homes and when parents are negative.

Linkage and association analyses have both been utilized in the search for genes affecting RD (see Smith et al. 2010 for a review). Segregation analyses have supported the existence of several genes of major effect, rather than many genes of small effect and linkage and association studies have localized the effects of genes to at least nine chromosomal regions (Smith 2010). In all cases, associations were found not in the DNA coding regions, which encode proteins, but in the noncoding regions, which regulate the degree to which the gene is expressed and therefore the amount of protein produced. Several of these are involved in neuronal and axonal migration in the brain.

1.6.7 Sex Differences

1.6.7.1 Mean Differences

Conventional wisdom tells us that girls perform better on tests of verbal ability and boys perform better on quantitative and spatial measures. These were the conclusions of a landmark review of the literature on sex differences by Maccoby and Jacklin (1974). Many studies focusing on sex differences have been conducted since that time. With the advent of meta-analytic procedures, the computation of effect sizes that quantify differences objectively has shown that the evidence for sex differences in childhood cognitive abilities is relatively weak. In general, few mean sex differences have been found for general or specific cognitive abilities (Halpern 2000), and results vary by study. For example, in some studies girls perform slightly better than boys on tests of verbal ability and memory (Cardon 1994; Halpern 2000; Hyde and Linn 1988), and sometimes on tests of math and quantitative reasoning (Halpern 2000; Wadsworth 1994) while at the same time males are overrepresented in the upper tail of the distribution of math scores. More recent evidence indicates, however, that, at least during the school years, there is little or no gender difference in performance on the National Assessment of Educational Progress (Hyde et al. 2008). Males may excel at verbal analogies while being overrepresented in the lower tail of the distribution of verbal scores (Halpern 2000). Some studies indicate that boys perform better on measures of visuospatial ability (Halpern 2000; Cardon 1994), and on nonverbal subtests of the WISC or WAIS (Wadsworth 1994).

Haworth et al. (2009b) analyzed data from six different samples from four countries and noted a small, but significant, mean sex difference in general cognitive ability in four of the six samples, with males scoring higher than females. Similarly,

Petrill et al. (2009) examined sex differences in math ability in 10-year-old twins scoring at or above the 85th percentile and found small mean differences favoring boys.

In a recent analysis of data from 788 children aged 5–16 years, for seven cognitive domains, sex differences were found for oral language, spatial abilities, and visual and tactile perceptual tasks, with boys performing better on all but the tactile perceptual task (Ardila et al. 2011). However, these differences accounted for only 1–3 % of the variance of any measure. Thus, the authors concluded that "gender differences during cognitive development are minimal, appear in only a small number of tests, and account for only a low percentage of the score variance" (p. 984).

Exploring gender differences in reading performance of subjects with and without RD, Hawke et al. (2009) noted that samples of poor readers typically include more males than females. Although mean reading scores differed only slightly by sex, males had much larger variances. This finding was consistent with that of Reynolds et al. (1996) based on data from the Virginia Twin Study of Adolescent Behavioral Development. The authors suggested that the larger variances for males may explain the greater number of males in samples of reading-disabled children and the higher gender ratios in more severely affected samples.

1.6.7.2 Differential Etiology

Although there may be small mean sex differences among measures of cognitive abilities and achievement, there may or may not be differences in etiologies of cognitive abilities as a function of sex. Baker et al. (1994) found no difference in etiology of general cognitive ability as a function of sex in 3–7-year-old adoptive and nonadoptive children participating in the CAP. They did, however, find small differences in etiologies of specific cognitive abilities: heritability for memory was higher among boys at all three ages, whereas that for spatial ability was higher among girls at ages 4 and 7. However, sample sizes were small and power to detect sex differences in etiologies low.

Haworth et al. (2009a) examined data from children from four countries for both quantitative sex differences (differences in the magnitude of genetic influences) and qualitative sex differences (i.e., different genetic factors, reflected in the DZ opposite-sex genetic correlation, which will be significantly less than 0.5 in the presence of qualitative differences). Although they found small mean differences, they found no significant qualitative sex differences, and only a small difference in the magnitude of genetic influences.

Although Petrill et al. (2009) found small mean sex differences in his sample of 10-year-old twins with high math ability, there was no evidence of differential etiology as a function of sex. This result was consistent with the findings of Kovas et al. (2007) for low math ability.

Exploring sex differences among subjects with low reading ability, Harlaar et al. (2005) noted a significant difference in the magnitude of h_g^2 estimates for 7-year-old TEDS subjects scoring at or below the 5th percentile, with higher heritability for males than for females ($h_g^2 = 0.72$ and 0.37, respectively). Although the quantitative gender difference was not significant in the lower 10% of the distribution, evidence was obtained for *qualitative* gender differences in this group, suggesting that different genetic factors may influence the development of RD in boys and girls.

Wadsworth and DeFries (2005) extended the DF basic model to test for gender differences in the magnitude of genetic influences on RD among 8–20-year-olds participating in the CLDRC. In contrast to the findings of Harlaar et al. (2005), h_g^2 estimates were somewhat higher for females than for males (0.63 and 0.53, respectively), but the difference was nonsignificant ($p > 0.3$). A test for qualitative gender differences was also nonsignificant.

Although sex differences may exist for some measures of specific cognitive abilities and achievement, different studies have obtained conflicting results, most likely due to differences in subjects' ages, definition of the phenotype, and measures used. However, it is important to note that the number of tasks in which mean sex differences are substantial and/or significant is small. Males and females are remarkably similar in their performance on measures of cognitive abilities and there is greater variability within sex than between sexes (Halpern 2000). Furthermore, there is little support for a differential etiology as a function of sex, suggesting that the same genetic influences contribute to the variation in performance of both males and females along the full continuum of scores.

1.7 Conclusions

The study of the behavioral genetics of cognition in children has grown exponentially in recent years and our understanding of the etiologies of cognitive abilities and their relations to achievement has also greatly increased. For example, results of recent studies have demonstrated that:

1. Heritabilities for cognitive abilities in childhood and adolescence are substantial, accounting for up to 50% of the variance.
2. The contribution from genetic influences increases from childhood to early adulthood, whereas that from shared environmental influences decreases (see chaps. 4 and 5 for developmental patterns in middle and later adulthood).
3. Etiology at the extremes of ability appears to follow the same patterns of heritability as ability within the normal range.
4. There is substantial genetic overlap between cognitive abilities and measures of academic achievement, but there are also independent influences.
5. Although linkage and association studies have not located any genes with large effect, GCTA analyses have demonstrated that SNP sets account for most of the genetic variation estimated from twin and adoption studies.
6. There is little evidence for sex differences in general cognitive ability; sex differences in specific cognitive abilities exist for very few tasks and account for a very small proportion of the variance.

Although much more research in quantitative genetics is warranted, such findings will guide molecular genetic studies in the search for specific genes and chromosomal regions responsible for individual differences in cognitive abilities and disabilities.

1.8 Future Directions

With the rapid pace of recent developments in both quantitative and molecular genetics, future research regarding the genetic and environmental etiologies of individual differences in cognitive abilities and academic achievement during childhood and adolescence will almost certainly accelerate. Such research during these important developmental epochs holds great promise for facilitating a better understanding of developmental genetic issues. Moreover, behavioral genetics research can tell us as much about the environment as it does about genetics (Plomin et al. 2013). Our research questions are no longer limited to "whether" and "how much?" but now include "what?" and "how?" Thus, in order to exploit recent developments in quantitative and molecular genetics, commensurate advances in our measurement and analysis of the relevant environments will be necessary. The resulting improvement in our understanding of salient environmental influences could greatly facilitate analyses of the interplay between genetic and environmental influences, including both genotype-environmental correlations and interactions.

Finally, individual differences in behavior develop on a landscape of environmental influences, which could potentially be optimized for each individual. Thus, advances in our understanding of the genetic and environmental etiologies of individual differences in cognitive abilities and achievements has considerable relevance for both basic research in child development and applied educational issues.

Acknowledgments This work was supported in part by grants HD027802 and HD010333 from The Eunice Kennedy Shriver National Institute of Child Health and Human Development and by grant MH063207 from the National Institute of Mental Health.

References

Ardila, A., Rosselli, M., Matute, E., & Inozemtseva, O. (2011). Gender differences in cognitive development. *Dev Psychol, 47*(4), 984–990. doi:10.1037/a0023819

Arnold, L. E., Abikoff, H. B., Cantwell, D. P., Conners, C. K., Elliott, G., Greenhill, L. L., et al. (1997). National institute of mental health collaborative multimodal treatment study of children with ADHD (the MTA). Design challenges and choices. *Archives of General Psychiatry, 54*(9), 865–870.

Astrom, R. L., Wadsworth, S. J., & DeFries, J. C. (2007). Etiology of the stability of reading difficulties: the longitudinal twin study of reading disabilities. *Twin Research and Human Genetics, 10*(3), 434–439. doi:10.1375/twin.10.3.434

Astrom, R. L., Wadsworth, S. J., Olson, R. K., Willcutt, E. G., & DeFries, J. C. (2011). DeFries-Fulker analysis of longitudinal reading performance data from twin pairs ascertained for reading difficulties and from their nontwin siblings. *Behav Genet, 41*(5), 660–667. doi:10.1007/s10519-011-9445-6

Baker, L. A., Ho, H., & Reynolds, C. (1994). Sex differences in genetic and environmental influences for cognitive abilities. R. P. J. C. DeFries & D. W. Fulker (Ed.), *Nature and nurture during middle childhood* (pp. 181–200). Cambridge: Blackwell.

Bereiter, C. (1969). The future of individual differences. *Harvard Educational Review, 39,* 310–318.

Binet, A., & Simon, T. (1905). Upon the necessity of establishing a scientific diagnosis of the inferior states of intelligence. *L'Annee Psychologique, 11,* 163–191.

Binet, A., & Simon, T. (1916). New methods for the diagnosis of the intellectual level of subnormals. In the development of intelligence in children. New York: Williams & Wilkins.

Bishop, D. V. (2001). Genetic influences on language impairment and literacy problems in children: Same or different? *Journal of Child Psychology and Psychiatry and Allied Disciplines, 42*(2), 189–198.

Bishop, E. G., Cherny, S. S., Corley, R., Plomin, R., DeFries, J. C., & Hewitt, J. K. (2003). Development genetic analysis of general cognitive ability from 1 to 12 years in a sample of adoptees, biological siblings, and twins. *Intelligence, 31*(1), 31–49.

Bouchard, T. J., & McGue, M. (1981). Familial studies of intelligence: A review. *Science, 212,* 1055–1059.

Brant, A. M., Haberstick, B. C., Corley, R. P. Wadsworth, S. J., DeFries, J. C., & Hewitt, J. K. (2009). The developmental etiology of high IQ. *Behavior Genetics, 39*(4), 393–405. doi:10.1007/s10519-009-9268-x

Butcher, L. M., Davis, O. S., Craig, I. W., & Plomin, R. (2008). Genome-wide quantitative trait locus association scan of general cognitive ability using pooled DNA and 500K single nucleotide polymorphism microarrays. *Genes Brain Behav, 7*(4), 435–446. doi:10.1111/j.1601-183X.2007.00368.x

Cardon, L. R. (1994). Specific cognitive abilities. R. P. J. C. DeFries & D. W. Fulker (Ed.), *Nature and nurture during middle childhood* (pp. 57–76). Oxford: Blackwell.

Cardon, L. R., DiLalla, L. F., Plomin, R., DeFries, J. C., & Fulker, D. W. (1990). Genetic correlations between reading performance and IQ in the Colorado Adoption Project. *Intelligence, 14,* 245–257.

Cardon, L. R., Fulker, D. W., DeFries, J. C., & Plomin, R. (1992a). Continuity and change in general cognitive ability from 1 to 7 Years of Age. *Developmental Psychology, 28*(1), 64–73.

Cardon, L. R., Fulker, D. W., DeFries, J. C., & Plomin, R. (1992b). Multivariate genetic analysis of specific cognitive abilities in the Colorado Adoption Project at Age 7. *Intelligence, 16*(3-4), 383–400.

Cardon, L. R. Smith, S. D., Fulker, D. W., Kimberling, W. J., Pennington, B. F., & DeFries, J. C. (1994). Quantitative trait locus for reading disability on chromosome 6. *Science, 266,* 276–279.

Carroll, J. B. (1993). *Human cognitive abilities: A survey of factor analytic studies.* New York: Cambridge.

Cattell, R. B. (1963). Theory of fluid and crystallized intelligence: A critical experiment. *Journal of Educational Psychology, 54,* 1–22.

Chabris, C. F., Hebert, B. M., Benjamin, D. J., Beauchamp, J. P., Cesarini, D., van der Loos, M. J. H. M., et al. (in press). Most reported genetic associations with general intelligence are probably false positives. *Psychological Science, 23*(11), 1314–1323

Cherny, S. S., & Cardon, L. R. (1994). General cognitive ability. J. C. DeFries, R. Plomin, & D. W. Fulker (Eds.), *Nature and nurture during middle childhood* (pp. 46–56). Oxford: Blackwell.

Cherny, S. S., Fulker, D. W., & Hewitt, J. K. (1997). Cognitive development from infancy to middle childhood. R. J. S. E. L. Grigorenko (Ed.), *Intelligence: Heredity and environment* (pp. 463–482). New York: Cambridge University Press.

Connolly, A. J., Nachtman, W., & Pritchett, E. M. (1976). *Key Math Diagnostic Arithmetic Test.* Circle Pines: American Guidance Service.

Davies, G., Tenesa, A., Payton, A., Yang, J., Harris, S. E., Liewald, D., et al. (2011). Genome-wide association studies establish that human intelligence is highly heritable and polygenic. *Mol Psychiatry, 16*(10), 996–1005. doi:10.1038/mp.2011.85

Davis, O. S., Butcher, L. M., Docherty, S. J., Meaburn, E. L., Curtis, C. J., Simpson, M. A., et al. (2010). A three-stage genome-wide association study of general cognitive ability: hunting the small effects. *Behav Genet, 40*(6), 759–767. doi:10.1007/s10519-010-9350-4

Deary, I. J., Johnson, W., & Houlihan, L. M. (2009). Genetic foundations of human intelligence. *Hum Genet, 126*(1), 215–232. doi:10.1007/s00439-009-0655-4

Deary, I. J., Yang, J., Davies, G., Harris, S. E., Tenesa, A., Liewald, D., et al. (2012). Genetic contributions to stability and change in intelligence from childhood to old age. *Nature, 482*(7384), 212–215. doi:10.1038/nature10781

DeFries, J. C. (1985). Colorado reading project. In D. B. Gray & J. F. Kavanagh (Eds.), *Biobehavioral measures of dyslexia* (pp. 107–122). Parkton, MD: York Press. (Reprinted from: NOT IN FILE).

DeFries, J. C., Filipek, P. A., Fulker, D. W., Olson, R. K., Pennington, B. F., Smith, S. D., & Wise, B. W. (1997). Colorado learning disabilities research center. *Learning Disabilities: A Multidisciplinary Journal, 8,* 7–19.

DeFries, J. C., & Fulker, D. W. (1985). Multiple regression analysis of twin data. *Behavior Genetics, 15,* 467–473.

DeFries, J. C., & Fulker, D. W. (1988). Multiple regression analysis of twin data: Etiology of deviant scores versus individual differences. *Acta Geneticae Medicae et Gemellologiae: Twin Research, 37,* 205–216.

DeFries, J. C., Johnson, R. C., Kuse, A. R., McClearn, G. E., Polovina, J., Vandenberg, S. G., & Wilson, J. R. (1979). Familial resemblance for specific cognitive abilities. *Behavior Genetics, 9*(1), 23–43.

DeFries, J. C., Knopik, V. S., & Wadsworth, S. J. (1999). Colorado twin study of reading disability. In D. D. Duane (Ed.), *Reading and attention disorders: Neurobiological correlates* (pp. 17–41). Baltimore: York Press.

DeFries, J. C., Olson, R. K., Pennington, B. F., & Smith, S. D. (1991). Colorado reading project: Past, present, and future. *Learning Disabilities: A Multidisciplinary Journal, 2,* 37–46.

DeFries, J. C., Plomin, R., & Fulker, D. W. (1994). *Nature and nurture during middle childhood.* Oxford: Blackwell Publishers.

Docherty, S. J., Davis, O. S. P., Kovas, Y., Meaburn, E. L., Dale, P. S., Petrill, S. A., et al. (2010). A genome-wide association study identifies multiple loci associated with mathematics ability and disability. *Genes Brain and Behavior, 9*(2), 234–247. doi:10.1111/j.1601-183X.2009.00553.x

Docherty, S. J., Kovas, Y., & Plomin, R. (2011). Gene-environment interaction in the etiology of mathematical ability using SNP sets. *Behavior Genetics, 41,* 141–154.

Doyle, A. E., Ferreira, M. A., Sklar, P. B., Lasky-Su, J., Petty, C., Fusillo, S. J., et al. (2008). Multivariate genome-wide linkage scan of neurocognitive traits and ADHD symptoms: Suggestive linkage to 3q13. *American Journal of Medical Genetics Part B (Neuropsychiatric Genetics), 147B*(8), 1399–1411. doi:10.1002/ajmg.b.30868

Dunn, L. M., & Markwardt, F. C. (1970). *Examiner's manual: Peabody individual achievement test.* Circle Pines: American Guidance Service.

Eichler, E. E., Flint, J., Gibson, G., Kong, A., Leal, S. M., Moore, J. H., & Nadeau, J. H. (2010). Missing heritability and strategies for finding the underlying causes of complex disease. *Nat Rev Genet, 11*(6), 446–450. doi:10.1038/nrg2809

Emde, R. N., & Hewitt, J. K. (2001). *Infancy to early childhood: Genetic and environmental influences on developmental change.* Oxford: Oxford University Press.

Erlenmeyer-Kimling, L., & Jarvik, L. F. (1963). Genetics and intelligence: A review. *Science, 142,* 1477–1479.

Friend, A., DeFries, J. C., & Olson, R. K. (2008). Parental education moderates genetic influences on reading disability. *Psychological Science, 19*(11), 1124–1130. doi:10.1111/j.1467-9280.2008.02213.x

Friend, A., DeFries, J. C., Olson, R. K., Pennington, B. F., Harlaar, N., Byrne, B., et al. (2009). Heritability of high reading ability and its interaction with parental education. *Behavior Genetics, 39*(4), 427–436. doi:10.1007/s10519-009-9263-2

Fulker, D. W., & Cherny, S. S. (1995). Genetic and environmental influences on cognition during childhood. *Population Research and Policy Review, 14,* 283–300.
Fulker, D. W., Cherny, S. S., & Cardon, L. R. (1993). Continuity and change in cognitive development. In R. P. G. E. McClearn (Ed.), *Nature, nurture and psychology* (pp. 77–98). Washington, DC: American Psychological Association.
Fulker, D. W., DeFries, J. C., & Plomin, R. (1988). Genetic influence on general mental ability increases between infancy and middle childhood. *Nature, 336,* 767–769.
Galton, F. (1869). *Hereditary genius: An inquiry into its laws and consequences.* London: Macmillan.
Gardner, H. (2006). *Multiple intelligences: New Horizons.* New York: Basic Books.
Gill, C. E., Jardine, R., & Martin, N. G. (1985). Further evidence for genetic influences on educational achievement. *British Journal of Educational Psychology, 55,* 240–250.
Gottfredson, L. S. (1997). Why g matters: The complexity of everyday life. *Intelligence, 23*(1), 79–132.
Gottfredson, L. S., & Deary, I. J. (2004). Intelligence predicts health and longevity, but why? *Current Directions in Psychological Science, 13*(1), 1–4.
Haier, A. J. (2011). Biological basis of intelligence. R. J. Sternberg & S. B. Kaufman (Eds.), *The Cambridge handbook of intelligence* (pp. 351–370). New York: Cambridge University Press.
Hallgren, B. (1950). Specific dyslexia (Congenital word-blindness): A clinical and genetic study. *Acta Psychiatrica et Neurologica Scandinavica, 65* (Supplement), 1–287.
Halpern, D. F. (2000). *Sex differences in cognitive abilities* (3rd ed.) Mahwah: Lawrence Earlbaum Associates.
Harlaar, N., Dale, P. S., & Plomin, R. (2007). From learning to read to reading to learn: Substantial and stable genetic influence. *Child Development, 78*(1), 116–131.
Harlaar, N., Spinath, F. M., Dale, P. S., & Plomin, R. (2005). Genetic influences on early word recognition abilities and disabilities: a study of 7-year-old twins. *Journal of Child Psychology and Psychiatry, 46*(4), 373–384. doi:10.1111/j.1469-7610.2004.00358.x
Hawke, J. L., Olson, R. K., Willcutt, E. G., Wadsworth, S. J., & DeFries, J. C. (2009). Gender ratios for reading difficulties. *Dyslexia, 15*(3), 239–242. doi:10.1002/dys.389
Hawke, J. L., Wadsworth, S. J., & DeFries, J. C. (2006). Genetic influences on reading difficulties in boys and girls: The Colorado Twin Study. *Dyslexia: An International Journal of Research and Practice, 12,* 21–29.
Hawke, J. L., Wadsworth, S. J., Olson, R. K., & DeFries, J. C. (2007). Etiology of reading difficulties as a function of gender and severity. *Reading and Writing, 20*(1–2), 13–25.
Haworth, C. M., Wright, M. J., Martin, N. W., Martin, N. G., Boomsma, D. I., Bartels, M., et al. (2009a). A twin study of the genetics of high cognitive ability selected from 11,000 twin pairs in six studies from four countries. *Behavior Genetics, 39*(4), 359–370. doi:10.1007/s10519-009-9262-3
Haworth, C. M. A., Dale, P. S., & Plomin, R. (2009b). Generalist genes and high cognitive ability. *Behavior Genetics, 39,* 437–445.
Haworth, C. M., Wright, M. J., Luciano, M., Martin, N. G., de Geus, E. J., van Beijsterveldt, C. E., et al. (2010). The heritability of general cognitive ability increases linearly from childhood to young adulthood. *Mol Psychiatry, 15*(11), 1112–1120. doi: mp200955[pii]10.1038/mp.2009.55
Hayiou-Thomas, M. E., Dale, P. S., & Plomin, R. (2012). The etiology of variation in language skills changes with development: a longitudinal twin study of language from 2 to 12 years. *Developmental Science, 15*(2), 233–249.
Hyde, J. S., Lindberg, S. M., Linn, M. C., Ellis, A. B., & Williams, C. C. (2008). Gender similarities characterize math performance. *Science, 321,* 494–495.
Hyde, J. S., & Linn, M. C. (1988). Gender differences in verbal ability: A meta analysis. *Psychological Bulletin, 104*(1), 53–69.
Iacono, W. G., Carlson, S. R., Taylor, J. T., Elkins, I. J., & McGue, M. (1999). Behavioral disinhibition and the development of substance use disorders: Findings from the Minnesota Twin Family Study. *Development and Psychopathology, 11,* 869–900.
Jensen, A. R. (1967). Estimation of the limits of heritability of traits by comparison of monozygotic and dizygotic twins. *Proceedings of the National Academy of Sciences, 58,* 149–157.

Jensen, A. R. (1969a). How much can we boost IQ and scholastic achievement? *Harvard Educational Review, 39,* 1–123.

Jensen, A. R. (1969b). Reducing the heredity-environment uncertainty. *Harvard Educational Review, 39,* 449–483.

Jensen, A. R. (1972). *Genetics and education*. New York: Harper and Row.

Kagan, J. S. (1969). Inadequate evidence and illogical conclusions. *Harvard Educational Review, 39,* 274–277.

Kirkpatrick, R. M., McGue, M., & Iacono, W. (2009). Shared environmental contributions to high cognitive ability. *Behavior Genetics, 39,* 406–416.

Koeppen-Schomerus, G., Spinath, F. M., & Plomin, R. (2003). Twins and non-twin siblings: different estimates of shared environmental influence in early childhood. *Twin Research, 6,* 97–105.

Kovas, Y., Haworth, C. M., Petrill, S. A., & Plomin, R. (2007). Mathematical ability of 10-year-old boys and girls: genetic and environmental etiology of typical and low performance. *Journal of Learning Disabilities, 40*(6), 554–567. doi:10.1177/00222194070400060601

Kutalik, Z., Whittaker, J., Waterworth, D., Beckmann, J. S., & Bergmann, S. (2011). Novel method to estimate the phenotypic variation explained by genome-wide association studies reveals large fraction of the missing heritability. *Genet Epidemiol, 35*(5), 341–349. doi:10.1002/gepi.20582

Light, J. G., & DeFries, J. C. (1995). Comorbidity of reading and mathematics disabilities: genetic and environmental etiologies. *Journal of Learning Disabilities, 28*(2), 96–106.

Loehlin, J. C., & Nichols, R. C. (1976). *Heredity, environment, and personality: A study of 850 sets of twins*. Austin: University of Texas Press.

Loo, S. K., Shtir, C., Doyle, A. E., Mick, E., McGough, J. J., McCracken, J., et al. (2012). Genome-wide association study of intelligence: Additive effects of novel brain expressed genes. *Journal of the American Academy of Child and Adolescent Psychiatry, 51*(4), 432–440.

Luciano, M., Wright, M. J., Duffy, D. L., Wainwright, M. A., Zhu, G., Evans, D. M., et al. (2006). Genome-wide scan of IQ finds significant linkage to a quantitative trait locus on 2q. *Behav Genet, 36*(1), 45–55. doi:10.1007/s10519-005-9003-1

Luo, D. S., Thompson, L. A., & Detterman, D. K. (2003). Phenotypic and behavioral genetic covariation between elemental cognitive components and scholastic measures. *Behavior Genetics, 33*(3), 221–246.

Maccoby, E. E., & Jacklin, C. N. (1974). *The psychology of sex differences*. Stanford: Stanford University Press.

Martin, N. G. (1975). The inheritance of scholastic abilities in a sample of twins. II. Genetical analysis of examination results. *Annals of Human Genetics, 39,* 219–229.

McArdle, J. J. (2006). Latent curve analyses of longitudinal twin data using a mixed-effects biometric approach. *Twin Research and Human Genetics, 9*(3), 343–359.

McArdle, J. J., Prescott, C. A., Hamagami, F., & Horn, J. L. (1998). A contemporary method for developmental-genetic analyses of age changes in intellectual abilities. *Developmental Neuropsychology, 14,* 69–114.

McGrew, K. S. (Ed.). (2005). *The Cattell-Horn-Carroll theory of cognitive abilities: Past, present, and future*. New York: Guilford Press.

McGue, M., Bouchard, T. J. J., Iacono, W. G., & Lykken, D. T. (1993). Behavioral genetics of cognitive ability: A life-span perspective. R. P. G. E. McClearn (Ed.), *Nature, Nurture, and Psychology* (pp. 59–76). Washington D.C.: American Psychological Association Press.

Meaburn, E. L., Harlaar, N., Craig, I. W., Schalkwyk, L. C., & Plomin, R. (2008). Quantitative trait locus association scan of early reading disability and ability using pooled DNA and 100K SNP microarrays in a sample of 5760 children. *Mol Psychiatry, 13*(7), 729–740. doi:10.1038/sj.mp.4002063

Medland, S. E., Wright, M. J., Geffen, G. M., Hay, D. A., Levy, F., Martin, N. G., & Duffy, D. L. (2003). Special twin environments, genetic influences and their effects on the handedness of twins and their siblings. *Twin Research, 6,* 119–130.

Neale, M. C., & Cardon, L. R. (1992). *Methodology for genetic studies of twins and families*. Dordrecht: Kluwer Academic Press.

Neale, M. C., & Maes, H. H. M. (2002). *Methodology for genetic studies of twins and families*. Dordrecht: Kluwer Academic Publishers.

Neale, M. C., & McArdle, J. J. (2000). Structured latent growth curves for twin data. *Twin Research, 3,* 165–177.

Petrill, S., Plomin, R., DeFries, J. C., & Hewitt, J. K. (Ed.), (2003). *Nature, nurture, and the transition to adolescence*. New York: Oxford University Press.

Petrill, S. A., & Deater-Deckard, K. (2004). The heritability of general cognitive ability: A within-family adoption design. *Intelligence, 32*(4), 403–409. doi101016/j.intell.2004.05.001

Petrill, S. A., Kovas, Y., Hart, S. A., Thompson, L. A., & Plomin, R. (2009). The genetic and environmental etiology of high math performance in 10-year-old Twins. *Behav Genet 39,* 371–379.

Petrill, S. A., Lipton, P. A., Hewitt, J. K., Plomin, R., Cherny, S. S., Corley, R., & DeFries, J. C. (2004). Genetic and environmental contributions to general cognitive ability through the first 16 years of life. *Developmental Psychology, 40,* 805–812.

Petrill, S. A., Saudino, K., Cherny, S. S., Emde, R. N., Fulker, D. W., Hewitt, J. K., & Plomin, R. (1998). Exploring the genetic and environmental etiology of high general cognitive ability in fourteen- to thirty-six-month-old twins. *Child Development, 69*(1), 68–74.

Petrill, S. A., Saudino, K., Cherny, S. S., Emde, R. N., Hewitt, J. K., Fulker, D. W., & Plomin, R. (1997). Exploring the genetic etiology of low general cognitive ability from 14 to 36 months. *Developmental Psychology, 33*(3), 544–548.

Plomin, R. (2012). Genetics: How intelligence changes with age. *Nature, 482*(7384), 165–166. doi:10.1038/482165a

Plomin, R., & DeFries, J. C. (1980). Genetics and intelligence: Recent data. *Intelligence, 4,* 15–24.

Plomin, R., & DeFries, J. C. (1985). *Origins of individual differences in infancy: The Colorado adoption project*. New York: Academic Press.

Plomin, R., & DeFries, J. C. (1998). The genetics of cognitive abilities and disabilities. *Scientific American, 278*(5), 62–69.

Plomin, R., DeFries, J. C., & Fulker, D. W. (1988). *Nature and nurture during infancy and early childhood*. New York: Cambridge University Press.

Plomin, R., DeFries, J. C., Knopik, V. S., & Neiderhiser, J. M. (2013). *Behavioral genetics* (6th ed.). New York: Worth Publishers.

Plomin, R., Fulker, D. W., Corley, R., & DeFries, J. C. (1997). Nature, nurture, and cognitive development from 1 to 16 years: A parent-offspring adoption study. *Psychological Science, 8*(6), 442–447.

Plomin, R., & Haworth, C. M. A. (2009). Genetics of high cognitive abilities. *Behavior Genetics, 39,* 347–349.

Plomin, R., Hill, L., Craig, I. W., McGuffin, P., Purcell, S., Sham, P., et al. (2001). A genome-wide scan of 1842 DNA markers for allelic associations with general cognitive ability: a five-stage design using DNA pooling and extreme selected groups. *Behavior Genetics, 31*(6), 497–509.

Plomin, R., & Kovas, Y. (2005). Generalist genes and learning disabilities. *Psychological Bulletin, 131*(4), 592–617. doi:10.1037/0033-2909.131.4.592

Plomin, R. A. D., DeFries, J. C. (1979). Multivariate behavior genetic analysis of twin data on scholastic abilities. *Behavior Genetics, 9,* 505–517.

Posthuma, D., Luciano, M., Geus, E. J., Wright, M. J., Slagboom, P. E., Montgomery, G. W., et al. (2005). A genomewide scan for intelligence identifies quantitative trait loci on 2q and 6p. *Am J Hum Genet, 77*(2), 318–326. doi:10.1086/432647

Reynolds, C. A., Hewitt, J. K., Erickson, M. T., Silberg, J. L., Rutter, M., Simonoff, E., et al. (1996). The genetics of children's oral reading performance. *Journal of Child Psychology and Psychiatry and Allied Disciplines, 37*(4), 425–434.

Risch, N., & Merikangas, K. (1996). The future of genetic studies of complex human diseases. *Science, 273*(5281), 1516–1517.

Rose, L. T., & Fischer, K. W. (2011).Intelligence in childhood. R. J. Sternberg & S. B. Kaufman (Eds.), *The Cambridge handbook of intelligence* (pp. 144–173). New York: Cambridge University Press.

Sattler, J. M. (1992). *Assessment of children*. San Diego: Jerome M. Sattler.

Smith, S. D. (2010). Learning disabilities. In J. Nurnberger & W. Berrettini (Eds.), *Psychiatric Genetics*. Cambridge: Cambridge University Press.

Snyderman, M., & Rothman, S. (1987). Survey of expert opinion on intelligence and aptitude testing. *American Psychologist, 42,* 137–144.

Spearman, C. E. (1923). *The nature of intelligence and the principles of cognition.* London: Macmillan.

Spearman, C. E. (1927). *The abilities of man.* New York: Macmillan.

Sternberg, R. (2012). Intelligence. *Dialogues in Clinical Neuroscience, 14,* 19–27.

Sternberg, R. J. (1985). *Beyond IQ: A triarchic theory of human intelligence.* New York: Cambridge University Press.

Sternberg, R. J., & Kauffman, S. B. (2011). *The cambridge handbook of intelligence.* New York: Cambridge University Press.

Stevenson, J., Graham, P., Fredman, G., & McLoughlin, V. (1987). A twin study of genetic influences on reading and spelling ability and disability. *Journal of Child Psychology and Psychiatry, 28*(2), 229–247.

Terman, L. M., & Merrill, M. A. (1973). *Stanford-Binet Intelligence Scale: 1972 Norms edition.* Boston: Houghton Mifflin.

Thomas, C. J. (1905). Congenital "word-blindness" and its treatment. *Ophthalmoscope, 3,* 380–385.

Thompson, L. A., Detterman, D. K., & Plomin, R. (1991). Association between cognitive abilities and scholastic achievement: Genetic overlap, but environmental differences. *Psychological Science, 2,* 158–165.

Thorndike, E. L. (1927). *The measurement of intelligence.* New York: Bureau of Publications, Teachers College, Columbia University.

Thurstone, L. L. (1938). Primary mental abilities. *Psychometric Monographs, Vol. 1.* ix + 121.

Thurstone, L. L., & Thurstone, T. G. (1941). Factorial studies of intelligence. *Psychometric Monographs, 2,* 1–94.

van Soelen, I. L., Brouwer, R. M., van Leeuwen, M., Kahn, R. S., Hulshoff Pol, H. E., & Boomsma, D. I. (2011). Heritability of verbal and performance intelligence in a pediatric longitudinal sample. *Twin Research and Human Genetics, 14*(2), 119–128. doi:10.1375/twin.14.2.119

Wadsworth, S. J. (1994). School achievement. In R. P. J. C. DeFries & D. W. Fulker (Ed.), *Nature and nurture during middle childhood* (pp. 86–101). Oxford: Blackwell.

Wadsworth, S. J., Corley, R. P., Hewitt, J. K. Plomin, R., & DeFries, J. C. (2002). Parent-offspring resemblance for reading performance at 7, 12, and 16 years of age in the Colorado Adoption Project. *Journal of Child Psychology and Psychiatry and Allied Disciplines, 43,* 769–774.

Wadsworth, S. J., Corley, R. P., Plomin, R., Hewitt, J. K., & DeFries, J. C. (2006). Genetic and environmental influences on continuity and change in reading achievement in the Colorado Adoption Project. In A. H. M. Ripke (Ed.), *Developmental contexts of middle childhood: Bridges to adolescence and adulthood* (pp. 87–106). New York: Cambridge University Press.

Wadsworth, S. J., & DeFries, J. C. (2005). Genetic etiology of reading difficulties in boys and girls. *Twin Research and Human Genetics, 8*(6), 594–601. doi:10.1375/183242705774860196

Wadsworth, S. J., DeFries, J. C., Fulker, D. W., & Plomin, R. (1995a). Cognitive ability and academic achievement in the Colorado Adoption Project: A multivariate genetic analysis of parent-offspring and sibling data. *Behavior Genetics, 25*(1), 1–15.

Wadsworth, S. J., DeFries, J. C., Fulker, D. W., Olson, R. K., & Pennington, B. F. (1995b). Reading performance and verbal short-term memory: A twin study of reciprocal causation. *Intelligence, 20,* 145–167.

Wadsworth, S. J., Olson, R. K., & DeFries, J. C. (2010). Differential genetic etiology of reading difficulties as a function of IQ: An update. *Behavior Genetics, 40*(6), 751–758.

Wadsworth, S. J., Olson, R. K., Willcutt, E. G., & DeFries, J. (2011). Multiple regression analysis of reading performance data from twin pairs with reading difficulties and nontwin siblings: The augmented model. *Twin Research and Human Genetics, 15,* 116–119.

Wadsworth, S. J., Olson, R. K., Willcutt, E. G., & DeFries, J. C. (2012). Multiple regression analysis of reading performance data from twin pairs with reading difficulties and nontwin siblings: The augmented model. *Twin Research and Human Genetics, 15*(1), 116–119.

Wasserman, J. D., & Tulsky, D. S. (2005). A history of intelligence assessment. D. P. Flanagan & P. L. Harrison (Eds.), *Contemporary intellectual assessment: Theories, tests and issues* (2nd ed., pp. 3–22). New York: Guilford Press.

Wechsler, D. (1958). *The measurement and appraisal of adult intelligence*. Baltimore: Williams and Wilkins.

Wechsler, D. (1974). *Manual for the Wechsler Intelligence scale for children, revised*. New York: The Psychological Corporation.

Wechsler, D. (1981). *Examiner's manual: Wechsler adult intelligence scale-revised*. New York: The Psychological Corporation.

Wechsler, D. (1991). *Manual for the Wechsler intelligence scale for children* (3rd ed.). San Antonio: The Psychological Corporation.

Wechsler, D. (2002). *Examiner's manual: The Wechsler individual achievement test—2nd edition*. San Antonio: The Psychological Corporation.

Wechsler, D. (2003). *Manual for the Wechsler Intelligence scale for children* (4th ed.). San Antonio: The Psychological Corporation.

Willis, J. O., Dumont, R., & Kaufman, A. S. (2011). Factor-analytic models of intelligence. R. J. Sternberg & S. B. Kaufman (Eds.), *The Cambridge handbook of intelligence* (pp. 39–57). New York: Cambridge University Press.

Wilson, R. S. (1983). The Louisville Twin Study: developmental synchronies in behavior. *Child Development, 54*(2), 298–316.

Woodcock, R. W., McGrew, K. S., & Mather, N. (2001). *Woodcock-Johnson III tests of achievement*. Itasca: Riverside Publishing.

Yang, J., Lee, S. H., Goddard, M. E., & Visscher, P. M. (2011). GCTA: a tool for genome-wide complex trait analysis. *American Journal of Human Genetics, 88*(1), 76–82. doi:101016/j.ajhg.2010.11.011

Zuk, O., Hechte, E., Sunyaev, S. R., & Lander, E. S. (2012). The mystery of missing heritability: Genetic interactions create phantom heritability. *Proceedings of the National Academy of Sciences U S A, 109*(4), 1193–1198.

Chapter 2
Interactions Between Socioeconomic Status and Components of Variation in Cognitive Ability

Eric Turkheimer and Erin E. Horn

In 2003, our lab published a paper demonstrating that the heritability of intelligence in 7-year-old children from the National Collaborative Perinatal Project was moderated by parental socioeconomic status (SES; Turkheimer et al. 2003). Among children raised in poor homes, identical (MZ) twins were no more correlated than fraternal (DZ) twins, heritability was close to zero, and the family environment accounted for more than half the variance; in children raised in middle-class or better homes, heritability was substantial and the effect of family environment approached zero. This finding was itself a "replication": the effect had been first reported by Sandra Scarr in the 1970s, was met with some methodological resistance at the time (Eaves and Jinks 1972), was placed on the back burner for 20 years, and then decisively replicated by Rowe et al. (1999). We have suggested that the effect be referred to as the "Scarr-Rowe" interaction (Turkheimer et al. 2009). It is not possible to be disinterested when discussing replications of one's own work, but we will do our best to do so in the current chapter. We reviewed the existing evidence in detail as recently as 2009 (Turkheimer et al. 2009), and somewhat more briefly in 2012 (Nisbett et al. 2012). We will do so again in this chapter, with emphasis on the most recent evidence. We will also dig a little deeper into the phenomenon itself.

E. Turkheimer (✉) · E. E. Horn
Department of Psychology, University of Virginia, PO Box 400400,
Charlottesville, VA 22904-4400, USA
e-mail: turkheimer@virginia.edu

E. E. Horn
e-mail: horn@virginia.edu

D. Finkel, C. A. Reynolds (eds.), *Behavior Genetics of Cognition Across the Lifespan*,
Advances in Behavior Genetics 1, DOI 10.1007/978-1-4614-7447-0_2,

2.1 Methodological Considerations

Heritabilities, let alone interactions with heritabilities, are complex quantities. A (broad) heritability is a standardized variance component, the proportion of phenotypic variance accounted for by variation in genotype. In its simplest form, heritability is computed as twice the difference between the intraclass correlations (ICCs) for MZ and DZ twins. ICCs, in turn, are ratios of within-pair and between-pair variances. So heritabilities are computed from zygosity-based differences in ratios of within- and between-pair variances, and interactions with heritabilites are about how differences between ratios vary as a function of linear or nonlinear changes in an exogenous variable.

With that in mind, an increase in heritability as a function of SES could result if MZ twin correlations increase with SES, if DZ correlations decrease, or both. Any such changes in correlations could occur because within-pair variances decrease or between-pair variances increase with SES in MZ pairs, because of the converse in DZ pairs, or because of some combination of the two. In addition, any model of variance changes in biometric components as a function of SES will also imply a model of the total phenotypic variance. Unpacking these effects is probably of greater developmental importance than focusing on the end product of standardized heritability coefficients, which are extremely hard to interpret substantively under the best of circumstances (Turkheimer and Harden, in press).

2.2 National Collaborative Perinatal Project (NCPP)

We will begin by reanalyzing some of the data we reported from the NCPP. Details of the sample can be found in our original report (Turkheimer et al. 2003). There were 114 MZ pairs and 205 DZ pairs, with abbreviated Wechsler Intelligence Scale for Children (WISC) scores obtained at 7 years of age. SES was measured as a weighted composite of parental income, education, and occupational level, and was the same for both members of the pair. We reported the quadratic effects of SES on the unstandardized and standardized variances of components attributable to the additive effect of genotype (A), environmental effects shared among siblings (C), and environmental effects unshared among siblings (E). A model including interactions between SES and each of the three ACE terms fit significantly better than a model with main effects only for Full Scale IQ (FSIQ) and Performance IQ (PIQ) but not for Verbal IQ (VIQ). For both standardized and unstandardized models of FSIQ and PIQ, the A term was near zero for children raised in the poorest homes, while the C term accounted for almost all the variation. In the most affluent homes, the situation was reversed, and the trend lines for the A and C components crossed at a level of SES corresponding to lower-middle-class homes.

We will use a slightly different, and less parameterized, model here. For Twin *i* in Pair *j* and Zygosity Group *k*, an IQ score y_{ijk} can be represented as a pair mean b_{0jk}, plus within-pair variance around the pair mean σ_k^2:

$$y_{ijk} = b_{0jk} + \sigma_k^2$$

The pair intercepts b_{0jk} can then be modeled as a population mean β_0 and between-pair variance τ_k^2, the variance of the pair means around the population mean. A classical twin model involves reparameterizing the within- and between-pair variances in MZ and DZ twins as the familiar ACE components: additive genetic (A), shared (C), and nonshared environmental (E) effects. In the MZ twins, the within-pair variance is equal to

$$\sigma_{MZ}^2 = E$$

and the between-pair variance is equal to

$$\tau_{MZ}^2 = A + C$$

In the DZ pairs, the within-pair variance equals

$$\sigma_{DZ}^2 = \frac{1}{2}A + E$$

And the between-pair variance equals

$$\tau_{DZ}^2 = \frac{1}{2}A + C$$

With the additional constraint that the total phenotypic variances are equal in the MZ and DZ pairs, these equations can be solved for the ACE variances. Expressing the ACE variances as proportions of their sum gives the familiar standardized ACE coefficients.

A biometric interaction model entails expressing the within- and between-pair variances, or the standardized or unstandardized ACE variances derived from them, as some function of a moderator variable, in this case, SES. We prefer an exponential function rather than a quadratic one as a model of the variances. Exponential models share with quadratic models the desirable property of being positive, but have the additional advantage of being monotonic uniformly increasing or decreasing with respect to the moderator. Quadratic models of variances are by definition parabolic with respect to the moderator, and once again, biometric interaction

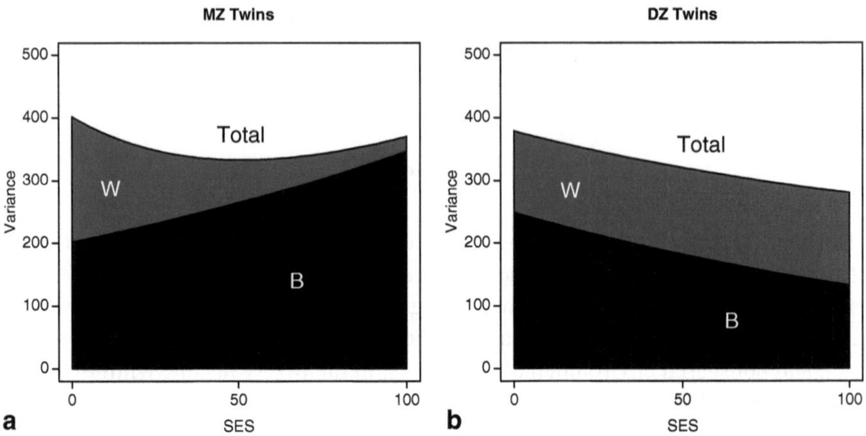

Fig. 2.1 Unstandardized between-(*B*), within-(*W*), and total variance in IQ for **a** identical (*MZ*) and **b** fraternal (*DZ*) twin pairs in the National Collaborative Perinatal Project (NCPP). *SES* socioeconomic status

models are difficult enough to explain without having to account for why a biometric variance first increases, and then decreases, as a function of SES.

In its least parameterized form, the model results in eight parameters, describing an intercept and log-linear effect of SES on the between- and within-pair variances in MZ and DZ twin pairs. The intercept tests for the magnitude of the between- and within-pair variances at SES=0; the log-linear slopes test the interactions between the variances and familial SES. We fit the model in Mplus 6.12 (Muthén and Muthén 2011), using full-information maximum likelihood estimation. As was the case in the original report, the individual interaction terms did not reach statistical significance, but the omnibus test of a model with all four interactions fit significantly better than a main-effects model ($p=.033$).

Since all the results we will report in this paper have been reported before, and because our main goal is comparative description of results across studies, we will focus on graphical presentation. The results for the between and within variances are illustrated in Fig. 2.1a and b. In this figure, which we will use extensively, the between and within or ACE variances are plotted as an exponential function of SES and stacked on top of each other, so the model for the total variance is evident as well as the model for the individual components. In this instance, we can see that the largest effects are for the MZ pairs, for whom differences within pairs decrease as a function of SES, while differences between pairs increase. To a lesser extent, the converse is true for the DZ pairs.

The ratio of between-pair variance to total phenotypic variance defines an ICC that can also vary as a function of SES. Figure 2.2 shows the result. As would be expected on the basis of the between and within variances, the ICC increases dramatically for the MZ pairs and decreases somewhat less dramatically for the DZ pairs. Finally, the ICCs can be combined with some identifying assumptions (equal

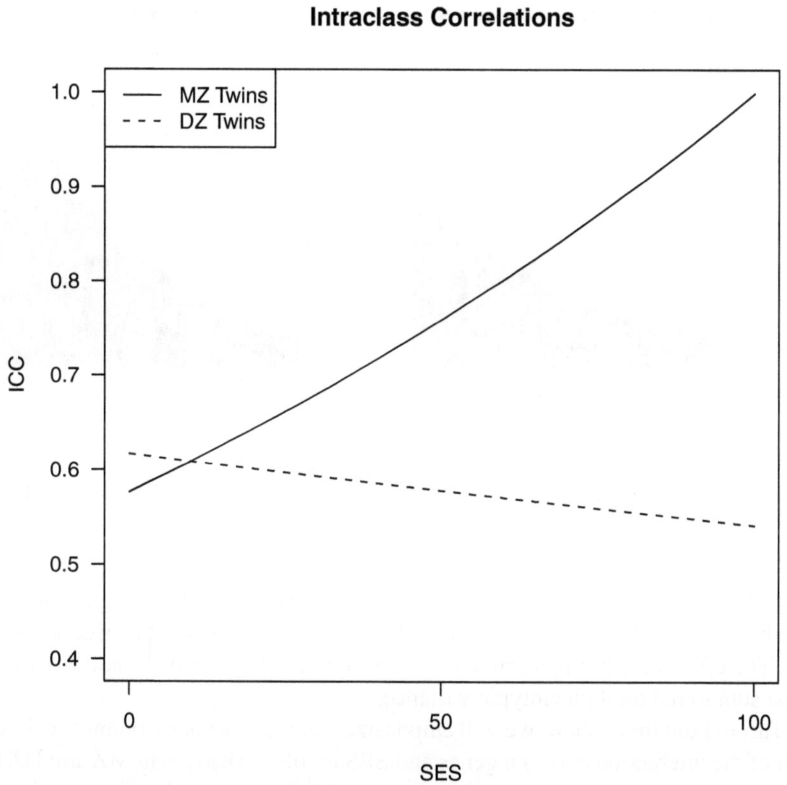

Fig. 2.2 Intraclass correlations (*ICCs*) of IQ for identical (*MZ*) and fraternal (*DZ*) twin pairs in the National Collaborative Perinatal Project (NCPP). *SES* socioeconomic status

phenotypic variance in MZ and DZ pairs, equal environmental similarity in MZ and DZ pairs, twice as much genetic similarity in MZ pairs) to parameterize the model in terms of either unstandardized or standardized variances attributed to additive genetic (A), familial environmental (C), and unique environmental (E) variances. As illustrated in Fig. 2.3a and b, these results are similar to those we originally reported. Since the total phenotypic variance has been set equal in the MZ and DZ pairs in the unstandardized ACE model illustrated in Fig. 2.3a, this is a good place to observe that the model of the total variance decreases slightly as a function of increasing SES.

We conclude that as familial SES increases, MZ twins tend to become more similar to each other, against a background of increases in the magnitude of differences between pairs. Differences within DZ pairs do not change substantially as a function of SES, whereas the magnitude of differences between pairs decreases. These between- and within-pair relations with SES can also be expressed as an increase in A variance with increasing SES, a corresponding decrease in C and E variance, or both. As will be the case in many of the studies we will review here, this

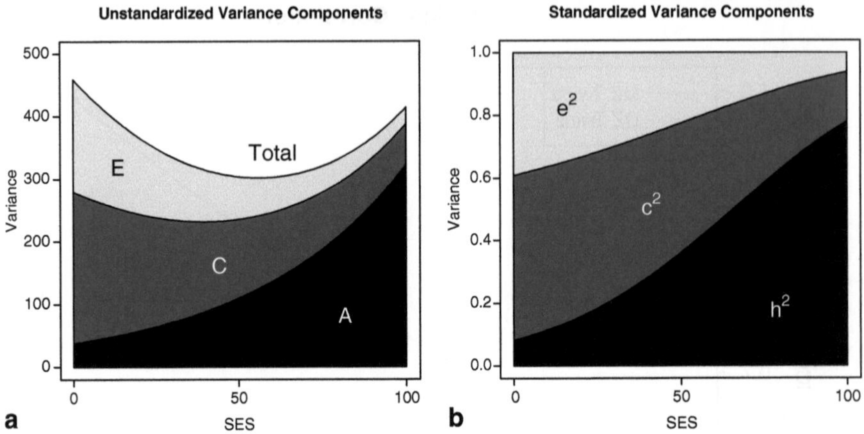

Fig. 2.3 **a** Unstandardized and **b** standardized additive genetic (*A*; h^2), shared environmental (*C*; c^2), and nonshared environmental (*E*; e^2) sources of variance in IQ in the National Collaborative Perinatal Project (NCPP). *SES* socioeconomic status

study lacks the statistical power to discriminate between changes in genetic effects as a function of SES in one direction and changes in environmental effects in the other. The effects are highly correlated, because ultimately the A, C, and E variances must sum to the total phenotypic variance.

Throughout this review, we will emphasize that the most basic biometric description of the interaction between genes and SES involves changes in MZ and DZ twin similarities and differences as a function of SES. Parameterizing these differences in terms of ACE parameters, although useful in many circumstances, involves making identifying assumptions that are sometimes viewed as controversial (Charney 2012), and introduces a level of abstraction about "genes" and "environment" that comes in between the observed data (involving changes in the between- and within-pair variances in MZ and DZ twins) and their interpretation. Standardizing the analysis to produce heritability coefficients introduces yet another layer of complexity to a phenomenon that is already difficult to understand.

Most of the studies we will review will not provide enough information to permit reinterpretation in terms of variation within and between pairs or unstandardized variance components. We will, nevertheless, focus our review on the following characteristics of study results, wherever possible:

1. Does phenotypic variance increase or decrease as a function of SES?
2. Do between- and within-twin pair variances increase or decrease as a function of SES?
3. Do MZ and DZ correlations increase or decrease as a function of SES?
4. Do standardized and unstandardized ACE components increase or decrease as a function of SES?

2.3 Early Studies

The original Scarr (Scarr-Salapatek 1971) study analyzed 635 Black and 357 White twin pairs from the Philadelphia school system who had been administered aptitude and achievement tests between the 2nd and 12th grades. Zygosity was unknown, so DZ twin correlations were estimated from the correlations in the opposite sex pairs, and MZ correlations were estimated from the difference between the same and opposite sex pairs. Tables 6 and 7 in Scarr-Salapatek (1971) report estimated ICCs for verbal and nonverbal aptitude scores broken down by race, SES, and estimated zygosity. In all four comparisons (Verbal and Nonverbal in Blacks and Whites), estimated MZ correlations were substantially lower in the lower SES group relative to the higher SES group (mean difference in correlation = .225), whereas in the DZ twins there was no difference in correlation between the high and low SES groups (mean difference in correlation = −.011). Scarr-Salapatek also reported the variances of Verbal, Nonverbal, and Total Scores for Low, Medium, and High SES Blacks and Whites. Across six (two races by three tests) comparisons, variances were higher in all six high SES groups compared to the low SES groups (mean difference = 13 %). Although Scarr-Salapatek also reported mean squares between and within twin pairs, due to the uncertainty about zygosity the data are presented in a way that does not allow computation of between- and within-pair or ACE variances for MZ and DZ twins. However, such computations are possible for Fischbein (1980), a description of which follows.

Fischbein (1980) analyzed a sample of 94 MZ and 229 DZ Swedish pairs who were administered a verbal and inductive reasoning test at age 12. Similar to Scarr-Salapatek (1971), MZ twin correlations increased as a function of increasing SES (Verbal Test r = .661, .678, and .755 for low, middle, and high SES groups, respectively; Inductive Test r = .439, .615, and .697). DZ twin correlations either decreased or were unchanged as a function of increasing SES (Verbal Test r = .519, .436, and .374; Inductive Test r = .332, .318, and .216). Fischbein (1980) reported what he called "variances" between and within pairs, although they are actually not variances but mean squares, of the kind that were commonly reported when variance components were computed using classical repeated measures analysis of variance rather than the random effects models that are more prevalent today (Shrout and Fleiss 1979). It is possible to compute the between- and within-pair variances from the mean squares Fischbein (1980) reported, and from those one can compute the ACE variances. These variances are reported in Table 2.1.The largest effects are on the between-pair variances, which decrease as a function of SES in the MZ twins and increase in the DZ twins. The pattern for the within-pair variances is less systematic, but the ACE variances follow a clear pattern: A variances increase sharply with increasing SES, while both C and E variances decrease.

van den Oord and Rowe (1998) analyzed data from 3,266 sibling, half-sibling, and cousin pairs drawn from the National Longitudinal Survey of Youth (NLSY). Children were administered the Mathematics, Reading Recognition, and Reading Comprehension tests at an average age of 9.5 years. Extensive data were available on their rearing environments. The data were analyzed using multilevel models;

Table 2.1 Between, within, and ACE variances derived from mean squares reported by Fischbein (1980)

Variance components	Low SES	Mid SES	High SES
Verbal test scores			
MZ twins			
B	21.67	25.90	26.57
W	11.13	12.20	8.61
DZ twins			
B	20.86	15.68	11.31
W	20.15	20.27	18.92
A	9.83	18.19	25.70
C	13.89	7.81	−.27
E	13.18	11.73	7.34
Inductive reasoning test scores			
MZ twins			
B	9.10	23.12	24.87
W	11.62	14.49	10.79
DZ twins			
B	12.47	13.20	23.12
W	25.11	21.46	14.49
A	10.10	16.93	29.68
C	3.23	5.46	−6.88
E	15.83	13.76	8.73

SES socioeconomic status, *MZ* identical, *DZ* fraternal, *B* between, *W* within, *A* additive genetic, *C* shared environmental, *E* nonshared environmental

ICCs were not reported. Relations between environmental variables and pair similarity were mostly small and nonsignificant. Nevertheless, the effects that were significant were in the direction of smaller nonshared environmental effects in better environments, which is consistent with the pattern observed in the Scarr-Salapatek (1971), Fischbein (1980), and Turkheimer et al. (2003) studies: higher MZ twin correlations with increasing SES.

2.4 Recent Studies

2.4.1 The National Longitudinal Study of Adolescent Health (Add Health)

Rowe et al. (1999) reported an analysis from the first wave of the Add Health Study, which included 1,909 sibling pairs: MZ twins, DZ twins, full-siblings, half-siblings, and cousins reared together. Participants were administered the Peabody Picture Vocabulary Test at a mean age of 16. Parental education was used as a measure of SES. Rowe et al. (1999) employed DeFries-Fulker (DF) analysis (DeFries and Fulker 1985), a method for biometric decomposition in which pairs are double entered, and then one member of a pair is predicted from the other member, zygosity, and the interaction of zygosity with the other member. When correctly parameterized, the regression coefficient for the other member of the twin pair is equal to the standardized estimate of the variance attributable to the shared environment, and the coefficient for the interaction is equal to the heritability. An extended version of the DF model (LaBuda and DeFries 1990) also includes a two-way interaction between the other twin member and a moderator variable (parental education) and a three-way interaction among zygosity, the other member, and the moderator. The coefficients for these terms estimate linear changes in the shared environmental and additive genetic terms, respectively, as a function of the moderator. Although the DF analysis reported by Rowe et al. (1999) does not provide enough information for us to compute between- or within-pair variances or unstandardized ACE components, their Table 6 provides pair correlations for Low and High maternal education groups separately for three groups: high relatedness (MZ twins), moderate relatedness (DZ twins and full siblings), and low relatedness (half-siblings and cousins). MZ twin correlations were lower in the low maternal education group ($r=.55$) than in the high maternal education group ($r=.75$); for the least related pairs, the effect was reversed (Low Education $r=.32$, High Education $r=.10$); the correlations in the group of DZ twins and full-siblings did not differ by maternal education (Low $r=.32$, High $r=.37$).

We have access to the Add Health data and can compute the other parameters directly. We fit a model using Mplus similar to the one we employed for the NCPP data, in which the between- and within-pair variances were estimated for the six relationship groups (we added the unrelated pairs), and then parameterized in terms of their unstandardized ACE coefficients and modeled as an exponential function of parental education. Results are illustrated in Fig. 2.4a–c. To simplify the descriptive analysis, we followed Rowe et al.'s procedure of analyzing the data in three groups of genetic relatedness: high (MZ twins), moderate (DZ twins and full siblings), and low (half-siblings, cousins, and unrelated siblings). The between-pair variance of the MZ twins increases modestly with increasing parental education, while the between-pair variance of the low relatedness group decreases. Neither effect is significant. The opposite pattern obtains within pairs, with the MZ twins becoming more similar with increasing parental education (within-pair variance decreases) while the least related pairs become more different. Once again, the individual effects are not significant. Figure 2.5 illustrates the changes in the ICC, with the high relatedness group becoming more similar and the low relatedness group less so with increasing SES. The ACE parameterization, illustrated in the left panel of Fig. 2.6a, shows that the additive genetic variance increases as a function of increasing SES ($p=.071$), while the shared environmental variance decreases ($p=.022$). The standardized components, in the right panel of Fig. 2.6b, show a similar pattern.

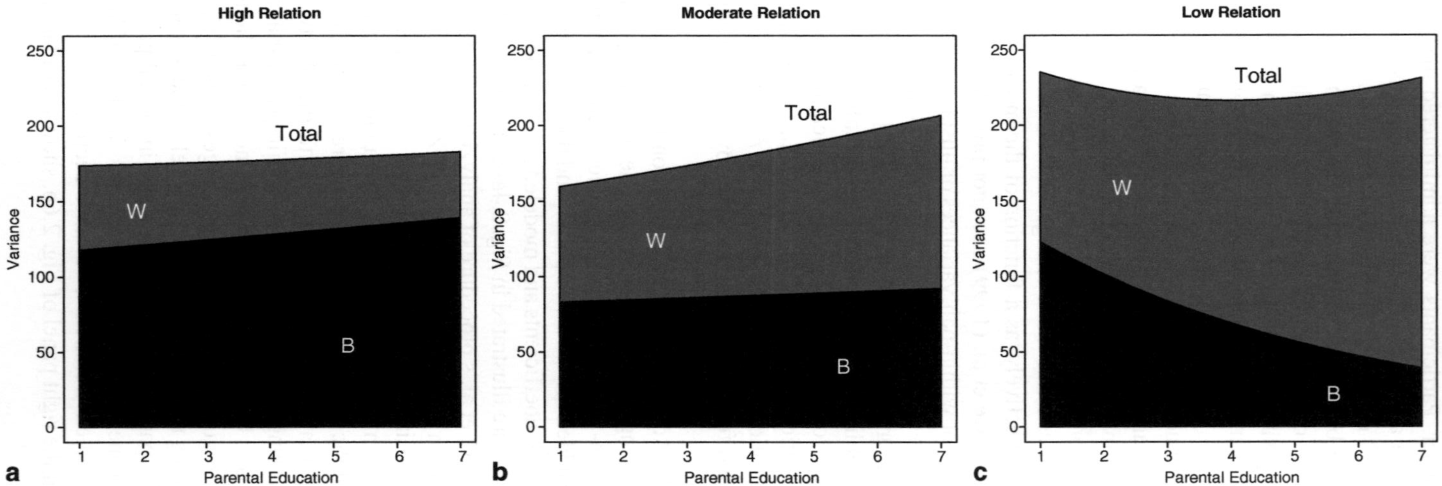

Fig. 2.4 Unstandardized between-(*B*), within-(*W*), and total variance in IQ for **a** highly related (monozygotic twins), **b** moderately related (dizygotic twins and full siblings), and **c** distantly related (half siblings, cousins, and unrelated siblings) sibling pairs in the National Longitudinal Study of Adolescent Health (Add Health). Parental education codes: *1* 8th grade or less, *2* did not graduate high school, *3* high school diploma or trade school, *4* trade school after high school, *5* some college, *6* college degree, *7* postgraduate education

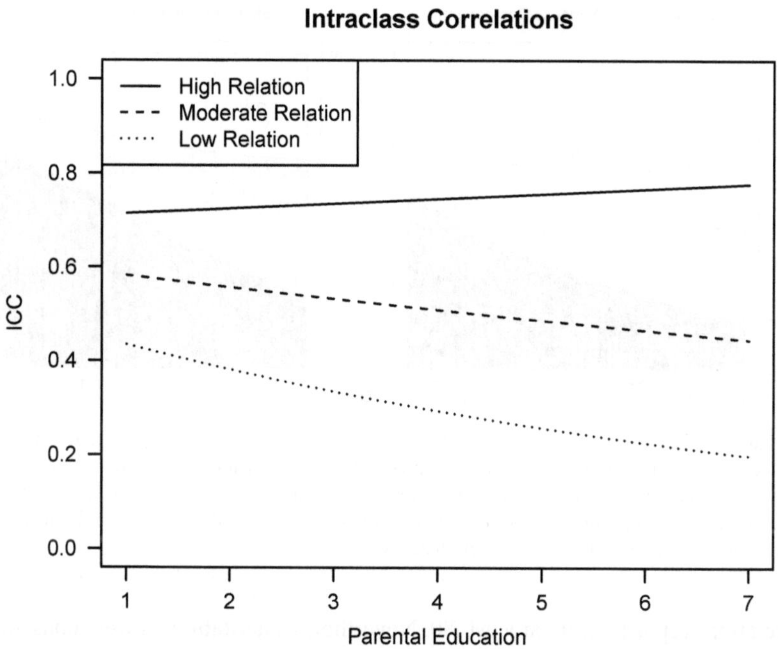

Fig. 2.5 Intraclass correlations (*ICCs*) of IQ for highly related (monozygotic twins), moderately related (dizygotic twins and full siblings), and distantly related (half siblings, cousins, and unrelated siblings) sibling pairs in the National Longitudinal Study of Adolescent Health (Add Health). Parental education codes: *1* 8th grade or less, *2* did not graduate high school, *3* high school diploma or trade school, *4* trade school after high school, *5* some college, *6* college degree, *7* postgraduate education

2.4.2 The Twins Early Development Study (TEDS)

The next study to be published on this topic, which has not been widely discussed, was the first report of G × E interactions from the TEDS (Koeppen-Schomerus et al. 2000), which at the time of its publication consisted of 1,134 MZ and 1,089 DZ twin pairs recruited in England and Wales. At 24 months, the parents of the children were asked to report on the children's verbal and nonverbal abilities using the MacArthur Communicative Development Inventory and the Parent Report of Children's Cognitive Abilities. The investigators were interested in interactions between the biometric components of ability and gestational risk, defined in terms of high- (25–31 weeks gestation), medium- (32–33 weeks), or low-risk (34 weeks or greater).

The use of parent report for the ability assessments of the children introduces some unusual properties into the results from the early childhood scores in TEDS. In particular, pair correlations were very high for both zygosities on both tests in all risk conditions, reaching as high as .98 and never falling below .69; 10 of the 12

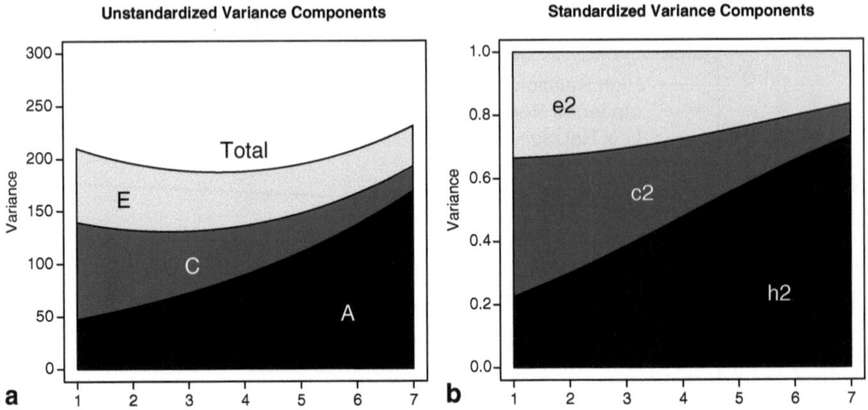

Fig. 2.6 **a** Unstandardized and **b** standardized additive genetic (*A*; h^2), shared environmental (*C*; c^2), and nonshared environmental (*E*; e^2) sources of variance in IQ in the National Longitudinal Study of Adolescent Health (Add Health). Parental education codes: *1* 8th grade or less; *2* did not graduate high school; *3* high school diploma or trade school; *4* trade school after high school; *5* some college; *6* college degree; *7* postgraduate education

correlations reported were at least .80. Nevertheless, heritabilities were consistently lower in the high-risk compared to the two lower-risk groups. Between- and within-pair variances were not reported, but based on the correlations that were reported, the effect appears to be arising mostly from increases in the similarity of DZ twins under higher risk conditions.

Asbury et al. (2005) reported a second analysis from the TEDS sample, examining interactions between a variety of environmental indicators and parent-administered Verbal and Non-Verbal ability tests when the children were 4 years of age. Results were mostly negative, and some were in the opposite direction than the Rowe et al. (1999) and Turkheimer et al. (2003) reports (i.e., heritabilities were, if anything, higher in the more environmentally disadvantaged twins). MZ and DZ twin correlations were not reported, but the parent-administered tests appear to display some of the unusual properties of the 2-year tests reported in the previous TEDS report: twin correlations were very high, and correlations with the environmental indicators were very low, mostly less than .3.

A much more comprehensive analysis of the TEDS sample was published by Hanscombe et al. (2012). By the time of this report, the twins had been tested eight times, at 2, 3, 4, 7, 9, 10, 12, and 14 years of age. Children at ages 2, 3, and 4 were tested via parent interview, as described in Asbury et al. (2005). At ages 7 and 9, families were mailed a test booklet administered by the parents, containing subtests from the WISC-III UK. At ages 10 and up, children were administered items from the multiple choice version of the WISC. Analyses were conducted separately at each age of testing, using measures of SES collected at registration in the study, at age 7, and at age 9. This created a total of 18 individual analyses (nine using the registration SES, five using the 7-year-old SES, and four using the 9-year-old SES). Results were quite diverse, but several consistencies can be noted. The phenotypic variance of ability decreased (not necessarily significantly) as a function of increasing SES in 15 of 18 analyses. The unstandardized variance attributable to the shared environment decreased as a function of increasing SES in 13 of 18. The authors concluded that shared environmental variance increases in poor environments, and that this effect may explain why heritability appears to have been reduced in poor environments: a constant genetic variance is being compared to a denominator that increases as SES declines. They also note that interactions with A and C are very highly correlated, so other studies without the enormous TEDS sample may have lacked the statistical power to discriminate between them.

2.4.3 Nagoshi and Johnson (2005)

The first study to be conceived explicitly as an attempted replication of Turkheimer et al. (2003) was Nagoshi and Johnson (2005), using the Hawaii Family Study. The analysis was not based on twins, however; instead they studied parent–child correlations for cognitive ability, and found that the correlations did not differ by familial SES. In fact, the study was not actually a replication of Turkheimer et al. (2003). As stated in the title of the paper, the Nagoshi and Johnson study pertained to the familiality of ability, not its heritability. Parent–child correlations comprise both shared environmental and genetic effects, which previous reports of the interaction had shown to be moderated by SES in opposite directions. The model from the NCPP analysis would have predicted exactly the null results described by Nagoshi and Johnson (2005), because as SES increases, the increasing A and decreasing C components of familiality would be expected to cancel each other out. Nevertheless, the study is still sometimes cited as a failure to replicate (e.g., Rushton and Jensen 2010).

2.4.4 Vietnam Era Twin Study (VETSA)

Kremen et al. (2005) reported on a sample of 176 MZ and 168 DZ adult twin pairs from the VETSA (see Chap. 4 for more information on VETSA). Twins were administered the reading subtest of the Wide Range Achievement Test, Version 3. Results showed a clear decrease in phenotypic variance with increasing parental education. Based on the reported variance (a single variance was reported for MZ and DZ twins at each third of the distribution of parental education) and ICCs, between- and within-pair variances appear to have decreased as a function of increasing parental education in both MZ and DZ twins. The models were not decisive as to whether this difference was attributable to proportional decreases in genetic and environmental terms, or specifically to decreases in shared and nonshared environmental variance with increasing parental education. The authors preferred the latter explanation. Grant et al. (2010) reported interactions between the Armed

Forces Qualification Test scores and parental education in the full sample of 3,203 twin pairs from the Vietnam Era Twin Study. Neither the total phenotypic variance nor any of the ACE parameters showed significant effects of parental education.

2.4.5 Two Studies in the Netherlands

van der Sluis et al. (2008) examined interactions between FSIQ scores and several environmental indicators in a sample of 385 young adult (ages 20–35) and 370 older adult (ages 36–69) twins in the Netherlands. Results were uniformly nonsignificant, and examination of the tables in the report does not suggest any systematic effects on phenotypic variance as a function of socioeconomic SES environment. Among the older participants in the sample, there was some tendency for shared and nonshared environmental terms to be greater among the more affluent participants or those with better educated parents. Bartels et al. (2009) describe a large study of 3,659 twin 12-year-old twin from the Netherlands Twin Register. The phenotypic variance of cognitive ability decreased with increasing parental education (variance equal to 83.0, 66.9, and 50.5 in the low, middle, and high parental education groups, respectively; the authors describe it as a ceiling effect). Neither MZ nor DZ correlations for cognitive ability showed substantial variation across the same environmental groups; standardized ACE parameters were constant across parental education groups as well.

2.4.6 National Merit Twin Study

Harden et al. (2007) described interactions between the common variance in five subtests from the National Merit Scholarship Qualifying Test (NMSQT) and measures of parental education and income. The study included 509 MZ pairs and 330 DZ pairs. Most participants were 17 years old. Methodologically, this was the first study to investigate interactions with a latent variable describing common variance among a set of measured ability scores. Significant interactions were found between SES and the genetic component of ability, such that the proportion of genetic variance increased with increasing parental income and education; the interaction with parental education was in the same direction but did not reach significance, but the magnitudes of the two interactions were not significantly different from each other. In a follow-up study, Tucker-Drob and Harden (2012) showed that the interaction was mediated by intellectual interest: the genetically mediated relationship between intellectual interest and ability was much stronger in children raised in high SES homes, and there was no interaction between SES and the portion of ability variance that is independent of academic interest.

Once again, we have access to the NMSQT data, so we can examine the results more closely. Figure 2.7a and b are plots of the between- and within-pair variances as a function of parental income in the MZ and DZ twins. The within-pair variance

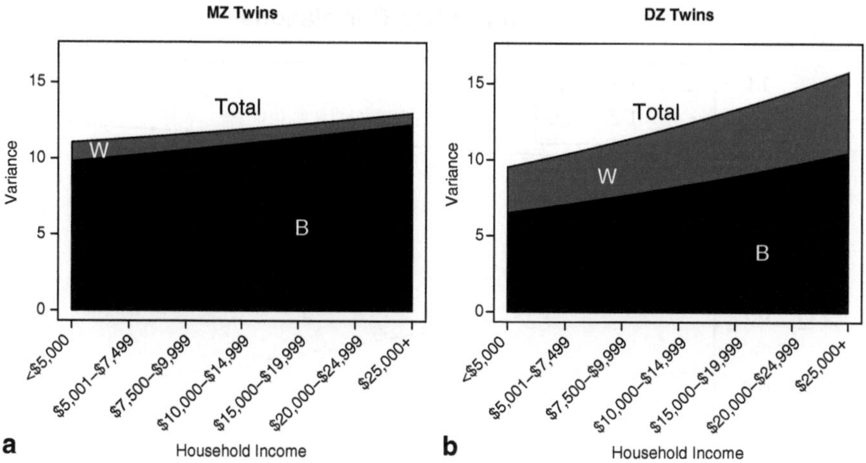

Fig. 2.7 Unstandardized between-(*B*), within-(*W*), and total variance in IQ for **a** identical (*MZ*) and **b** fraternal (*DZ*) twin pairs in the National Merit Twin Study

decreases slightly in the MZ pairs and increases in the DZ pairs; the between-pair variance increases for both. Similar results are obtained when parental education is used in place of household income as a moderator. The ICCs, shown in Fig. 2.8, increase as a function of increasing family income in the MZ twins, and decrease in the DZ twins. Results of an ACE parameterization of the model are illustrated in Fig. 2.9. As reported by Harden et al. (2007), the A variance increases as a function of increasing household income, in both unstandardized and standardized models. The phenotypic variance of the ability score increases as a function of SES.

2.4.7 *Early Childhood Longitudinal Study (ECLS)*

The ECLS comprises approximately 750 (the National Center for Education Statistics requires that reported sample sizes be rounded to the nearest 50) twin pairs from a nationally representative sample of children born in 2001. Cognitive data were collected when the children were approximately 10 months, 2, 4, and 5 years old. At 10 months and 2 years of age, children were administered the mental and motor scales of the Bayley Scales of Infant Development—Research Edition. SES was computed as a composite of parental education, occupation, and income. Tucker-Drob et al. (2011) examined interactions of SES with biometric components of ability at 10 months and with change in ability between 10 months and 2 years. They did not find significant interactions between SES and the ability scores at 10 months (there was no genetic variance at all at 10 months) but did find an interaction with change between 10 months and 2 years, whereby genetic variance increased as a function of increasing SES. A second analysis (Rhemtulla and Tucker-

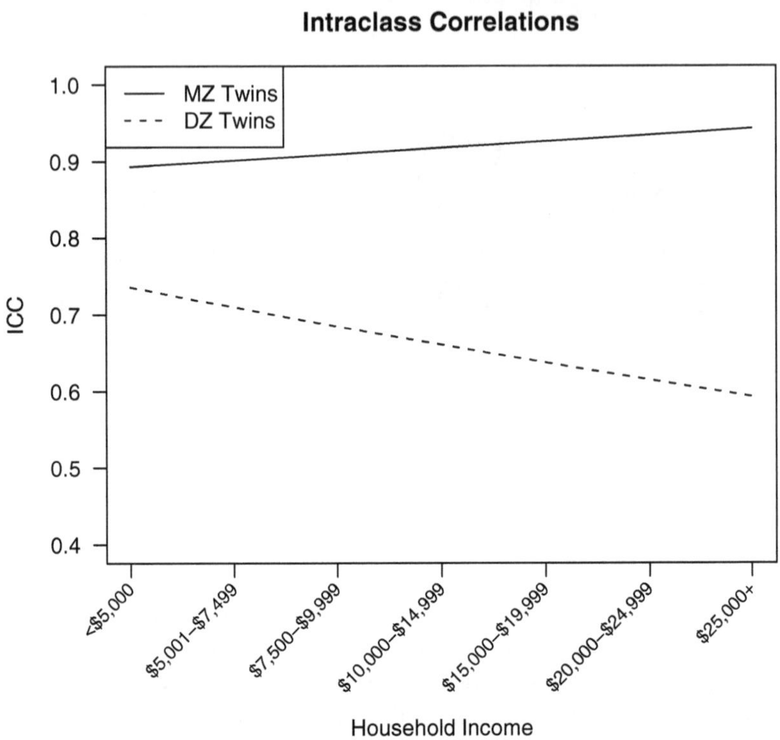

Fig. 2.8 Intraclass correlations (*ICCs*) of IQ for identical (*MZ*) and fraternal (*DZ*) twin pairs in the National Merit Twin Study

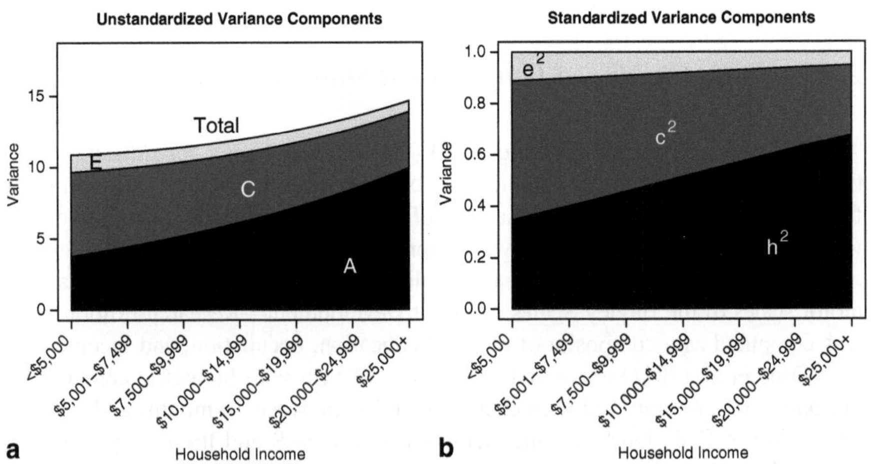

Fig. 2.9 **a** Unstandardized and **b** standardized additive genetic (A; h^2), shared environmental (C; c^2), and nonshared environmental (E; e^2) sources of variance in IQ in the National Merit Twin Study

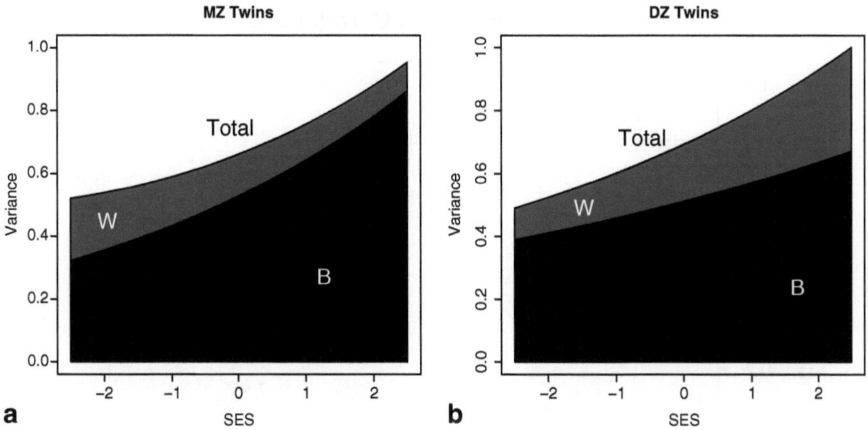

Fig. 2.10 Unstandardized between-(*B*), within-(*W*), and total variance in 4-year Math Score for **a** identical (*MZ*) and **b** fraternal (*DZ*) twin pairs in the Early Childhood Longitudinal Study (ECLS). *SES* socioeconomic status

Drob 2012) showed a similar interaction between SES and preschool mathematics, but not reading scores, at 4 years of age.

An examination of the between- and within-pair variance plots in Fig. 2.10a and b shows a tendency for between- and within-pair variances to increase at higher levels of SES. Figure 2.11 shows that the ICCs increase for MZ twins and decrease for DZ twins. The unstandardized and standardized ACE parameterizations in Fig. 2.12a and b show that the effect is manifest as a strong increase in the genetic variance, and small decreases in the environmental variances, as a function of increasing SES.

2.5 Molecular Genetic Studies

Several studies have recently begun to examine whether associations between specific genes and ability are moderated by SES, although the search has been somewhat limited because of the difficulty of finding main effects of genetic associations with ability (Chabris et al. 2012). Enoch et al. (2009) examined associations among the COMT Val158Met genotype, education (of the participants, not their parents), and ability scores in a sample of 328 middle-aged Plains American Indians. They found that Met allele carrier scores on memory and executive function measures improved with increased years of education, while the scores of Val/Val carriers improved only marginally. In the high-education group, Met allele carriers outperformed the Val/Val carriers, but in individuals with fewer than 11 years of education the opposite was true.

Keltikangas-Järvinen et al. (2010) studied a population-based sample of 982 Finnish students aged 9, 12, and 15. Participants were genotyped for the T102C polymorphism of the serotonin receptor 2A. Although the polymorphism did not have a significant main effect on school achievement, there was a significant interaction

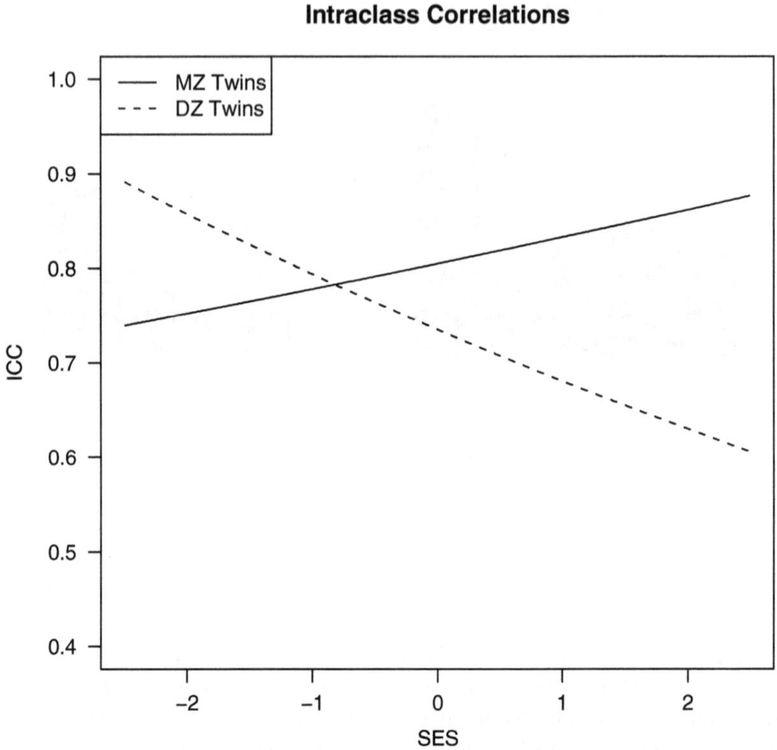

Fig. 2.11 Intraclass correlations (*ICCs*) of IQ for identical (*MZ*) and fraternal (*DZ*) twin pairs in the Early Childhood Longitudinal Study (ECLS). *SES* socioeconomic status

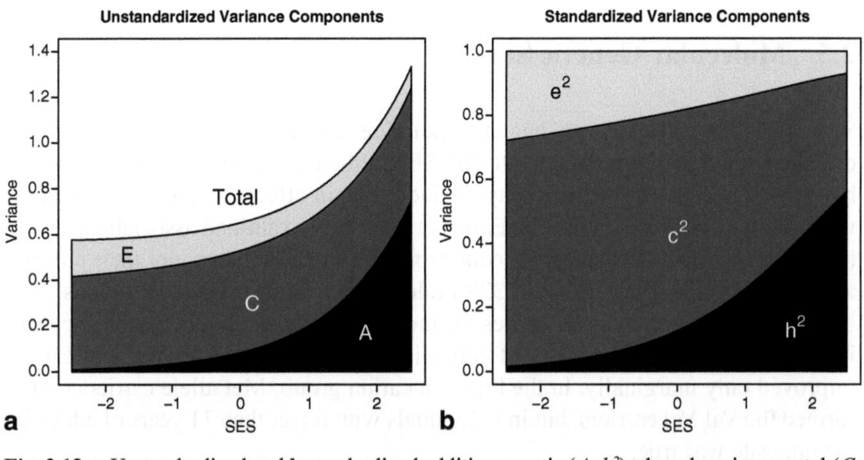

Fig. 2.12 **a** Unstandardized and **b** standardized additive genetic (A; h^2), shared environmental (C; c^2), and nonshared environmental (E; e^2) sources of variance in 4-year Math Score in the Early Childhood Longitudinal Study (ECLS). *SES* socioeconomic status

between the polymorphism and parental education. Among children whose mothers had more than a secondary education, children with the T/T genotype had higher grade point averages that children with either the T/C or C/C genotypes; among children whose mothers were less educated, the T/T children's GPAs were slightly lower than the other groups.

2.6 Studies of Reading

Renewed interest in population-level G × E interactions in cognitive ability has led to investigations of several related phenotypes of interest. One active area of research has involved reading ability and disability. Friend et al. (2008) studied a sample of 545 twin pairs between the ages of 8 and 20 who were selected for school history of reading disability and low performance on measures of reading ability. Thirteen percent of the parents in the sample had fewer than 12 years of education, 23 % had a high school diploma only, and the remaining 64 % had at least some college. Parental education was correlated .09 with the multivariate reading disability score. DF analysis suggested there was a significant correlation between parental education and the magnitude of the standardized genetic effect, and univariate analyses following a median split showed a heritability of .71 in the high parental education group, and .49 in the low parental education group. McGrath et al. (2007) showed that in a sib-pair linkage design of a variety of carefully constructed phonological and language measures, two linkage peaks on chromosomes 6 and 15 showed significant interactions with continuous measures of the home language and literacy environment. Linkage effects were stronger in sib-pairs reared in relatively enriched environments.

Taylor et al. (2010) studied the interaction between teacher quality and reading scores obtained from statewide testing in Florida in a sample of 280 MZ and 526 DZ twin pairs in the 1st and 2nd grades. The investigators obtained teacher quality ratings based on improvements in test scores in the teachers' classrooms. Results showed a significant positive correlation between teacher quality and the heritability of reading ability. It should be noted, however, that this study was unusual in that the moderator potentially varied within twin pairs, unlike typical measures of SES or parental education that vary at the level of families. van der Sluis et al. (2012) have recently shown that studies of this kind are subject to Type I errors unless some additional procedures are followed.

2.7 Some Other Methods and Phenotypes

A recent study suggests that the interaction may operate on the level of the cerebral cortex. Chiang et al. (2010) assessed cortical white matter integrity using diffusion tensor imaging in 705 twins and their siblings. They demonstrated that white matter integrity was highly heritable, and moreover that there were significant G × E interactions, such that heritabilities were higher among twins of higher SES and higher IQ scores. Yeo et al. (2011) reported correlations between scores on the Wechsler Abbreviated Scale of Intelligence and the total length of rare copy number variations in a sample of 74 individuals with alcohol dependence. The total length of the rare deletions was correlated with intelligence at $r = -.30$, and the correlation was higher in the Anglo/White group than in the non-White group.

2.8 Conclusions

2.8.1 Replicability of the Scarr-Rowe Interaction

We have considered the original instances of the Scarr-Rowe interaction, the replication of it that we provided in 2003, and our own and others' attempts at replication since that time. Within the domain of ability test scores in American youth, the interaction has replicated reasonably well. It has been detected, in more or less the same form, in the original Scarr-Salapatek (1971) Philadelphia Study, then in four large American datasets: the NCPP (Turkheimer et al. 2003), the Adolescent Health Project (Rowe et al. 1999), the National Merit Twin Study (Harden et al. 2007; Tucker-Drob and Harden 2012), and the ECLS (Tucker-Drob et al. 2011; Rhemtulla and Tucker-Drob 2012). In addition, the molecular genetic studies that have been reported to date appear to support the interaction, in that the effects of measured genes are stronger in children born to more highly educated parents.

There have been two notable failures to replicate in American samples: the van den Oord and Rowe (1998) study from the NLSY, and the Grant et al. (2010) analysis of the Vietnam Era Twin Study. Of these exceptions, the NLSY data are in need of further analysis. Much more is known about the kinships in the NLSY than was available in the 1990s, and the children have been retested several times since then. All one can note about the VETSA sample is that they were no longer of school age (average age of 19.6 years, with approximately 20 % of the sample older than 21), although there was still a main effect for the shared environment. We would have predicted that this would have been a dataset where the effect should have appeared: American, large, not too severely restricted at lower levels of SES. As of now, it stands as a significant outlier.

Whether the Scarr-Rowe interaction is found among European children is a complex question. A few years ago we would have said no, based on the early reports from the TEDS study and the two analyses of Dutch data, but the most recent comprehensive report of the TEDS sample is much more favorable to the hypothesis. The conclusion of the TEDS investigators that the effect appears to consist more of a reduction of shared environmental variance in favorable environments, as opposed to an increase in the genetic variance, is not at all contradictory of the original hypotheses about moderation of standardized heritability. Some other reports are

Table 2.2 Gene by socioeconomic status (SES) interactions of reanalyzed datasets

Variance components	Dataset			
	NCPP	Add Health	National Merit Twin Study	ECLS
MZ between	↑	↑	↑	↑
MZ within	↓*	↓	↓	↓
DZ between	↓	↑	↑	↑
DZ within	↑	↑	↑	↑
χ^2 (df), p	15.774 (4), .003	5.400 (6), .494	8.290 (4), .082	7.487 (4), .112
MZ ICC	↑*	↑	↑	↑
DZ ICC	↓	↓	↓	↓*
χ^2 (df), p	21.017 (2), .000	5.137 (3), .162	3.228 (2), .199	5.949 (2), .051
A variance	↑*	↑	↑*	↑
C variance	↓	↓*	↓	↓
E variance	↓*	↓	↓	↓
Total variance	↓	↓	↑	↑
χ^2 (df), p	25.40 (3), .000	6.433 (3), .092	7.043 (3), .071	4.048 (3), .256
A standardized	↑	↑	↑	↑
C standardized	↓	↓	↓	↑
E standardized	↓	↓	↓	↓

↑ indicates increasing variance with increasing SES, ↓ indicates decreasing variance with increasing SES

NCPP National Collaborative Perinatal Project, *Add Health* National Longitudinal Study of Adolescent Health, *ECLS* Early Childhood Longitudinal Study, *MZ* identical, *DZ* fraternal, *ICC* intraclass correlation, *A* additive genetic, *C* shared environmental, *E* nonshared environmental, *df* degrees of freedom

*$p<.05$ (statistically significant estimates)

consistent with the shared environmental hypothesis and some are not, but few of them have the statistical power available in the TEDS study, and discriminating A from C interactions is probably not within the reach of any study with fewer than several thousand twin pairs. We are eager to see multivariate longitudinal models of the TEDS data that are currently in preparation.

2.8.2 *Between- and Within-Pair Variances*

As we have emphasized throughout the chapter, there is more to be learned from examining interactions of SES with the unstandardized between- and within-pair components of standardized heritabilities than can be gleaned from the heritabilities themselves. These results are summarized in Table 2.2 for the studies we were able to reanalyze, and Table 2.3 for the studies that we reviewed based on published reports. In all four of the datasets we have reanalyzed here, the within-pair variance of MZ twin pairs increases at lower levels of SES: poverty appears to have the effect

Table 2.3 Summary of gene by socioeconomic status (SES) interactions in reviewed literature

Variance components	Dataset									
	Scarr-Salapatek (1971)	Fischbein (1980)	van den Oord and Rowe (1998)	Koeppen-Schomerus et al. (2000)	Asbury et al. (2005)	Hanscombe et al. (2012)	Kremen et al. (2005)	Grant et al. (2010)	van der Sluis et al. (2008)	Bartels et al. (2009)
MZ between		↓								
MZ within		↓								
DZ between		↑								
DZ within		↓/=								
MZ correlation	↑	↑		↑/=						=
DZ correlation	=	↓/=		↓						=
A variance		↑	↓			↓	↓/=	=	=	
C variance		↓	↓			↓	↓	=	↑/=	
E variance		↓	↓			↓	↓	=	↑/=	
Total variance						↓	↓	=	=	↓
A standardized		↑		↑	↑/=	↑	↑/=	=		=
C standardized		↓		↓		↓	↓/=	=		=
E standardized		↓		↓		↓	↓/=	=		=

↑ indicates increasing variance with increasing SES, ↓ indicates decreasing variance with increasing SES, = indicates no variance changes across levels of SES
MZ identical, *DZ* fraternal, *A* additive genetic, *C* shared environmental, *E* nonshared environmental

of making MZ twins more different from each other. There appears to be a less consistent effect on within-pair variances in DZ pairs, but if anything they decrease in low SES environments. The between-pair variance of MZ pairs decreases in poor environments, whereas for DZ pairs there is little evidence of a systematic pattern.

An ICC between twins is a ratio of between-pair variance to the phenotypic variance, which is the sum of the between- and within-pair variances. In all four of the reanalyzed studies, the MZ twin correlation increases with increasing SES, and the DZ twin correlation decreases. There is no consistent pattern to whether one changes more strongly than the other. Finally, ACE variances are the product of the ICC and the total phenotypic variance, and it is here that the greatest degree of consistency can be observed across studies. In all four of the American datasets, A increases while C and E decrease as a function of increasing SES. In the ECLS, the sharp increase in genetic variance at the upper end of the SES distribution produces a similar increase in the phenotypic variance, but for the other studies there is no large and systematic effect on the total variance.

2.8.3 Differences in Environments and Abilities

Although the Scarr-Rowe interaction is commonly described as an interaction between SES and IQ, our review has suggested that analyses have not been limited to these particular constructs. Environmental variables have often included parental education and income, which may also be included in socioeconomic composite variables. The TEDS sample includes a wide variety of more detailed measures of the home environment, although few of them produced significant interactions with the heritability of IQ. Similarly, the ability measures employed in studies have varied considerably, from the Wechsler IQs included in the Turkheimer et al. (2003) study (which also included seven WISC subtests, VIQ, and PIQ) to a wide variety of other ability, achievement and school-based scores.

This diversity of environmental and ability measures offers both opportunities and dangers. It would, of course, be important if one could narrow down the kinds of abilities or environmental measures for which SES by heritability interactions are likely to occur. Such investigations will be hampered, however, by the limitations of statistical power available in most studies, compounded by the high correlations that are common among multiple measures of cognitive ability or the family environment. It should also be noted, however, that there are significant risks attendant to situations where there are multiple means of testing hypotheses in a context of limited statistical power. As has been amply documented in the recent literature on gene x environment interactions using measured DNA, it is all too easy for well-intentioned researchers to capitalize on the multiple testing opportunities that are available when several ability tests are crossed with several environmental measures (Hewitt 2012). The advantage of omnibus constructs like SES and IQ is that they focus attention on the common variance rather than on the idiosyncratic variance in individual measures. To the extent researchers choose to investigate individual measures in an exploratory way, it will be important to report all results, both positive and negative. At some point in the near future, a meta-analysis of this literature will be an important contribution.

2.8.4 *Hypotheses About Mechanism*

Why does the Scarr-Rowe interaction occur? We will conclude by offering a speculative proposal in the hope that it forms the basis for future investigation. Other than the Scarr-Rowe interaction, the best known moderator of the heritability of cognitive ability is age: heritability increases throughout childhood at a pace that accelerates in late adolescence and then continues steadily through middle adulthood; environmental variance then reasserts itself at the end of life (see Chaps. 4 and 5 for more detail). It is surprising how little is known about why heritability changes with age, starting with whether it occurs because of increases in MZ twin correlations, decreases in DZ twin correlations, or both.

We propose that these two moderation effects on the heritability of ability are related to each other, indeed that they are two manifestations of the same phenomenon. Beam and Turkheimer (2013) have recently shown that age-related increases in heritability can be produced by accumulating within-pair correlations between phenotypes and environment, which we call rPE. If the member of a twin pair with greater phenotypic ability is subsequently exposed to more stimulating environments, pairs of siblings will diverge over time at a rate inversely proportional to their genetic relatedness. MZ twins, for whom phenotypic differences can only arise in the nonshared environment, have little differential phenotype to provide energy for a divergence process based on rPE. DZ twins, however, are different genetically, providing an initial basis for a developmental process that exposes them to differential environments, leading to an accelerating course of divergence. One could easily expect this process to accelerate in adolescence, leading to the steep acceleration in heritability that is observed at that age. We propose that the Scarr-Rowe interaction is the result of an interruption of the normal developmental process of accumulating rPE. For sibling pairs reared in deprived environments, it does not matter how much phenotypic ability one displays, because there are no favorable niches to seek out. In the absence of favorable environments to select, DZ twins and other less genetically related kinds of sibling pairs are not induced to differentiate more rapidly than MZ pairs.

The analyses we reported here are at least broadly consistent with the hypothesis. In the Add Health (Fig. 2.4), National Merit (Fig. 2.7), and ECLS-B (Fig. 2.10) datasets, the interactions appear to be driven largely by increases in within-pair variances with increasing SES in the DZ pairs, as our hypothesis would predict. In the NCPP data (Fig. 2.1), it must be noted, the effect appears to arise largely in the MZ pairs, who appear to become more similar (smaller within-pair variances) in more affluent environments. We are currently working on more detailed modeling of the predictions of the model. Definitive testing of our hypothesis that the Scarr-Rowe interaction and increases in heritability with age are actually the same phenomenon will require an integration of developmental research designs, in which biometric variances are computed repeatedly over development, with socioeconomic designs that attempt to show that biometric parameters vary as a function of changes in SES. Specifically, we hypothesize that not only cross-sectional heritability, but the rate of increase in heritability across development, will be lower among children raised in poverty. Several datasets are available to test this hypothesis. In the United States, the NLSY includes longitudinal assessment of siblings and half-siblings, with a substantial proportion from lower SES families. These data have not been examined for interactions with SES since the van den Oord and Rowe (1998) analysis, 15 years ago. In the ECLS, it has been demonstrated that heritability of ability is zero in infancy and increases substantially by kindergarten, and also that interactions between heritability and SES become larger in this time span (Tucker-Drob et al. 2011). The TEDS data should allow for similar analyses, although we remain uncertain whether modern European samples include a sufficient proportion of children raised in poverty to manifest the interaction reliably.

The idea that small initial differences in phenotype have the potential to interact with environmental selection processes has a long history in developmental behavioral genetics, often as part of discussions of active or evocative GE correlation (Towers et al. 2003; see Chap. 6). A central role for niche-picking has sometimes been proposed in either a hereditarian (Scarr and McCartney 1983) or an environmentalist (Bronfenbrenner and Ceci 1994) perspective, but as we have argued for some time, the most important consequence of complex developmental dynamics is that they can lead to large differences in outcome that are not easily attributable to either genes or environment:

> The idea of P × E interactions does not make sense in strictly cross-sectional models because it would involve an interaction between a dependent (P) and an independent (E) variable, but in developmental models it makes perfect sense to postulate that the effect of an environmental event depends on the phenotype of the organism at the time the event occurs; indeed, this model appears much more plausible than the idea that environmental effects are somehow mediated directly by the genotype. If intelligent children evoke more complex linguistic interactions with their caregivers, it is observable phenotypic aspects of their behavior, not their genotype, that is having an effect on surrounding adults. This phenotype is in turn the cumulative result of developmental interactions between the child's genotype and previous environmental events. (Turkheimer and Waldron 2000, pp. 91–92)

Increases in heritability throughout the lifespan and the Scarr-Rowe interaction both highlight that simple linear, additive models cannot do justice to the developmental dynamics of genetic and environmental effects. Heritabilities are almost never a scientific end in themselves, and even moderation of heritability is unlikely to provide useful insight unless it can be translated into meaningful developmental models. It is also the case that constructs like SES that are commonly discussed as "environmental" often contain substantial genetic variance (Plomin and Bergemann 1991). Our proposal is that the key developmental dynamic of genetic influence across the lifespan involves the constant selection of new environments, driven by the current state of the organism, which is itself the product of the developmental process up to that point in time.

References

Asbury, K., Wachs, T. D., & Plomin, R. (2005). Environmental moderators of genetic influence on verbal and nonverbal abilities in early childhood. *Intelligence, 33,* 643–661. doi:10.1016/j.intell.2005.03.008.

Bartels, M., van Beijsterveldt, C. E. M., & Boomsma, D. I. (2009). Breastfeeding, maternal education and cognitive function: A prospective study in twins. *Behavior Genetics, 39,* 616–622. doi:10.1007/s10519-009-9293-9.

Beam, C. R., & Turkheimer, E. (2013). Phenotype-environment correlations in longitudinal twin models. *Development and Psychopathology, 25,* 7–16.

Bronfenbrenner, U., & Ceci, S. J. (1994). Nature-nuture reconceptualized in developmental perspective: A bioecological model. *Psychological Review, 101*, 568–586. doi:10.1037/0033-295X.101.4.568.

Chabris, C. F., Hebert, B. M., Benjamin, D. J., Beauchamp, J. P., Cesarini, D., van der Loos, M. J. H. M., et al. (2012). Most reported genetic associations with general intelligence are probably false positives. *Psychological Science, 23,* 1314–1323. http://economics.cornell.edu/dbenjamin/IQ-SNPs-PsychSci-20111205-accepted.pdf.

Charney, E. (2012). Behavior genetics and post genomics. *Behavioral and Brain Sciences, 35,* 331–358. http://issuu.com/echar/docs/evan_charney__behavior_genetics_and_post-genomics.

Chiang, M.-C., McMahon, K. L., de Zubicaray, G. I., Martin, N. G., Hickie, I., Toga, A. W., Wright, M. J., & Thompson, P. M. (2010). Genetics of white matter development: A DTI study of 705 twins and their siblings aged 12 to 29. *NeuroImage, 54,* 2308–2317. doi:10.1016/j.neuroimage.2010.10.015.

DeFries, J., & Fulker, D. (1985). Multiple regression analysis of twin data. *Behavior Genetics, 15,* 467–473. doi:10.1007/BF01066239.

Eaves, L. J., & Jinks, J. L. (1972). Insignificance of evidence for differences in heritability of IQ between races and social classes. *Nature, 240,* 84–88.

Enoch, M.-A., Wahed, J. F., Harris, C. R., Albaugh, B., & Goldman, D. (2009). COMT Val 158Met and cognition: Main effects and interaction with educational attainment. *Genes, Brain, and Behavior, 8,* 36–42. doi:10.1111/j.1601-183X.2008.00441.x.

Fischbein, S. (1980). IQ and social class. *Intelligence, 4,* 51–63.

Friend, A., DeFries, J. C., & Olson, R. K. (2008). Parental education moderates genetic influences on reading disability. *Psychological Science, 19,* 1124–1130. doi:10.1111/j.1467-9280.2008.02213.x.

Grant, M. D., Kremen, W. S., Jacobson, K. C., Franz, C., Xian, H., Eisen, S. A., Toomey, R., Murray, R. E., & Lyons, M. J. (2010). Does parental education have a moderating effect on the genetic and environmental influences of general cognitive ability in early adulthood? *Behavior Genetics, 40,* 438–446. doi:10.1007/s10519-010-9351-3.

Hanscombe, K. B., Trzaskowski, M., Haworth, C. M., Davis, O. S., Dale, P. S., & Plomin, R. (2012). Socioeconomic status (SES) and children's intelligence (IQ): In a UK-representative sample SES moderates the environmental, not genetic, effect on IQ. *PLoS One, 7,* 1–16. doi:10.1371/journal.pone.0030320.

Harden, K. P., Turkheimer, E., & Loehlin, J. C. (2007). Genotype by environment interaction in adolescents' cognitive aptitude. *Behavior Genetics, 32,* 273–283. doi:10.1007/s10519-006-9113-4.

Hewitt, J. K. (2012). Editorial policy on candidate gene association and candidate gene-by-environment interaction studies of complex traits. *Behavior Genetics, 42,* 1–2. doi:10.1007/s10519-011-9504-z.

Keltikangas-Järvinen, L., Jokela, M., Hintsanen, M., Salo, J., Hintsa, T., Alatupa, S., & Lehtimäki, T. (2010). Does genetic background moderate the association between parental education and school achievement? *Genes, Brain and Behavior, 9,* 318–324. doi:10.1111/j.1601-183X.2009.00561.x.

Koeppen-Schomerus, G., Eley, T. C., Wolke, D., Gringras, P., & Plomin, R. (2000). The interaction of prematurity with genetic and environmental influences on cognitive development in twins. *The Journal of Pediatrics, 137,* 527–533. doi:10.1067/mpd.2000.108445.

Kremen, W. S., Jacobson, K. C., Xian, H., Eisen, S. A., Waterman, B., Toomey, R., Neale, M. C., Tsuang, M. T., & Lyons, M. J. (2005). Heritability of word recognition in middle-aged men varies as a function of parental education. *Behavior Genetics, 35,* 417–433. doi:10.1007/s10519-004-3876-2.

LaBuda, M. C., & DeFries, J. C. (1990). Genetic etiology of reading disability: Evidence from a twin study. In G. T. Pavlidis (Ed.), *Perspectives on dyslexia* (Vol. 1, pp. 47–76). New York: Wiley.

McGrath, L. M., Pennington, B. F., Willcut, R. G., Boada, R., Shriberg, L. D., & Smith, S. D. (2007). Gene × environment interactions in speech sound disorder predict language and preliteracy outcomes. *Development and Psychopathology, 19,* 1047–1072. doi:10.1017/S0954579407000533.

Muthén, L. K., & Muthén, B. O. (2011). *Mplus V. 6.12* [Computer software]. Los Angeles, CA: Authors.

Nagoshi, C. T., & Johnson, R. C. (2005). Socioeconomic status does not moderate the familiality of cognitive abilities in the Hawaii Family Study of Cognition. *Journal of Biosocial Science, 37,* 773–781. doi: 10.1017/S0021932004007023.

Nisbett, R. E., Aronson, J., Blair, C., Dickens, W., Flynn, J., Halpern, D. F., & Turkheimer, E. (2012). Intelligence: New findings and theoretical developments. *American Psychologist, 67,* 130–159. doi:10.1037/a0026699.

Plomin, R., & Bergeman, C. S. (1991). The nature of nurture: Genetic influence on "environmental" measures. *Behavioral and Brain Sciences, 14,* 373–386. doi:http://dx.doi.org/10.1017/S0140525X00070278.

Rhemtulla, M., & Tucker-Drob, E. M. (2012). Gene-by-socioeconomic interaction in school readiness. *Behavior Genetics, 42,* 549–558. doi:10.1007/s10519-012-9527-0.

Rowe, D. C., Jacobson, K. C., & van den Oord, E. J. C. G. (1999). Genetic and environmental influences on vocabulary IQ: Parental education level as a moderator. *Child Development, 70,* 1151–1162. doi:10.1111/1467-8624.00084.

Rushton, J. P., & Jensen, A. R. (2010). Race and IQ: A theory-based review of the research in Richard Nisbett's *Intelligence and How to Get It. The Open Psychology Journal, 3,* 9–25. doi: 10.2174/1874350101003010009.

Scarr, S., & McCartney, K. (1983). How people make their own environments: A theory of genotype greater than environment effects. *Child Development, 54,* 424–435. http://www.jstor.org/stable/1129703.

Scarr-Salapatek, S. (1971). Race, social class, and IQ. *Science, 174,* 1285–1295. doi:10.1126/science.174.4016.1285.

Shrout, P. E., & Fleiss, J. L. (1979). Intraclass correlations: Uses in assessing rater reliability. *Psychological Bulletin, 86,* 420–428. doi:10.1037/0033-2909.86.2.420.

Taylor, J., Roehrig, A. D., Hensler, B. S., Connor, C. M., & Schatschneider, C. (2010). Teacher quality moderates the genetic effects on early reading. *Science, 328,* 512–514. doi:10.1126/science.1186149.

Towers, H., Spotts, E. L., & Neiderhiser, J. M. (2003). Genetic and environmental influences on parenting and marital relationships. *Marriage & Family Review, 33,* 11–29. doi:10.1300/J002v33n01_03.

Tucker-Drob, E. M., & Harden, K. P. (2012). Intellectual interest mediates gene × socioeconomic interaction on adolescent academic achievement. *Child Development, 83,* 743–757. doi:10.1111/j.1467-8624.2011.01721.x.

Tucker-Drob, E. M., Rhemtulla, M., Harden, K. P., Turkheimer, E., & Fask, D. (2011). Emergence of a gene x socioeconomic status interaction on infant mental ability between 10 months and 2 years. *Psychological Science, 22,* 125–133. doi:10.1177/0956797610392926.

Turkheimer, E., Harden, K. P., D'Onofrio, B., & Gottesman, I. I. (2009). The Scarr-Rowe interaction between measured socioeconomic status and the heritability of cognitive ability. In K. McCartney, & R. A. Weinberg (Eds.) *Experience and development: A festschrift in honor of Sandra Wood Scarr* (pp. 81–97). New York: Psychology Press.

Turkheimer, E., & Harden, K. P. (in press). Behavior genetic research methods: Testing quasi-causal hypotheses using multivariate twin data. In H. T. Reis & C. M. Judd (Eds.), *Handbook of Research Methods in Personality and Social Psychology*. Cambridge: Cambridge University Press.

Turkheimer, E., & Waldron, M. (2000). Nonshared environment: A theoretical, methodological, and quantitative review. *Psychological Bulletin, 126,* 78–108. doi:10.1037//0033-2909.1261.78.

Turkheimer, E., Haley, A., Waldron, M., D'Onofrio, B., & Gottesman, I. I. (2003). Socioeconomic status modifies heritability of IQ in young children. *Psychological Science, 14,* 623–628.

van den Oord, E. J. C. G., & Rowe, D. C. (1998). An examination of genotype-environment interactions for academic achievement in an U.S. national longitudinal survey. *Intelligence, 25,* 205–228. doi:10.1016/S0160-2896(97)90043–X.

van der Sluis, S., Willemsen, G., de Geus, E. J. C., Boomsma, D. I., & Posthuma, D. (2008). Gene-environment interaction in adults' IQ scores: Measures of past and present environment. *Behavior Genetics, 38,* 348–360. doi:10.1007/s10519-008-9212-5.

van der Sluis, S., Posthuma, D., & Dolan, C. V. (2012). A note on false positives and power in G × E modelling of twin data. *Behavior Genetics, 42,* 170–186. doi:10.1007/s10519-011-9480-3.

Yeo, R. A., Gangestad, S. W., Liu, J., Calhoun, V. D., & Hutchinson, K. E. (2011). Rare copy number deletions predict individual variation in intelligence. *PLoS One, 6,* 1–8. doi:10.1371/journal.pone.0016339.

Chapter 3
Genetic and Environmental Influences on Intellectual Disability in Childhood

Michèle Carlier and Pierre L. Roubertoux

Scientific literature on atypical development is so vast that a systematic review could not fit in some 30 pages; therefore, we had to make choices. First, we have limited our presentation to intellectual disability (ID), leaving aside behavioral and psychiatric disorders. After defining ID, the main causes are presented (genetic and environmental) with special emphasis on gene–environment correlations and/or interactions. We then selected two genetic disorders linked to ID (Phenylketonuria and Fragile X) to present both the research methodologies and the type of findings, before discussing the contribution of cross-syndrome comparisons. To uncover a causal link between genetic events and a behavioral phenotype, it is often essential to use model organisms. The advantage of such models, plus the requirements and limitations involved in their use, are presented before concluding the chapter.

3.1 Definition and Epidemiology

3.1.1 Definition

The history of non-normative development in cognition is probably as long as the story of the human species, although the dating of biological evidence on developmental disabilities is relatively recent. Czarnetzki et al. (2003) studied osseous

M. Carlier (✉)
Aix-Marseille Université, CNRS UMR 7290 Psychologie Cognitive, Fédération de Recherche 3C - Comportement Cerveau Cognition, case D, Bât 9, 3 Place Victor Hugo, 13331 Marseille Cedex 3, Marseille, France
e-mail: michele.carlier@univ-amu.fr

P. L. Roubertoux
Aix-Marseille Université, Génétique et Neurosciences, INSERM U 910, Génétique Médicale, Génomique Fonctionnelle, 27 bvd Jean Moulin, 13385 Marseille Cedex 5, Marseille, France
e-mail: pierre.roubertoux@univ-amu.fr

D. Finkel, C. A. Reynolds (eds.), *Behavior Genetics of Cognition Across the Lifespan*, Advances in Behavior Genetics 1, DOI 10.1007/978-1-4614-7447-0_3,

remains (7,063 specimens) from different points in Europe; one skeleton dated as being 2,550 years old was diagnosed as a woman with signs of Trisomy 21 (T21) who had died at the age of 18 or 20. "Psychomotor retardation" was described in the Book of Hearts in Eber's papyrus written in 1,600 BC (Okasha 1999). It is beyond the scope of this chapter to give an overview of the main views on intellectual disabilities from ancient times to the present, but one point is worth noting: the acceptance of disabled persons by society has varied at different periods of time and from country to country. Misconceptions on the nature and causes of non-normative development have had dramatic consequences for persons with intellectual disabilities (Roubertoux 2004; Smith 2006), and while considerable progress has been made, a recent survey of American college students has shown that misconceptions about cognitive and adaptive behavior of people with mild intellectual disabilities are still relatively frequent (Musso et al. 2012) and that an implicit negative stereotype is still alive (Enea-Drapeau et al. 2012). In modern Western societies, the late eighteenth century and early nineteenth century marked a major milestone in the care of these persons. Jean-Marie Gaspard Itard (1774–1838) was probably the first person in a western country to attempt to stimulate and educate an "incurable" child—the wild boy of Aveyron discovered in 1798. His observation of the child's behavior (Itard 1801) is a model of humanistic thinking and contains many accurate insights into developmental psychology. Later, Edouard Seguin (1812–1880) developed a systematic method of education for "idiots" and "imbéciles" at Bicêtre Hospital (Paris). Despite the very good appraisal of his method by the Académie royale des sciences (Royal Academy of Science, December, 1843), Seguin came into conflict with the hospital administration; he left the hospital, set up a private center and published a book with the first description of children with T21 (Seguin 1846). He later migrated to the United States and in 1876 became the first president of the Association of Medical Officers of American Institutions for Idiotic and Feeble-Minded Persons. The organization has changed its name several times over the past 100 years, going from the "American Association on Mental Deficiency," (AAMD), to the "American Association on Mental Retardation" (AAMR) and finally, in 2006, to the "American Association on Intellectual and Developmental Disabilities" (AAIDD). Changes in the nomenclature and the choice of the term "intellectual disabilities" to replace "mental retardation" are in line with changes in biological and social knowledge of non-normative development, but debate on the pros and cons is still continuing—see Fisch (2011); and for a historical presentation of the definitions of mental retardation in the AAID, see Leonard and Wen (2002).

Four classifications system are widely used to describe ID. In the 11th edition of the *AAIDD Definition Manual on Intellectual Disability*, ID is characterized as a "limitation both in intellectual functioning and adaptive behavior as expressed in conception, social, and practical adaptive skills" (2010, p. 1). The disability is developmental if it occurs before the age of 18. Five assumptions need to be validated for the definition to apply: (1) consideration of the context of the community environment where the person lives, (2) consideration of cultural and linguistic diversity, (3) a valid assessment of not only limitations but also strengths, (4) a profile of support needed, and (5) the expectation that the person will improve.

The definition in the International Classification of Diseases and Related Health Problems (10th revised edition: ICD-10 version 2010) of the World Health Organization (WHO) is close to the AAID definition, but the term "mental retardation" is still used. The classification considers mental retardation as "impairment of skills manifested during the developmental period, skills, which contribute to the overall level of intelligence, i.e., cognitive, language, motor, and social abilities. Retardation can occur with or without any other mental or physical condition." Assuming improvement "as a result of training and rehabilitation" is also a key point, but the focus on social adaptation is less significant than in the AAIDD definition as there is no requirement for an assessment of social adaptation ("Degrees of mental retardation are conventionally estimated by standardized intelligence tests. These can be supplemented by scales assessing social adaptation in a given environment.")

All 191 WHO Member States officially endorsed the International Classification of Functioning, Disability, and Health, known more commonly as the ICF (WHO, 2001; 2002). It is a classification of health and health-related domains on three levels: the level of the body or body part, the level of the whole person, and the level of the whole person in a social context. The classification uses four lists: body function (mental function, sensory functions, pain, etc.), body structure (structures of the nervous system, structures of the cardiovascular, immunological and respiratory systems, etc.), activity and participation (learning and applying knowledge, communication, etc.), and environmental factors (support and relationships, the natural environment and human-made changes to the environment). Impairment, disability, and handicap are distinct concepts. Impairment is defined as "any loss or abnormality of a psychological, physiological or anatomical structure or function" (e.g., blindness or mental retardation). Disability is a "restriction or lack (resulting from an impairment) of ability to perform an activity in the manner or within the range considered normal for a human being" and describes a functional limitation or restriction of activity caused by impairment (e.g., difficulty in seeing or speaking). Handicap is defined as a "disadvantage for a given individual, resulting from an impairment or disability that limits or prevents the fulfillment of a role that is normal (depending on age, sex and social and cultural factors) for that individual." The term is also a classification of "circumstances in which disabled people are likely to find themselves," and finally "Such disadvantages affect the interaction of the person with a specific environment and culture" (e.g., being confined to home or unable to use public transport). This "ecological" perspective is a significant step forward from the classical medical and functional models (although these are still useful for certain cases), which focus on the personal deficits and limitations of the individual and fail to take into account the social, economic, and attitudinal barriers faced by persons with disability (Fuchs et al. 2007).

The fourth well-known classification of ID is from the American Psychiatric Association's Diagnostic and Statistical Manual (DSM-IV). The 5th edition is scheduled for May 2013, and proposed revisions are already available on the American Psychiatric Association Web site (code name: A 00-01). It is relevant to quote the definition here: "Intellectual Developmental Disorder is a disorder that includes both a current intellectual deficit and a deficit in adaptive functioning with onset during the developmental period. All three of the following criteria must be met."

The criterion for a diagnosis of ID is the same in all classifications: approximately, two standard deviations below the mean (i.e., an IQ score below 70 in the most commonly used scales). The WHO classification still has subdivisions for mild, moderate, severe, and profound deficiency with approximate IQ ranges of, respectively, 50–69, 35–49, 29–34, and under 20. It may be noted that these subdivisions are now outdated. Most of the psychological instruments currently available for assessing general cognitive levels (Wechsler, Stanford-Binet, and Kaufman scales) have floor effects for persons with disability and generally fail to detect valid differences in the lower ranges (Youngstrom et al. 2003; Carlier and Ayoun 2007; Heissl et al. 2009; AAIDD 2010). Some studies, have overcome this difficulty by considering only the raw scores (Hessl et al. 2009; Couzens et al. 2011) or by using a scale developed for children younger than the subjects of the study (Chabrol et al. 2005). Another solution is to develop a specific assessment battery for individuals with ID (Edgin et al. 2010), but such types of procedures are not relevant for epidemiological studies where the use of standardized scores is mandatory.

3.1.2 Epidemiology

The use of the ICF-based indicators has been more widely adopted in some countries (Australia, Canada, India, Italy, Japan, Mexico, and the Netherlands) than in others (WHO Web site, January 30, 2012). The WHO Web site refers users to the United Nations Statistics Division (DISTAT) website for epidemiological data. The demographic yearbook recently published by the United Nations (2011) does not contain data on handicaps, but some items should be of interest to the reader of this chapter (items on live births, infant births, and fetal deaths). An earlier DISTAT report (1990) included a warning for interpreting prevalence rate comparisons across countries because of differences in the concepts and methods used to identify persons with disabilities, and two interesting comments were made. First, for reports on severe impairments (blind, deaf, leg amputated, mentally retarded, etc.), rather than mild-to-moderate impairments, male/female ratios of disabled subjects were greater than 1.0, indicating a predominance of disabled males for severe impairments; and second, a large proportion of surveys found that, on the average, disabled persons are less educated, have lower socioeconomic status, and are more likely to reside in rural and poor areas compared to able-bodied persons.

Studies using ICF criteria can be found in Canadian statistics, and data extracted from a national survey (Participation and Activity Limitation Survey, 2006) are presented in Table 3.1. In children, disability rates are higher for boys than for girls, in particular for the types of disability linked to ID, i.e., affecting development, communication, learning, and memory. Reported rates of learning and developmental disabilities among young people are also higher for boys, but the difference is smaller; the reverse is found for emotional/psychological disability and pain. One limitation of the PALS (as for other surveys developed with the WHO) was the use of self-reporting to identify disability (or reporting by parents or guardians for children (14 and under)).

Table 3.1 Disability type by gender and age for children and youths with disabilities (percentage of the population)—from *Disability in Canada: A 2006 Profile.* (Human Resources and Skills Development Canada 2011)

Disability type	Ages					
	Children				Youths	
	Under 5 years (Overall: 1.7)		5–14 (Overall: 4.6)		15–19 (Overall: 4.6)	
	Boys	Girls	Boys	Girls	Boys	Girls
Chronic condition	1.4	0.9	3.8	2.2	–	–
Developmental	1.3	0.8	1.9	0.9	1.4	0.8
Hearing	0.2	0.2	0.6	0.4	0.5	0.5
Seeing	0.2	0.2	0.5	0.3	0.5	0.6
Communication	–	–	2.8	1.3	1.6	1.1
Emotional/psychological	–	–	2.1	1.0	0.8	1.1
Learning	–	–	4.1	2.2	3.3	2.1
Agility	–	–	1.3	0.6	1.3	1.4
Mobility	–	–	0.6	0.6	1.6	1.8
Memory					1.1	0.7
Pain					1.6	2.5
Other	–	–	0.2	0.2	2.2	2.3

More than one disability type could be identified for each survey respondent. The numbers of specific disability types differed depending on the survey respondent's age (from 4 to 11)

Fuchs et al. (2007) also used ICF and reported epidemiological data from a survey in the Canadian province of Manitoba, describing the population of children with disabilities cared for by the child welfare system in Manitoba during the 2004–2005 fiscal year (n = 1,869). One-third of children in care were found to have a disability. Boys are overrepresented (60 %), as are First Nations children (68.7 %). The number of children with disabilities increases until the age of 13 years and then declines. Disabilities were classified into six main categories: intellectual (75.1 % of the children affected), mental health (45.8 %), medical (22 %), physical (18 %), sensory (5 %), and learning (3 %). Children often had more than one disability (58.1 %) and the most common combination was intellectual and mental health. Approximately 49 % of disabilities had no known cause. Fetal alcohol spectrum disorder (FASD) was diagnosed in 34.2 % of the children and for 51.6 % maternal substance abuse was considered, or suspected, to be the cause of the disability.

McDermott et al. (2007) noted that the distinction made by epidemiologists between incidence (the risk of developing a condition within a specified period of time) and prevalence (the total number of cases in the population at a given time) is difficult to use for ID as this may vary for the same person at different periods of her/his life (prenatally, at birth, at school age). Reviewing data (mostly published before 2000), the authors reported a prevalence of ID at 10–20 per 1,000, but lower and higher estimates could also be found depending on the populations surveyed and methods used (nationality and age of the population, national registry or not, cross-sectional data on children in mainstream public schools, data from special education schools, etc.). Such inconsistency in data collected may be largely attributable to the classifications system revisions. In practice, many epidemiological studies do not take adaptive behavior into account and the sole criterion used to estimate the prevalence of ID is IQ. The few studies to include adaptive behavior as one of the criteria suggests that the prevalence of ID would go down from 2 to 1% when it is included (Leonard and Wen 2002).

A number of consistencies can be found in the literature. First, the prevalence of mild-to-moderate ID is higher than severe ID. Second, age-specific prevalence rates increase with age, peaking at about 10–14 years. This trend could reflect differences in case of ascertainment, the ability of adults with ID to adapt to the demands of society with the passage of time, IQ changes, or differentials in mortality between people with ID and the general population (Leonard and Wen 2002). Third, males are more likely to have ID than females, especially in the younger age groups. Some biological factors may be put forward to explain the higher proportion of males (see below). Fourth, social, economic, cultural, and ethnic factors influence the prevalence of ID. A higher prevalence of moderate ID was consistently found in groups with low socioeconomic status, and with certain ethnic groups (e.g., Afro-American children, indigenous Australians, Canadian Aboriginals). Many variables could explain these differences (Leonard and Wen 2002), for example, social, demographic, economic and cultural factors, prenatal and/or postnatal biological factors, plus probable interactions between these factors.

3.2 Main Causes of ID

ID has many different causes. The AAIDD 2010 proposes a multifactorial approach with four types of factor: biomedical (genetic disorders, nutrition), social (social and family interaction, child abuse), behavioral (e.g., activities causing injury, and maternal substance abuse), and educational (availability of educational support); the last factor is outside the scope of this chapter, therefore, it will not be discussed here. Another categorization can be made according to the timing of the risk factors—prenatal, perinatal, and postnatal.

3.2.1 Genetic Factors

There is no doubt that genetic factors are of primary importance in the etiology of ID. How many genes are involved? In the database Online Mendelian Inheritance in Man (OMIM), 1,883 items can be found when using "mental retardation" as the key words (but only 181 with the more recent term of "intellectual disability"). The number of entries in the catalog has been increasing since the beginning of the century with the explosion of genetic information (McKusick 2007). Certain disorders occur relatively frequently, while others are very rare (Billuart et al. 1998; Chabrol et al. 2005, for example) and more difficult to detect. It is frequently assumed that in approximately half of ID cases, there is no known cause, but more and more requests are being made to screen for genetic defects in cases of moderate-to-severe ID.

Most numerical chromosomal anomalies are lost by miscarriage. Trisomy 13 and 18 are found among live births but also have a high rate of fetal death. Many fetuses with T21 can survive and consequently the syndrome is the most common genetic disorder involving ID. An extra copy of one chromosome is relatively easy to detect, but more sophisticated techniques are needed to detect balanced or unbalanced chromosomal rearrangements. de Vries et al. (2001) developed a checklist to help preselect cases for subtelomere testing, which included: (1) family history of ID, (2) prenatal onset growth retardation, (3) postnatal growth abnormalities, (4) ≥2 dysmorphic facial features, and (5) one or more dysmorphic nonfacial features. Following these recommendations, Popp et al. (2002) selected 30 patients with unexplained developmental delay; using conventional cytogenetics and multiplex FRSH telomere integrity assay, chromosomal aberrations were detected in 4 of the 30 patients (13.3 %). All were young children (under 3 years). The authors observed that facial dysmorphism is more difficult to detect in younger children; it is, however, an important criterion in the decision to carry out genetic screening. de Vries et al. (2003) reviewed 20 studies including 2,500 persons with ID of unknown cause. In 125 patients (4.8 %), a telomeric defect was detected. One year later, Koolen et al. (2004) confirmed the high probability of finding subtelomeric rearrangements in patients with unexplained ID and reported an aberration in 14 of 210 patients (6.7 %: 10 deletions and 4 duplications). Once a telomeric defect is found, a key question still has to be solved: can the defect be considered as the cause of the ID? It is important to establish whether the defect has been observed in other patients with ID, and examples can be found in recently published data. Manolakos et al. (2010) used array-CGH in a cohort of 82 Greek children (mean age 4.9 years) with unexplained ID (normal karyotype), dysmorphic facial features, and congenital malformations, and detected 13 patients (15.8 %) with cryptic chromosomal imbalances: 6 patients with duplications, 5 patients with deletions, 1 with triplication and 1 with two duplications. In 3 out of the 13 patients, the chromosomal rearrangements occurred *de novo* and were said to be the putative cause of the ID. As the other aberrations had been inherited from a healthy parent, the authors concluded that they were probably benign. After sequencing the exomes of 10 case-parents trios, Vissers et al. (2010) identified unique nonsynonymous *de novo* mutations in nine genes. Three genes do not seem to play a role in ID, but the other six genes are linked to ID. The authors concluded that *de novo* mutations are a major cause of unexplained ID.

Autosomal single mutations with either dominant or recessive or X-linked modes of transmission and short deletions are known to be linked to ID. Severe dominant forms of ID are not transmitted as it is unlikely that the patients will reproduce. According to Ropers (2008), little is known about the prevalence of dominant ID, but such cases are probably not so rare, given the high proportion of apparently relevant *de novo* copy number variants (CNV). Autosomal recessive forms of ID (AR-ID) due to mutations are probably common, although the often-quoted estimate of up to

25 % of unexplained cases of ID has not been confirmed by recent epidemiological data. Cumulating the data, Ropers (2008) concluded that 22 loci for nonsyndromic AR-ID have been found. To date, only six AR-ID genes[1] have been identified, but Rogers predicted "their number will soon explode." (p. 244). And, he was right. Performing homozygosity mapping in a cohort of 136 consanguineous families mainly from Iran, he and his coauthors (Najmabadi et al. 2011) announced 3 years later, that they had discovered 50 novel AR-ID genes. However, the causal links between the new genes and ID have yet to be confirmed. In the same year, Abou Jamra et al. (2011) used the same strategy with 64 Syrian consanguineous families with nonspecific ID and uncovered 11 novel loci. On the basis of the number of ID genes on the X chromosome, Schuurs-Hoeijmakers et al. (2011) estimated that there are approximately 2,000 AR-ID genes (11 % of the autosomal protein-coding genes). To overcome the difficulties encountered by colleagues who used large (but very rare) consanguineous families, these authors performed homozygosity mapping in outbred families with multiple ID-affected siblings. In 10 families, they found 21 homozygous regions shared by affected siblings. The regions overlap neither with the six nonsyndromic AR-ID genes nor with the syndromic AR-ID genes. The authors concluded that homozygosity mapping in outbred families may help identify novel AR-ID genes.

With more boys with ID among institutionalized children and the disproportionate number of families with intellectually disabled boys, only, there has long been an argument for the sex linkage of ID, and this form of transmission is easier to detect. In a survey of all children with a very low IQ (30–50) born between 1955 and 1964 in the State of New South Wales (Australia), Turner and Turner (1974) estimated a prevalence rate from brother pairs in excess of 0.74/1,000 males, concluding that one in every five of the "mentally retarded" boys in the IQ range of the survey may have an X-chromosomal form of ID. Ropers (2008) reported more than 80 genes for X-linked ID identified after collecting data in large cohorts of families studied by international consortia (e.g., EuroMRX consortium; de Brouwer et al. 2007). The Fragile X syndrome may account for 25 % of X-linked ID. One year later, Géczet al. (2009) estimated that more than 90 different X-linked ID genes (~ 11 % of the X-chromosome genes) had been identified but that many more genes remain uncharacterized. In many cases, ID is not the sole disorder; there is a frequent co-occurrence of autism spectrum disorder, epilepsy, or behavioral and psychiatric problems.

3.2.2 Environmental Factors

Mwaniki et al. (2012) conducted a systematic review to estimate risks of long-term neurocognitive and other sequelae after intrauterine and neonatal insults such as preterm-birth complications, intrapartum-related factors (hypoxic ischemic encephalopathy, infections and in particular sepsis, meningitis, and neonatal tetanus),

[1] In some papers, the old label "mental retardation" is kept and the genes are labeled MR genes.

and other conditions such as jaundice and congenital infections (cytomegalovirus, toxoplasmosis, syphilis, and rubella). Of 28,212 papers identified by search, only 153 met their inclusion criteria (identifiable and well-defined neonatal insult, the use of standardized tests or controls in neurodevelopmental assessment, and less than 20% of survivors lost to follow-up). In all, 22,161 neonates were assessed. The overall median risk of at least one sequela in any domain was 39.4%. The most common impairments were learning difficulties and cognition or developmental delay (59%). Multiple impairments were frequent (e.g., cognitive impairment, motor impairment, and hearing and vision loss). Behavioral problems were relatively low (11%) but may have been underestimated (Thompson and Gillberg 2012).

Alcohol and drug use by pregnant woman are risk factors for the neonate causing intrauterine growth retardation, birth defects, altered behavior, and withdrawal syndromes. Fortunately, most adverse effects of prenatal drug exposure are rare or less than what might be expected, with the exception of alcohol exposure (Chiriboga 2003). The highly adverse effect of alcohol appears to have been known since ancient times (see, for example, the story of Samson's mother in the Bible, and Aristotle's warnings), although whether the teratological effects were directly known at the time is debatable (Warren and Hewitt 2009). The major features of fetal alcohol syndrome (FAS) can be divided into three categories: intrauterine and postnatal growth retardation, craniofacial dysmorphisms—e.g., small palpebral fissures, flattened philtrum, and thin upper lip—and evidence of central nervous system anomalies—decreased cranial size, structural brain anomalies, neurological hard or soft signs (Stratton et al. 1996). It is not easy to diagnose FAS at birth as facial dysmorphisms have also been reported in children exposed to substances other than alcohol (Chiriboga 2003). Fetal alcohol exposure is the leading known environmental cause of ID and a dose response is observed: the stigmata are present in proportion to the degree of exposure, but it is difficult to detect a threshold below which the risk does not exist; pregnant women are therefore advised to avoid drinking alcohol. Less severe outcomes for the child are categorized into FASD. The prevalence in the United States has been reported as 1–3 and 9.1 per 1,000 live births for FAS and FASD, respectively (Chudely et al. 2005). It may be higher in some countries and/or populations as can be seen in the following data. In a Canadian First Nations community, the prevalence of FAS and partial FAS was estimated at 55–101 per 1,000. In other Canadian communities, the rates were also very high, but it was very low in Saskatchewan: 0.51 per 1,000 in the period 1973–1977 (Chudley et al. 2005). In a recent central Italian population-based screening of children attending primary school, the prevalence of FAS was estimated at 4.0–12.0 per 1,000, and of FASD between 23.1 and 62.1 per 1,000 (May et al. 2011). In a group of 100 Israeli children under the age of 2 who were candidates for domestic adoption or in foster care, 15% either had FASD or were at risk of developing symptoms (Tenenbaum et al. 2011). In short, it is impossible to establish a prevalence level that applies to all populations.

Many postnatal environmental risk factors have been reported and some examples are presented here. Severe malnutrition during development can cause infant mortality, smaller physical size, and ID among the survivors. The relation between malnutrition and cognitive level is complex as both nutrition and intellectual factors are associated with a number of social factors (e.g., the caregivers who may themselves be ill or malnourished, the geographical conditions, and/or socioeconomic status). Data published by Ivanovic et al. (2000) are particularly valuable. The authors studied the long-term effects of severe undernutrition during the first year of life on brain development and IQ in two groups of poor Chilean high school graduates, one which had been undernourished and the other which had not suffered from undernutrition. The socioeconomic conditions were similar for both groups, except for the mean number of years of education of the mothers, which was 2 years less for the group that had been undernourished. The group of students who had suffered from undernutrition recorded a mean IQ lower than the other group (a difference of up to 24.5 points). Multiple regression analysis showed that maternal education and undernutrition explained 71.4% of the IQ variance. The effect of childhood iodine deficiency is considered to be the most common cause of preventable ID worldwide, with pregnant women and young children being particularly susceptible. The number of iodine-deficient countries is decreasing, but more than 200 million school-age children still have an insufficient iodine intake (Andersson et al. 2012). An association has also been established between the body burden of lead (in blood or tooth dentine) and a lower IQ, even after adjustments are made for other environmental factors (Taylor and Rogers 2005).

We do not want to end this short and nonexhaustive overview of environmental risk factors without mentioning acute and chronic psychological stress, physical abuse, exposure to family violence, and institutional deprivation. One example can be taken from the excellent English and Romania Adoptee (ERA) study—see Rutter et al. (2010)—and other chapters of the 2010 Monograph. In a follow-up of children who had suffered severe institutional deprivation in Romania during the Ceauşescu regime, the team found sound evidence that institutional deprivation does truly cause deprivation-specific psychological patterns (quasi-autistic patterns, disinhibited attachment, inattention/overactivity, and cognitive impairment).

3.2.3 Gene–Environment Interactions and/or Correlations

The etiology of ID is complex, and in practice it is often difficult to disentangle genetic and environmental risk factors when considering individual cases. The idea of a causal linear relationship between genes and behavior, or between the environment and behavior is now obsolete (Roubertoux and Carlier 2007, 2011). In almost all cases, the phenotype linked to a specific genetic disorder is highly variable. Causes may be biological (e.g., an epistastic effect or an interaction between genes) or environmental (e.g., environmental adversity or cultural transmission). De Smedt et al. (2007), to cite an example, reported that one of the factors contributing to the variability of the 22q11 deletion phenotype (Di George/Velo-Cardio-Facial syndrome) is the mode of transmission of the deletion (*de novo* vs. familial): children with familial deletion have a lower IQ than children with a *de novo* deletion. Studying a sample of 103 children, they found that the difference in IQ between the two groups of children could be attributed to the lower educational attainment level of the parents of children with familial inherited deletion.

Another example can be found with Fragile X syndrome. Maternal responsivity predicts language development in young children with Fragile X syndrome (Warren et al. 2010); it has been well established in the literature that maternal responsivity is dependent on the child's behavior (transactional model; Warren and Brady 2007). In disorders such as Fragile X (see below), not only does the child have a full mutation affecting his/her ability to communicate, but the mother also has a genetic defect (either a full mutation or premutation), which may, in turn, affect her own skills required for communicating with her child(ren). There is evidence supporting this hypothesis, showing that differential sensitivity to life stress is associated with CGG repeat length (Seltzer et al. 2011), which characterizes the mutation (see below).

Fetus vulnerability in FASD is an excellent illustration of gene–environment interaction. Less that 10% of women who drink during pregnancy have children with FAS. What is the reason for the differential vulnerability of fetuses? One explanation may be in differences in the genetic background of the fetus and of the mother. Alcohol dehydrogenase (ADH) is the principal enzyme catalyzing ethanol oxidation to acetaldehyde. Three functionally distinct polymorphisms exist for *ADH1B* with different binding affinity for alcohol and maximal turnover rates. Warren and Li (2005) reviewed the literature on human and animal studies and reported that the presence of either *ADH1B*2* or *ADH1B*3* alleles in the maternal and fetal genomes appears to afford protection from alcohol-derived teratogenesis; this is not the case when the *ADH1B*1* allele is present. Other candidate genes are highly probable (Lombard et al. 2007).

3.3 Neuronal and Behavioral Phenotypes of Genetic Developmental Disorders

As explained when introducing the chapter, we are not presenting all genetic diseases associated with ID. McKusick's team, in the database OMIM, is carrying out this useful and daunting task. We have chosen to give a brief review of current knowledge of two disorders (Phenylketonuria (PKU) and Fragile X) to show how knowledge has progressed in the field of ID linked to genetic disorders. We shall then discuss the contribution of a research method used extensively in this area: the cross-syndrome comparison.

3.3.1 A Single Gene Genetic Disorder: PKU

PKU (OMIN 261600) is an autosomal recessive genetic disorder caused by the *PAH* gene (located at 12q23.2), which encodes the phenylalanine (Phe) hydroxylase enzyme needed to metabolize Phe into tyrosine. Some 500 mutations of the gene have been identified. This locus heterogeneity explains part of the within-group biological phenotypic variability (Kayaalp et al. 1997). The mutated allelic form produces a deficiency of the enzyme Phe hydroxylase, thereby causing an accumulation of Phe, which in turn affects the biosynthesis of neurotransmitters (dopamine and norepinephrine). Elevated Phe and low tyrosine levels are thought to impair, *inter alia*, brain myelination. In Europe, PKU occurs in approximately 1/10,000–1/15,000 births, but regional differences in incidence have been reported (see Williams et al. 2008 for a qualitative review). As with any other autosomal recessive genetic disorder, parental consanguinity increases the prevalence of PKU. Most cases of untreated PKU are associated with growth failure, microcephaly, seizures, and severe ID caused by the accumulation of a toxic by-product of Phe metabolism. Interestingly, some untreated patients develop normally (Möller et al. 2003). Treatment is a low-Phe diet and dietary compliance is difficult in adolescence and adult life; even though new treatments have been developed, diet is still the best therapy (Giovannini et al. 2012). What are the outcomes for early-treated individuals with PKU? In our qualitative review on cognitive development, we concluded (Carlier and Ayoun 2007) that these individuals had a mean IQ close to 100 (i.e., normal) provided they followed the diet until at least 10 years of age. However, some lower scores, compared to controls, were found for executive functions. DeRoche and Welsh (2008) conducted a meta-analysis on neurocognitive outcomes of early-treated patients with "classical PKU" (blood Phe levels from 600 to 1,200 μmol/l) to establish whether a profile of deficits in intelligence and executive functions had emerged from empirical research published between 1980 and 2004. The meta-analysis covered 33 studies totaling 1,109 individuals with early-treated PKU, and 1,145 peer control individuals from 5 to 35 years of age. For intelligence tests, the effect size of the differences, with lower scores for patients, was small to moderate (from 0.20 to 0.42). Differences were greater for executive functions with an effect size of 0.79 for the Total executive function score, and up to 1.15 for cognitive flexibility. Of the different measurement tools, intelligence tests had effect sizes that were homogeneous across outcomes, but this was not the case for executive functions tasks. Studies published more recently have reported larger effect sizes for these functions than did earlier studies. This may be because of advances in the assessment of executive functions and/or the use of more sensitive tasks to detect subtle impairments of cognitive processes. The deficit observed in executive functions was consistent with the prefrontal model of PKU: as noted earlier, the mutation disrupts the normal synthesis of dopamine and norepinephrine. However, as DeRoche and Welsh (2008) observed, there is an overlap between neuropsychological tasks measuring "prefrontal processes" and "white matter integrity." The white matter hypothesis could, therefore, also be put forward. Neuroimaging data have provided considerable evidence of white matter abnormalities associated with early-treated PKU; see, for example, Anderson et al. (2007) for a qualitative review. DeRoche and Welsh (2008) published their own data, presenting new evidence of links between diffuse white matter damage and cognitive deficits, including attention and executive functions (citing planning ability, spatial organization, cognitive flexibility, and conceptual reasoning). There was one limitation to their study and that was the large mean difference in IQ between the PKU group and the control group (WISC-III; 91 vs. 104). Controlling for the effect of IQ (and not only for age, gender, and SES) would have provided interesting additional information. A study by Anderson et al. (2007) showed peripheral levels of Phe to be inversely correlated with cognitive performance, confirming earlier observations. With two groups matched for IQ, gender, and demographical variables, Banerjee et al. (2011) assessed executive strategic processing during verbal fluency performance in 32 children with PKU. The two groups of children had the same mean IQ (i.e., global cognitive level), but mean differences were found in phonemic fluency trials (word generation in a food/drink category; words beginning with S and F), and for a number of semantic or phonemic switches. On average, the performance of the PKU group was 0.6 standard deviation below the control group (i.e., a medium effect size), with a larger effect size for the older children (1.5). No significant correlations were found between any of the Phe and verbal fluency variables. The negative correlation between IQ and Phe level has been well-documented (Waisbren et al. 2007 for a meta-analysis), but less is known about more specific aspects of cognition. Viau et al. (2011) studied a sample of 55 patients and found that the correlations between cognitive tests and treatment variables were highly variable and depended on the cognitive variable under consideration. In addition to the Phe level, the ratio between Phe and tyrosine levels may potentially play a role in brain development, and therefore in cognitive processes (Sharman et al. 2010). Campistol et al. (2011) draw attention to the fact that the mild form of hyperphelylalaninemia, characterized by a plasma Phe concentration lower than 360 µmol/l, may also, to a lesser extent, have a negative impact on cognitive development.

Not only is a high Phe level a high-risk factor for patients, but also, in the event of pregnancy, for the fetus, with more neonatal sequelae in untreated pregnancies (see Prick et al. 2012 for data on a large cohort).

3.3.2 A Single Gene X-Linked Disorder: The Fragile X Syndrome

Fragile X syndrome (FXS, OMIM 300624) is the most frequent cause of ID because of mutations to a single gene, and also the most common monogenic cause of Autism Spectrum Disorder characterized by three areas of dysfunction before 3 years of age: atypical social behavior, deficits in verbal and nonverbal communication, and repetitive and highly restricted interests. The gene involved is *FMR-1* (Fragile X site Mental retardation–1) located at Xq27.3. The syndrome is usually caused by an expansion of cytosine guanine-guanine (CGG) repeats in the 5' untranslated region of exon 1 of the *FMR-1* gene. The normal size of CGG repetitions ranges from 5 to 54, the most common value being 30. In the event of CGG expansion being transmitted by the female, the number of repetitions increases and the woman transmits a premutation to her offspring; any of her female offspring are then likely to transmit an even larger number of repetitions to the next generation. When there are more than 200 repetitions, the offspring will carry the full mutation with hypermethylation of the CpG island in the promoter region of the *FMR1* gene, the consequence of which is that Fragile X mental retardation protein (FMRP) is not produced. Crawford et al. (2001) conducted a review of population-based studies and estimated the prevalence of the full mutation, which ranges from 1/3,717 to 1/8,918 males in the general Caucasian population. The prevalence of the premutation in Caucasian populations is ~1/1,000 for males and from 1/246 to 1/468 for females. FMRP is mainly expressed in the brain and gonads and has multiple functions in RNA metabolism, including mRNA decay, dendritic targeting of mRNAs, and protein synthesis (see De Rubeis and Bagni 2011; De Rubeis et al. 2012 for recent reviews). It is beyond the scope of the chapter to go deeply in the biology of FXS, but one point is worth noting: understanding the role of the FMRP in the development of the brain and adult neurogenesis is, of course, critical in the development of therapies. New perspectives are reviewed by Wang et al. (2012).

The facial characteristics are more pronounced in males than in females and are more apparent in male carriers after puberty (long narrow face, large ears, and prominent jaw and forehead). Motor and cognitive development is delayed in infancy and adolescence (Reiss and Dant 2003 for a thorough qualitative review); the mean adult IQ is in the range 42–55. The estimate clearly depends on the method of recruitment of patients and the psychological test used to measure IQ. As noted previously, intelligence scales have very limited discriminative power in low scores because floored scores are frequent. This issue can be illustrated once more as per the data reported by Hessl et al. (2009). They assessed a large sample of 217 school-age boys and girls with FXS. They found a mean IQ of 50 with high variability (standard deviation 19.5, range 40–123). The percentage of participants with floored standard scores in the subtests of Wechsler's WISC III scale ranged from 40% (picture completion) to 70% (arithmetic). In other words, the mean IQ does not make much sense. The rate of intellectual development during school age was measured in a longitudinal design where boys and girls with FXS and their unaffected siblings were assessed twice. During the time, between the first and second assessments (on average 3.89 years), the annual rate of intellectual development was approximately 2.2 times faster in the unaffected children compared to the children with FXS (Hall et al. 2008). This confirmed earlier data on longitudinal changes in IQ scores in children and adolescents with FXS (Fisch et al. 1996, 2002; Fisch 1997; Bailey et al. 1998) and highlights an important fact: in most cases, the IQ decline observed in children with ID cannot be attributed to cognitive regression but rather to a slower rate of development compared to normally developing children. Weaknesses in executive function, visual memory, visual–spatial relationships, arithmetic, and relatively less severe impairments in verbal skills were also reported in male individuals with FXS. Not only is there frequently Autistic Spectrum Disorder (between 15 and 25% according to Bailey et al. 2001; up to 35.1% in a sample of 37 boys with FXS—Hall et al. 2010), but also anxiety disorder and attention deficit hyperactivity disorder coexisting in males with FXS (Hagerman et al. 1985; see Gallagher and Hallahan 2012 for a clinical overview). The most prominent feature of FXS brain morphology is the dysgenesis of the dendritic spines that are longer and thinner than normal (Koukoui and Chaudhri 2007; De Rubeis and Bagni 2011; De Rubeis et al.

2012). Structural studies have pointed out a significantly enlarged caudate nucleus and a decrease in the size of the cerebellar vermis. Functional magnetic resonance imaging (MRI) studies have detected specific patterns of activation linked to cognitive and emotional tasks (Lightbody and Reiss 2009).

As females have two X chromosomes, the effects of the full mutation are less compared to males. Summarizing the published data, Jacquemont et al. (2007) considered that most of the females have an IQ in the 75–90 range, and about 25 % have an IQ less than 70. Interestingly, even though many females may have a normal intellectual level, other difficulties, in visuospatial processing and mathematics, have been reported, and, at the emotional level, hyperactivity, shyness, and anxiety (Lachiewicz and Dawnson 1994; Bennetto et al. 2001; Gallagher and Hallahan 2012).

A mild "fragile X phenotype" has been described in carriers of the permutation: the Fragile X Tremor/Ataxia syndrome (FXTAS, OMIM 300623). Older persons carrying a premutation are more likely to develop neurological disorders, with severe tremor and difficulty in walking and maintaining balance, and eventually Parkinsonism and cognitive decline (Hagerman and Hagerman 2007). In premutation females, the prevalence of premature ovarian failure is high (Schwartz et al. 1994; Jacquemont et al. 2007; Cornish et al. 2008). Subtle cognitive impairments were described in young premutated women (Goodrich-Hunsaker et al. 2011) and male premutation carriers (Hunter et al. 2012).

The discovery of CGG expansion in the *FMR-1* gene has provided an opportunity to study correlations between the characteristics of the mutation (the number of CGG repeats, the amount of protein produced, the methylation output ratio, and the activation ratio) and scores on cognitive tasks. Some authors have observed very high correlations, but recent publications suggest that the relationship is too small to make any individual prediction; in the study by Lightbody et al. (2006), for example, protein levels could explain only 7 % of IQ variance. In their follow-up study, Hall et al. (2008) concluded that the FMRP level accounted for only 5 % of the intellectual score at time 1, and 13 % at time 2. However, reference should be made to a recent report using the methylation status of FREE2 CpG sites to identify low-functioning full-mutation females (Godler et al. 2012). The study included 74 control females (<40 CGG repeats), 62 premutation females (55–200 CGG repeats), and 18 full-mutation females. Examiners blind to the DNA status of the participants assessed the participants with the Wechsler Intelligence test appropriate for the chronological age (WAIS III or WISC III). Data showed that methylation of FMR1 intronic CpG unit 10–12 was the most significant predictor of cognitive impairment (IQ<70) in full-mutation females. Specific impairment in arithmetic skills in high-functioning (IQ>70) full-mutation females was also linked to methylation in CpG unit 10–12. These authors did not observe a correlation between FMR1 activation ratio and FMRP production with any of the IQ measures and concluded that the activation ratio is probably a less reliable measure than methylation status of intronic CpG unit 10–12, given that methylation from random X inactivation varies among different tissues and organs. As interesting as this research may be, it needs to be cross-validated, preferably with more recent editions of the Wechsler's scales that are closer to contemporary approaches in cognitive and clinical psychology.

Table 3.2 Neuronal and behavioral phenotypes associated with four genetic diseases

Characteristics	Syndrome			
	Trisomy 21 (Down)	Williams-Beuren	Fragile X	Di George/ velocardiofacial
Neuronal	Reduced brain volume; reduced frontal and temporal lobes; major reduction of hippocampus	Reduced brain volume; anomalous sulcal patterning; primary dorsal stream impairment	Enlarged caudate nucleus and thalamus; decreased volume of cerebellar vermis, amygdala and superior temporal gyrus	Specific anomalies are uncommon; reduction of cerebral white matter; disturbance of the GABA-ergic nervous system?
IQ (mean)	<50	60–70	Boys: <55 Higher in girls	About 75
Language	Seriously impaired	Relatively strong	Relatively strong	Preserved
Spatial cognition	Relative strength	Seriously impaired	Impaired	Impaired
Behavioral or psychiatric disorders	Behavioral problems	Hypersensitivity to sound, attention deficit disorder	Autistic-type behavior, hyperactivity	Psychiatric disorders, Schizophrenia
Personality	Loving	Hypersociability	Anxiety	Shyness

3.3.3 Cross-Syndrome Comparisons

Table 3.2 gives an approximate picture of profiles of patients with one of four genetic syndromes: T21 (or Down), Williams-Beuren, Fragile X, and DiGeorge/Velocardio facial syndromes. Fragile X syndrome has already been described, but information is needed on the other three syndromes before commenting on the table. We must notice that the table shows large between-syndromes differences, but it hides the large within-syndrome variability for any phenotype.

Williams-Beuren syndrome (WBS, OMIM 194050) is caused by hemizygous contiguous gene deletion (1.5–1.8 Mb) on chromosome 7q11.23, which contains approximately 28 genes. The syndrome is rare. However, it has probably been underestimated: 1 in 20,000 live births was the prevalence reported in early publications but a more recent estimation is for up to 1 in 7,500 (Strømm et al. 2002). In addition to medical problems, the main characteristics are distinctive facial features and a specific psychological profile with a hypersocial personality and severe difficulties in visuospatial tasks (Morris and Mervis 2000; Eckert et al. 2006). Autistic disorders are more frequent than expected and suggest that the common WBS deletion can result in a continuum of social communication impairment, ranging from excessive talkativeness and overfriendliness to absence of verbal language and poor social relationships (Tordjman et al. 2012).

Velocardiofacial (OMIM 192430) and DiGeorge (OMIM 188400) syndromes are both caused by a 1.5–3.0 Mb hemizygous deletion on chromosome 22q11.2, a common deletion, which encompasses approximately 45 genes. Although there is a distinction between the two syndromes in the OMIM database, many papers considered VCFS/DGS as a single category and we have chosen to do this. The deletion occurs in approximately 1/4,000 live births; patients have a mean IQ in the borderline range, with frequent learning difficulties, autism and, in adulthood, psychiatric disorders—up to 33 % of the patients can develop schizophrenic disorders (Murphy et al. 1999; Raux et al. 2007; De Smedt et al. 2009; Campbell et al. 2010; Philip and Bassett 2011).

T21 or Down syndrome (OMIM 190685) remains the major genetic cause of ID. Jérôme Lejeune (Lejeune et al. 1959) reported that what was then called "mongolism" was caused by an extra copy of chromosome 21, now labeled HSA21 for Homosapiens autosome chromosome 21. Watanabe et al. (2004) recorded 283 genes encoding proteins on this chromosome. T21 is caused by a chromosomal imbalance involving HSA21. Although the cell carries three allelic forms, the genes of HSA21 show dosage-dependent difference in persons with T21, i.e., the genes are overexpressed to varying degrees, and some are not overexpressed at all (Reymond et al. 2002). Lyle et al. (2004), Kahlem et al. (2004), and Kahlem (2006) have shown that a number of genes were not overexpressed, and that the level of expression was tissue- and age-dependent. In addition to skeletal and medical abnormalities (Roubertoux and Kerdelhué 2006), ID is the main characteristic, but with large within-group differences (Chapman and Hesketh 2003; Carlier and Ayoun 2007). The cognitive behavioral phenotype includes deficits in speech and language production, and broad impairment of the memory domain (Chapman and Hesketh 2000; Vicari 2006; Carlier and Ayoun 2007; Menghini et al. 2011a). Roubertoux and Carlier (2009) summarized earlier studies and concluded that not only is the size of the brain structures generally smaller in persons with T21 (compared to normally developing persons), but the size of the hippocampus is also dramatically reduced (by more than 50 %). The challenge is now to determine which genes have an extra copy-causing ID and to describe the pathophysiological pathways of the brain and cognitive dysfunction involved. One methodology for shedding light on genotype–phenotype correlations is the use of mouse models (see Roubertoux and Carlier 2009, and the section below the part "Model organism of ID").

Studies of a single genetic disorder could pave the way to uncover causal mechanisms between the biological and/or environmental events and the patient's phenotype. Recent illustrations can be found in Menghini et al. (2011a, b) for T21 and WBS syndromes. A methodology commonly used has the disorder group matched to two separate typically developing control groups, one matched for chronological age and the second matched for mental age, and is based on a standardized test (Thomas et al. 2009). We could say, quoting Meyer-Lindenberg et al. (2006), that the study of neural or behavioral mechanisms in one specific disorder provides "a unique window to genetic influences on cognition and behavior".

Another approach is to compare the behavioral phenotypes of two or more disorder groups. Many studies have been conducted comparing persons with FXS and

T21 syndromes (58 items in EBSCO database) and comparing WBS and T21 (50 items in the same database).

The samples used are generally matched for different characteristics including mental age or IQ. As the mean IQ is lower in the T21 group than in the other groups (see Table 3.2), persons with higher cognitive scores are selected in the T21 group and persons with lower cognitive scores for the other group, which substantially restricts any meaningful comparison. Notwithstanding these limitations, the methodology demonstrates that behind differences in general cognitive levels, persons with different genetic disorders have very different neural and behavioral phenotypes, thus offering scope for causal gene to phenotype hypotheses (Walter et al. 2009).

It is also particularly helpful to compare behavioral and/or neural phenotypes in syndromes with identifiable genetic causes in a bid to identify the ways in which certain cognitive traits may influence one another. This strategy was adopted by McDuffie and Abbeduto (2009) when they compared language development in children with T21, FXS, and WBS; they concluded that the relation between language and cognition differs across the three syndromes. Annaz et al. (2009) used the cross-syndrome design to study the development of holistic face recognition in children with autism, T21, WBS, and typically developing persons. Atypical profiles were found in each group of patients, but every disorder group was atypical in a different way. The same strategy was chosen by Carlier et al. (2011) when seeking to establish whether atypical laterality observed in persons with ID was mainly due to ID and cognitive delay. They compared hand, foot, eye, and ear patterns of laterality in groups of patients with one genetic disorder (T21, WBS, and VCFG/DGS) and one group of typically developing persons. Their data showing the existence of a cognitive threshold, below which lateral preference is atypical, argue in favor of a causal link between cognition and laterality in persons with a low IQ.

3.4 Model Organisms of ID

Model organisms are part of the *translational* strategy, which includes not only cellular models, but also a pathophysiological investigation and a clinical approach. Translational strategy endeavors to decipher or confirm the role of a gene, and more precisely the genetic mechanisms that cause the disease, and then to propose remedial molecules. The need to identify the gene and subsequently the defective protein so as to discover a treatment is the rule even if, paradoxically, the first treatment of a genetic disease, PKU, considered the biochemical aspect of the disease only. There has never been a model organism of PKU.

Model organisms of diseases can be seen as a direct consequence of the Darwinian view of evolution. We use model organisms because species have a common ancestor and similar characteristics. Model organisms were first analogous. A species or strain is considered to be a model for a disease when the observed characteristics of the organism tally with the anatomical, physiological, and pathological criteria defining the disease (see Chap. 9 for more discussion of animal models of cognition).

Phenomenological similarities originally appeared to be satisfactory. Senescent rats have been presented as models of Alzheimer disease, an A/J mouse as a model for leukemia, and C57BR and C3H mouse strains that have been judged poor learners and therefore suggested as models of ID. McKinney (1977) and later Robbins and Sahakian (1979) proposed more stringent rules to improve the validity of model organisms. The advent of transgenic and "knock-out" mice technologies and the development of common tools for humans and other organisms such as MRI have led experts to reconsider and refine the criteria (Tordjman et al. 2012). We suggest the following criteria for a model organism of ID: (1) The disorder must have identical etiology in humans and the model organism. This criterion means that it is possible to modify the homologous gene in the model organism (living animal or cell line) to reproduce the genetic events occurring in humans, i.e., for genetic etiology; (2) the metabolic and cellular mechanisms must be the same; (3) brain structure volumes and neurotransmission mechanisms contributing to ID must tally; (4) the impaired intellectual processes must be comparable; and (5) the physiological mechanisms and intellectual processes must improve in the same way by using similar compounds.

3.4.1 Identical Human and Model Organism Etiology

The development of a model organism provides a tool to confirm a hypothesis on the role of a gene in the development of a disease. This requires preliminary examination of human genetic results, e.g., for the development of mice with an extra chromosomal copy, which may be involved in the ID and neurological disorders in T21. Lejeune et al. (1959) demonstrated that the syndrome is caused by an extra copy of the human chromosome 21 (HSA21). The estimated number of genes encompassed in the triplicated region is relatively small, thus making it feasible to adopt a genotype–phenotype correlation approach for the HSA21 genes and the cognitive characteristics observed in TRS21 (Hattori et al. 2000; Watanabe et al. 2004); there is only a small number of HSA 21 genes, approximately 300. A region between D21S17 and ETS2 has been reported as being associated with most of the Jackson signs, including ID (Delabar et al. 1993; Korenberg et al. 1994). Smith et al. (1995, 1997) developed a mouse model of trisomy in which extra fragments from the human D21S17 and ETS2 region were inserted into the mouse genome. The D21S17 and ETS2 regions being syntenic to MMU16, they created segmental trisomy for the region. As an extra copy of a chromosomal fragment including the *Dyrk1a* gene was known to generate cognitive disorders, Altafaj et al. (2001) developed a transgenic mouse overexpressing the *Dyrk1a* gene only, because it was suspected of playing a major role in cognitive disorders. The story of Fragile-X syndrome also shows that the development of a mouse model depended on knowledge of the disease. Oberlé et al. (1991) and Yu et al. (1991) simultaneously reported that the syndrome was the consequence of both the instability of a DNA-segment and abnormal methylation. Both repeats and hypermethylation shut down the transcription of *FMR1* with a loss of the FMR protein that contributes to synaptic functions. The two genetic events create a loss of function similar to the one generated by gene targeting. Oostra's *FMR1* Consortium (Bakker et al. 1994) developed a homologous *FMR1* knockout mouse model. FMR knockout mice present various cognitive disorders and brain dysfunctions generally associated with ID. However, the development of model organism cannot be simply the addition or deletion of a gene associated with a disease. The genetic mechanisms of the diseases are often more complex than a full null allele or an allele overexpression, as shown by Hutchinson-Gilford in the case of progeria syndrome, caused by a heterozygote point mutation (nucleotide 1824—C1824 to T1824) that causes a splicing event in the Lamine A (*LMNA*) gene (located at 1.q22). The mutation leads to the elimination of the 3' half of exon 11 (about 150 bp or 50 amino acid) resulting in a truncated form of prelamin A called progerin (De Sandre-Giovannoli et al. 2003). The relevant mouse model for Hutchinson-Gilford progeria (Osorio et al. 2011) carried a deletion of exon 11 in the homologous mouse *LMNA*.

A number of qualitative reviews have pointed out the limits of mouse models for medical genetics, but genetic engineering is progressing in this field. Until recently, gene targeting was performed using one of the many 129 stem cells transferred into C57BL/6 blastocysts, producing a heterogeneous genetic background that cannot compensate for an insufficient number of back crosses. The heterogeneous genetic background generates "noise" that interacts with the gene effect. It is now possible to generate targeted mice with stem cells that belong to the inbred strain that is used as host. More care is now given to the selection of the promoter in transgenic mice, and differences in gene expressions cannot be attributed to different efficiencies of the promoter. On the other hand, the discovery of regulatory sequences on the noncoding regions of the genome has added complications for gene-targeting technologies. Noncoding regions carry micro RNAs, (miRNA) sequences that regulate transcription factors. Thousands of miRNAs have been reported in mammals (Kozomara and Griffiths-Jones 2011). The miRNAs contribute to developmental diseases (Sayed and Abdellatif 2011) and also to neuronal and cognitive development (Hansen et al. 2010). The deletion of the full gene (intronic plus exonic sequences) cumulates the effect of the protein for which the gene codes and the effect of other proteins, which may be regulated by the miRNAs.

3.4.2 The Same Metabolic and Cellular Mechanisms

The identification of cellular mechanisms provides the opportunity to use cellular models, but the model developed is more a model of the cellular or metabolic conditions required for the onset of the disease rather than a model of the disease. There are several models (*Caenorhabditis elegans, Zebra fish, Drosophila,* and yeast) of neurological and developmental diseases. Mason and Giorgini (2011) reviewed the yeast cell model used to test for a number of mechanistic relationships between the abnormal expansion of a polyglutamine tract and huntingtin protein toxicity. Tauber et al. (2011) published the complex gene expression profiling in mutant yeast for huntingtin and the resulting chart is crucial for the analysis of the transcriptional consequences of huntingtin toxicity.

While the model may fit at the genetic level, it may fail at the metabolic level, as seen with the attempt to model Lesch-Nyhan syndrome in mice. In patients, the syndrome is characterized by cognitive disorders and self-mutilation. The cause of the syndrome is known and the development of model organisms could pave the way to treatments. Lesch-Nyhan disease is because of a mutation in the hypoxanthine phosphoribosyltransferase (*HPRT*) gene mapped at Xq26.2-q26.3. HPRT regulates the metabolism of purines. The mutation results in a lack of HPRT in Lesch-Nyhan syndrome inducing an abnormal purine metabolism (overproduction and overexcretion of purines), with patients having no or low levels of HPRT. Experiments targeting the homologous *Hprt* gene in the mouse did not generate self-injury behavior (Hooper et al. 1987; Kuehn et al. 1987). Purine metabolism is, in fact, different in humans and the mouse. The findings suggest that mice are protected against HPRT loss and that purine metabolism is less HPRT-dependent in the mouse than in humans. Nonmutant mice did not salvage circulation hypoxanthine. A second enzyme, adenine phosphoribosyltransferase (APRT), is involved in the purine salvage pathway in mice. Wu and Melton (1993) observed that the HPRT/APRT ratio was lower in mice than in humans. They then administered (9-ethyladenine), which inhibits APRT to a group of mice lacking HPRT. The *HPRT*-targeted mice given 9-ethyladenine displayed self-injurious behavior (Tordjman et al. 2007).

3.4.3 Brain Structure Volumes and Neurotransmission Mechanisms

Postmortem studies and different MRI techniques investigating humans provide opportunities for comparing brain structures and brain chemistry with the central nervous system of a model organism. Here, we must overcome prejudices. Small organisms may provide unexpected models. The *DYRK-1A* gene is a homolog of the *minibrain* gene, as described in *Drosophila* by Tejedor et al. (1995), Guimera et al. (1996), and Song et al. (1996). The mutation leading to the minibrain phenotype is associated with reduced mushroom bodies and learning deficits in drosophila (Heisenberg et al. 1985). The reduction of mushroom bodies can be paralleled with the small brain or small hippocampus that has been reported in persons with T21 and with mouse models of segmental trisomy.

The results obtained from MRI and other techniques used to visualize the brain and to estimate brain structures are paving the way for the development of model organisms. The reduction in the size of brain structures in T21 (Roubertoux and Carlier 2009) provides a framework for examining animal models, for example, the mouse. The methyl-CpG-binding protein 2 (*MECP2*) gene contributes to Rett syndrome. *Mecp2*-null mice present volumetric and metabolic brain abnormalities (Saywell et al. 2006) that are also present in patients with Rett diagnosis (Naidu et al. 2001). The study of the brain phenotype provides the means to refine the phenotypic comparison between the model organism and the patient. Conversely, examination of results obtained with a model organism should guide the clinician in decisions on the need for a brain examination.

The mouse provides valuable models of neurological diseases, but the mouse is not always relevant for studying the brain or a neurological phenotype. The neurochemical characteristics of the brain are quite similar in the mouse brain and the primate brain. In mammals, the different neurotransmitters and their receptors are controlled by orthologous genes and consequently the functions of the neurotransmitters and their receptors do not differ across the species. The neurotransmission system does not appear to be differentiated across the mammalian species. The anatomy of the brain, however, is specific to the species, even though there are similarities. Ergic systems are like the liquids used by all cars, but they do not produce the same result in a Trabant as they do in a Porsche. Some striking anatomical differences between model organisms and human must be pointed out. The mouse brain does not include a language center. While the prefrontal cortex does exist in rodents, it shows less differentiation than in primates. The cortical layers differ. The prefrontal cortex in rodents has connections that are more similar to the median cortex connections than to the prefrontal cortex of primates. The medullar organization of the motoneurons differs between rodents and primates. The corticospinal tract is a descending medullar way; dexterity depends on it being intact. The organization of the tract differs across the species and is consequently a factor in the selection of a model organism for the study of motor disorders. The number of fibers varies with the estimated number being greater in human (1,101,000) than in nonhuman primates (40,000) and rodents (137,000). Direct corticospinal connections are found with motoneurons in primates, including Human, but the connections are different in rodents, which have no direct connection between the corticospinal neurons and the cervical motoneurons innervating the limb muscles. The organization of the fibers in the spinal cord also differs. A large percentage of corticospinal fibers follow an ipsilateral descending medullar way, but in rodents most of the fibers are in the dorsal column, whereas in Primates most of these are in the lateral column (Courtine 2007). Differences in motor tracts disqualify the mouse from modeling motoneuron genetic disorders. Comparative studies of brain substrates of clinical signs should preclude any attempt to develop model organism of ID.

3.4.4 Intellectual Processes in the Model Organism

The prospect of mimicking intellectual processes affected in genetic disease comes up against one main difficulty: the model organisms have no access to language. This is a limiting factor as speech and language disorders are often features in pediatric symptomatology. To date, none of the different attempts to find substitutes for language have been satisfactory. Therefore, only non-speech-dependent intellectual processes can be considered in this section.

There are two approaches to modeling intellectual processes when using model organisms of ID. The first is to perform an exhaustive annotation of the cognition-related traits that can be observed in the organism. This interesting approach provides information about the phenotypes associated with a given gene or mutation. Schaevitz et al. (2010), for example, screened various behavioral traits related to cognition in a mouse model of Rett syndrome. Other studies have investigated ethological traits such as aggression against a conspecific or ultrasound production. The results may be relevant in an annotation perspective but not in a modeling approach. The second approach, which will be adopted here because it is better suited to the use of model organisms for diseases, consists in: (1) selecting the traits that characterize the syndrome from clinical observations or from psychological reports, and (2) creating conditions that could generate responses mimicking the traits in the model organism.

Several strategies are available. The testing of model organisms of T21 has the advantage of an abundance of publications on the syndrome. A general profile appears with relative strength in associative tasks, difficulty in responding by new strategies to new conditions, poor long-term memory, and attention difficulties (see Roubertoux and Kerdelhué 2006; Roubertoux and Carlier 2009). The psychological profile deduced from clinical studies has been used as framework for mouse models of T21 (Chabert et al. 2004). An exhaustive review of the studies (Sérégaza et al. 2006; Roubertoux and Carlier 2009) showed that most of them attempted to adjust the model organism to the human profile. An abundance of information on a syndrome is not enough to initiate work to develop a model organism. Much is known on the psychological profile of patients with WBS, but little has been done on the development of mouse models. Mouse models with the genes encompassed in the deleted region should enlighten the function of these genes.

A paper by Milner et al. (1998) that recalls the onset of cognitive psychology provides another strategy for exploring the intellectual functions. The key point in the approach is to validate every cognitive alteration by a specific brain structure dysfunction. Milner et al. (1998) proposed two main types of memory, declarative and nondeclarative, based on distinct brain systems. Performance scores for separate memory categories can be measured in rodents. Declarative aspects can be found: (1) in the reduction in the number of freezing episodes when the mouse is subjected to fear conditioning with changes in the context, (2) in nonrepeated visits of a reinforced arm in the radial maze, and (3) in reversal difficulties or reduced time in a virtual quadrant in the Morris water maze. The variables are the "ability to respond appropriately to stimuli through practice, as the result of conditioning or habit learning" (Milner et al. 1998, p. 450). Nondeclarative memory is comprised of three categories: (1) procedural memory, i.e., the formation of habits and acquisition of skills—reaching the platform under proximal cue conditions, (2) priming, and (3) associative, with classical conditioning measured as output; this could be described as an emotional or skeletal response. The modification of the performance when the conditioned stimulus is presented in the fear conditioning protocol illustrates classical conditioning with emotional response, whereas the operant schedule response illustrates classical conditioning with skeletal and muscular response.

A third strategy consists in transposing a protocol developed for humans to model organisms. The best known is the eyelid conditioning protocol used in psychiatric investigations and which was transposed to mice (Chen et al. 1999). An original strategy is to transpose a rat or mouse protocol to the human species, as was done by Foti et al. (2011) who created a radial maze for children to measure spatial memory.

The wisest approach in defining homologous processes across the species is to consider a *transversal homology* investigating the homology of the pathways between two levels of biological organization.

3.4.5 Similar Reactions to Treatments

The translational approach aims ultimately at offering treatment of genetic diseases. Model organism functions are seen as a means of providing preliminary screening of potentially curing compounds. Positive results achieved with a model organism can then be the starting point of a cure strategy for the clinician. Preliminary positive results are also required in many countries to initiate the legal procedure for prescribing a treatment. There are a number of examples of human and mouse models having the same reaction to a treatment. The best known is the reduction of anxiety induced by the benzodiazepine family, by 5-HT reuptake inhibitors and by 5-HT1A agonists. The effects work in the same direction in humans, as tested in clinical interviews, and in mice, tested in the elevated plus maze. Tuberous sclerosis complex, or Bourneville disease, is a mammalian target of rapamycin (mTOR) overactivation syndrome. Inhibition of mTOR in mice improves the neuronal and behavioral characteristics in the mouse model (Aarts et al. 2010). de Vries (2010) reported that "Molecularly-targeted treatments using mTOR inhibitors (such as rapamycin) are showing great promise for the physical and neurological phenotype of TSC. Preclinical and early-phase clinical studies of the cognitive and neurodevelopmental features of TSC suggest that some of the neuropsychiatric phenotypes might also be reversible, even in adults with the disorder." Partially identical results have been reported with immunization treatment for Alzheimer disease in a mouse model and patients. Schenk et al. (1999) developed a mouse model of Alzheimer disease presenting an overproduction of the predominant form found in the amyloid plaques of Alzheimer disease, the 42-amino acid form of the peptide (Abeta42). Mice immunized with Abeta42 at 6 weeks of age showed an improvement in learning and a reduction of the beta-amyloid-plaque formation. Aβ peptide immunotherapy approach in patients is associated with clearance of the beta-amyloid-plaque but it does not improve cognitive performance. Moreover, undesirable effects accompany the administration of the selected molecules used in Aβ peptide immunotherapy (Delrieu et al. 2012).

3.5 Conclusion and Perspectives. Genes Involved in Rare Genetic Diseases and Gene Contributing to the Normal Range of Variation

A major field of behavior genetic analysis is deciphering the genetic mechanisms underlying the nonpathological range of variation. Does understanding the genetic mechanisms of cognitive dysfunction help understanding the genetic mechanisms contributing to cognitive differences within the normal range of variation?

The specificity of the pathological processes, and particularly the development of compensatory processes associated with a single mutation, has been defended by the opponents of the pathological method. Things have changed with the possibility to target genes or parts of genes. We use daily pathological genetic events for understanding the nonpathological variation in experimental animal genetics. What is a targeted gene? It is an abnormality. We use the different annotations resulting from the observation of the disturbed mouse to predict the genes involved in the nonpathological range of variation. Human geneticists can adopt a similar strategy as long as they do not forget the developmental processes. Most of the genetic events contributing to the identification of genetic disorders of cognition are because of mutations generating truncated proteins or abnormal protein isophorms. Fragile-X syndrome is a "natural knock out" occurring in our species. It is equivalent to the "experimental knock out" produced in the mouse. We can thus use the results obtained with genetic disorders to infer the function of genes contributing to nonpathological variation. Mutations in nicotinic receptor A7 (CHRNA7), dopamine receptor 4 (DRD4), and dopamine transporter (DAT1) that induce severe cognitive dysfunctions may contribute to quantitative differences in attention. Catecholamin-O-methyltransferase (COMT) is associated with several brain pathologies but some well-identified allelic forms modulate episodic memory in the nonpathological population. We do not defend the idea that all the genes contributing to individual differences in normal variation can be found by the study of rare genetic disorders but genetic events at work within the normal range of variation and the rare cognitive disorders do overlap.

References

AAIDD. (2010). *Intellectual disability: definition, classification and system of supports/The AAIDD Ad Hoc Committee On terminology And Classification* (11th ed.). Washington, DC: American Association on Intellectual and Developmental Disabilities.

Aarts, H., Ruys, K. I., Veling, H., Renes, R. A., de Groot, J. H., van Nunen, A. M., et al. (2010). The art of anger: Reward context turns avoidance responses to anger-related objects into approach. *Psychological Sciences, 21*(10), 1406–1410.

Abou J. R., Wohlfart, S., Zweier, M., Uebe, S., Priebe, L., Ekici, A., et al. (2011). Homozygosity mapping in 64 Syrian consanguineous families with non-specific intellectual disability reveals 11 novel loci and high heterogeneity. *European Journal of Human Genetics, 19*(11), 1161–1166.

Altafaj, X., Dierssen, M., Baamonde, C., Marti, E., Visa, J., Guimera, J., et al. (2001). Neurodevelopmental delay, motor abnormalities and cognitive deficits in transgenic mice over-expressing Dyrk1A (minibrain), a murine model of Down's syndrome. *Human Molecular Genetics, 10*(18), 1915–1923.

Anderson, P. J., Wood, S. J., Francis, D. E., Coleman, L., Anderson, V., & Boneh, A. (2007). Are neuropsychological impairments in children with early-treated phenylketonuria (PKU) related to white matter abnormalities or elevated phenylalanine levels? *Developmental Neuropsychology, 32*(2), 645–668.

Andersson, M., Karumbudathan, V., & Zimmermann, M. B. (2012). Global iodine status in 2011 and trends other the past decade. *The Journal of Nutrition, 142*(4), 744–750.

Annaz, D., Karmiloff-Smith, A., Johnson, M. H., & Thomas, M. S. C. (2009). A cross-syndrome study of the development of holistic face recognition in children with autism, Down syndrome, and Williams syndrome. *Journal of Experimental Child Psychology, 102*(4), 456–486.

Bailey, D. B., Hatton, D. D., & Skinner, M. (1998). Early developmental trajectories of males with fragile X syndrome. *American Journal of Mental Retardation, 103*(1), 29–39.

Bailey, D. B., Hatton, D. D., Skinner, M., & Mesibov, G. (2001). Autistic behavior, FMR1 protein, and developmental trajectories in young males with Fragile X syndrome. *Journal of Autism and Developmental Disorders, 31*(2), 165–174.

Bakker, C. E., Verheij, C., Willemsen, R., Vanderhelm, R., Oerlemans, F., Vermey, M., et al. (1994). FMR1 knockout mice: a model to study fragile X mental retardation. *Cell, 78*(1), 23–33.

Banerjee, P., Grange, D. K., Steiner, R. D., & White, D. A. (2011). Executive strategic processing during verbal fluency performance in children with phenylketonuria. *Child Neuropsychology, 17*(2), 105–117.

Bennetto, L., Pennington, B. F., Porter, D., Taylor, A. K., & Hagerman, R. J. (2001). Profile of cognitive functioning in women with the fragile X mutation. *Neuropsychology, 15*(2), 290–299.

Billuart, P., Bienvenu, T., Ronce, N., des Portes, V., Vinet, M. C., Zemni, R., et al. (1998). Oligophrenin-1 encodes a rhoGAP protein involved in X-linked mental retardation. *Nature, 392*(6679), 923–926.

Campbell, L. E., Azuma, R., Ambery, F., Stevens, A., Smith, A., Morris, R. G., et al. (2010). Executive functions and memory abilities in children with 22q11.2 deletion syndrome. *The Australian and New Zealand Journal of Psychiatry, 44*(4), 364–71.

Campistol, J., Gassió, R., Artuch, R., & Vilaseca, M. A. (2011). Neurocognitive function in mild hyperphenylalaninemia. *Developmental Medicine & Child Neurology, 53*(5), 405–408.

Carlier, M., & Ayoun, C. (2007). *Déficiences intellectuelles et intégration sociale*. Wavre: Mardaga.

Carlier, M., Desplanches, A. G., Philip, N., Stefanini, S., Vicari, S., Volterra, V., et al. (2011). Laterality Preference and Cognition: Cross–Syndrome Comparison of Patients with Trisomy 21 (Down), del7q11.23 (Williams–Beuren) and del22q11.2 (DiGeorge or Velo–Cardio–Facial) syndromes (2011). *Behavior Genetics, 41*(3), 413–422.

Chabert, C., Jamon, M., Cherfouh, A., Duquenne, V., Smith, D. J., Rubin, E., et al. (2004). Functional analysis of genes implicated in Down syndrome: 1. Cognitive abilities in mice transpolygenic for Down syndrome chromosomal region-1 (DCR-1). *Behavior Genetics, 34*(6), 559–569.

Chabrol, B., Girard, N., N'Guyen, K., Gérard, A., Carlier, M., Villard, L., et al. (2005). Delineation of the clinical phenotype associated with OPHN–1 mutations based on the clinical and neuropsychological evaluation of three families. *American Journal of Medical Genetics. Part A, 138*(4), 314–317.

Chapman, R. S., & Hesketh, L. J. (2000). Behavioral phenotype of individuals with Down syndrome. *Mental Retardation and Developmental Disabilities Research Reviews, 6*(2), 84–95.

Chen, L., Bao, S., & Thompson, R. F. (1999). Bilateral lesions of the interpositus nucleus completely prevent eyeblink conditioning in Purkinje cell-degeneration mutant mice. *Behavioral Neuroscience, 113*(1), 204–10.

Chiriboga, C. A. (2003). Fetal alcohol and drug effects. *Neurologist, 9*(6), 267–279.

Chudley, A. E., Conry, J., Cook, J. L., Loock, C., Rosales, T., LeBlanc, N., et al. (2005). Fetal alcohol spectrum disorder: Canadian guidelines for diagnosis. *Canadian Medical Association Journal Journal de l'association médicale canadienne, 172*(5 Suppl), S1–S21.

Cornish, K., Turk, J., & Hagerman, R. (2008). The fragile X continuum: New advances and perspectives. *Journal of Intellectual Disability Research, 52*(6), 469–482.

Courtine, G., Bunge, M. B., Fawcet, J. W., Grossman, R. G., Kaas, J. H., Lemon, R., et al. (2007). Can experiments in nonhuman primates expedite the translation of treatments for spinal cord injury in humans? *Nature Medicine, 13*(5), 561–566.

Couzens, D. Cuskelly, M., & Haynes, M. (2011). Cognitive development and Down syndrome: Age-related change on the Stanford–Binet test (Fourth Edition). *American Journal of Intellectual Disability Deficiency, 116*(3), 181–204.

Crawford, D. C., Acuña, J. M., & Sherman, S. L. (2001). FMR1 and the fragileX syndrome: human genome epidemiology review. *Genetics in medicine: official journal of the American College of Medical Genetics, 3*(5), 359–371.

Czarnetzki, R., Blin, N., & Pusch, C. M. (2003). Down'syndrome in ancient Europe. The Lancet, 362(9388), 1000.

De Brouwer, A. P., Yntema, H. G., Kleefstra, T., Lugtenberg, D., Oudakker, A. R., de Vries, B. B., et al. (2007). Mutation frequencies of X-linked mental retardation genes in families from the EuroMRX consortium. *Human Mutation, 28*(2), 207–208.

De Rubeis, S., & Bagni, C. (2011). Regulation of molecular pathways in the Fragile X Syndrome: Insights into Autism Spectrum Disorders. *Journal of Neurodevelopmental Disorders, 3*(3), 257–269.

De Rubeis, S., Fernández, E., Buzzi, A., Di Marino, D., & Bagni, C. (2012). Molecular and cellular aspects of mental retardation in the Fragile X syndrome: From gene mutation/s to spine dysmorphogenesis. *Advances in Experimental Medicine and Biology, 970*, 517–51.

De Sandre-Giovannoli, A., Bernard, R., Cau, P., Navarro, C., Amiel, J., Boccaccio, I., et al. (2003). Lamin a truncation in Hutchinson-Gilford progeria. *Science, 300*(5628), 2055.

De Smedt, B., Devriendt, K., Fryns, J.–P., Vogels, A., Gewillig, M., & Swillen, A. (2007). Intellectual abilities in a large sample of children with velo-cardio-facial syndrome: An update. *Journal of Intellectual Disability Research, 51*(9), 666–670.

De Smedt, B., Swillen, A., Verschaffel, L., & Ghesquière, P. (2009). Mathematical learning disabilities in children with 22q11.2 deletion syndrome: a review. *Developmental Disabilities Research Reviews, 15*(1), 4–10.

de Vries, P. J. (2010). Targeted treatments for cognitive and neurodevelopmental disorders in tuberous sclerosis complex. *Neurotherapeutics, 7*(3), 275–282.

de Vries, B. B., White, S. M., Knight, S. J., Regan, R., Homfray, T., Young, I. D. et al. (2001). Clinical studies on submicroscopic subtelomeric rearrangements: A checklist. *Journal of Medical Genetics, 38*(3), 145–150.

de Vries, B. B., Winter, R., Schinzel, A., & van, R. avenswaaij–A. rts (2003). Telomeres: A diagnosis at the end of chromosomes. *Journal of Medical Genetics, 40*(6), 385–398.

Delabar, J.-M., Theophile, D., Rahmani, Z., Chettouh, Z., Blouin, J.-L., Prieur, M., Noel, B., & Sinet, P.-M. (1993). Molecular mapping of twenty-four features of Down syndrome on chromosome 21. *European Journal of Human Genetics, 1*(2), 114–124.

Delrieu, J., Ousset, P. J., Caillaud, C., & Vellas, B. (2012). Clinical trials in Alzheimer's disease': Immunotherapy approaches. *Journal of Neurochemistry, 120*(Suppl 1), 186–193.

DISTAT Department of International Economic and Social Affairs Statistical Office. (1990). Statistics on special population groups, Series Y, No 4 Disability Statistics Compendium. United Nations. http://unstats.un.org/unsd/demographic/sconcerns/disability/. Accessed 30 Jan 2012.

DeRoche, K., & Welsh, M. (2008). Twenty-five years of research on neurocognitive outcomes in early-treated phenylketonuria: Intelligence and executive function. *Developmental Neuropsychology, 33(*4), 474–504.

Eckert, M. A., Galaburda, A. M., Mills, D. L., Bellugi, U., Korenberg, J. R., & Reiss, A. L. (2006). The neurobiology of Williams syndrome: Cascading influences of visual system impairment? *Cellular and Molecular Life Science, 63*(16), 1867–1875.

Edgin, J. O., Mason, G. M., Allman, M. J., Capone, G. T., Deleon, I., Maslen, C., et al. (2010). Development and validation of the Arizona cognitive test battery for Down syndrome. *Journal of Neurodevelopmental Disorders, 2*(3), 149–164.

Enea–Drapeau, C., Carlier, M., & Huguet, P. (2012) Tracking subtle stereotypes of children with Trisomy 21: From facial-feature-based to implicit stereotyping. *PLoS ONE, 7*(4), e34369. doi:10.1371/journal.pone.0034369.

Fisch, G. S. (1997). Longitudinal assessment of cognitive behavioral deficits produced by the Fragile-X mutation. *International Review of Research in Mental Retardation, 21*, 221–247.

Fisch, G. S. (2011). Mental retardation or intellectual disability? Time for a change. *American Journal of Medical Genetics, 155A*(12), 2907–2908.

Fisch, G. S., Simenson, R. J., Tarleton, J., Chalifoux, M., Holden, J. A., Carpenter, N., et al. (1996). Longitudinal study of cognitive abilities and adaptive behavior levels in Fragile X males. *American Journal of Medical Genetics, 64*, 356–361.

Fisch, G. S., Simensen, R. J., & Schroer, R. J. (2002). Longitudinal changes in cognitive and adaptive behavior scores in children and adolescents with the Fragile X mutation or autism. *Journal of Autism and Developmental Disorders, 32*(2), 107–114.

Foti, F., Menghini, D., Petrosini, L., Valerio, G., Crinò, A., Vicari, S., et al. (2011). Spatial competences in Prader-Willi syndrome: A radial arm maze study. *Behavior Genetics, 41*(3), 445–456.

Fuchs, D., Burnside, L., Marchenski, S., Mudry, A. (2007). Children with disabilities involved with the child welfare system in Manitoba: Current and future challenges. In I. Brown, F. Chaze, D. Fuchs, J. Lafrance, S. McKay, & S. Thomas Prokop (Eds.), *Putting a human face on child welfare: Voices from the Prairies* (pp. 127–145). Prairie Child Welfare Consortium www.uregina.ca/spr/prairechild/index.html/. Centre of Excellence for Child Welfare www.cecw–cepb.ca.

Gallagher, A., & Hallahan, B. (2012). Fragile X-associated disorders: A clinical overview. *Journal of Neurology, 259*(3), 401–413.

Gécz, J., Shoubridge, C., & Corbett, M. (2009). The genetic landscape of intellectual disability arising from chromosome X. *Trends in Genetics, 25*(7), 309–316.

Giovannini, M., Verduci, E., Salvatici, E., Paci, S., & Riva, E. (2012). Phenylketonuria: nutritional advances and challenges. *Nutrition & Metabolism (London), 9*(1), 7.

Godler, D. E., Slater, H. R., Bui, Q. M., Storey, E., Ono, M. Y., Gehling, F., et al. (2012). Fragile X mental retardation 1 (FMR1) intron 1 methylation in blood predicts verbal cognitive impairment in female carriers of expanded FMR1 alleles: Evidence from a pilot study. *Clinical Chemistry, 58*(3), 590–598.

Goodrich–Hunsaker, N. J., Wong, L. M., McLennan, Y., Srivastava, S., Tassone, F., Harvey, D., et al. (2011). Young adult female fragile X premutation carriers show age-and genetically-modulated cognitive impairments. *Brain and Cognition, 75*(3), 255–260.

Guimera, J., Casas, C., Pucharcos, C., Solans, A., Domenech, A., Planas, A. M., et al. (1996). A human homologue of Drosophila minibrain (MBN) is expressed in the neuronal regions affected in Down syndrome and maps to the critical region. *Human Molecular Genetics, 5*(9), 1305–1310.

Hagerman, P. J., & Hagerman, R. J. (2007). Fragile X associated tremor/ataxia syndrome—an older face of the fragile X gene. *Nature Clinical Practice Neurology, 3*(2), 107–112.

Hagerman, R., Kemper, M., & Hudson, M. (1985). Learning disabilities and attentional problems in boys with the fragile X syndrome. *American Journal of Diseases of Children, 139*(7), 674–678.

Hall, S. S., Burns, D. D., Lightbody, A. A., & Reiss, A. L. (2008). Longitudinal changes in intellectual development in children with Fragile X syndrome. *Journal of Abnormal Child Psychology, 36*(6), 927–939.

Hall, S. S., Lightbody, A. A., Hirt, M., Rezvani, A., & Reiss, A. L. (2010). Autism in fragile X syndrome: a category mistake? *Journal of the American Academy of Child and Adolescent Psychiatry, 49*(9), 921–933.

Hansen, K. F., Sakamoto, K., Wayman, G. A., Impey, S., & Obrietan, K. (2010). Transgenic miR132 alters neuronal spine density and impairs novel object recognition memory. *PLoS One, 5*(11):e15497, doi:10.1371/journal.pone.0015497.

Hattori, M., Fujiyama, A., Taylor, T. D., Watanabe, H., Yada, T., Park, H. S., et al. (2000). The DNA sequence of human chromosome 21. *Nature, 405*(6784), 311–319.

Heisenberg, M., Borst, A., Wagner, S., & Byers, M. D. (1985). Drosophila mushroom body mutants are deficient in olfactory learning. *Journal of Neurogenetics, 2*(1), 1–30.

Hessl, D., Nguyen, D. V., Green, C., Chavez, A., Tassone, F., Hagerman, R. J., et al. (2009). A solution to limitations of cognitive testing in children with intellectual disabilities: the case of fragile X syndrome. *Journal of Neurodevelopmental Disorders, 1*(1), 33–45.

Hooper, M., Hardy, K., Handyside, A., Hunter, S., & Monk, M. (1987). HPRT-deficient (Lesch–Nyhan) mouse embryos derived from germline colonization by cultured cells. *Nature, 326*(6110), 295–298.

Human Resources and Skills Development Canada. (2011). Disability in Canada: A 2006 profile. http://www.hrsdc.gc.ca/eng/disability_issues/reports/disability_profile/2011/page01.shtml. Accessed Mar 2012.

Hunter, J. E., Sherman, S., Grigsby, J., Kogan, C., & Cornish, K. (2012). Capturing the fragile X premutation phenotypes: A collaborative effort across multiple cohorts. *Neuropsychology, 26*(2), 156–164.

Itard, J. (1801). Mémoire sur les premiers développements de Victor de l'Aveyron. Paris http://dx.doi.org/doi:10.1522/cla.itj.rap. English traduction (1802) An historical account of the discovery and Education of a Savage Man, or of the first developments, physical and moral, of the Young sauvage caught in the woods near Aveyron, in the year 1798. London: Richard Phillips.

Ivanovic, D. M., Leiva, B. P., Perez, H. T., Inzunza, N. B., Almagià, A. F., Toro, T. D., et al. (2000). Long-term effects of severe undernutrition during the first year of life on brain development and learning in Chilean high-school graduates. *Nutrition, 16*(11–12), 1056–1063.

Jacquemont, S., Hagerman, R. J., Hagerman, P. J., & Leehey, M. A. (2007). Fragile-X syndrome and fragile X-associated tremor/ataxia syndrome: Two faces of FMR1. *Lancet Neurology, 6*(1), 45–55.

Kahlem, P. (2006). Gene dosage effect on chromosome 21 transcriptome in trisomy 21: Implication in Down's syndrome cognitive disorders. *Behavior Genetics, 36*(3), 416–428.

Kahlem, P., Sultan, M., Herwig, R., Steinfath, M., Balzereit, D., Eppens, B., et al. (2004). Transcript level alterations reflect gene dosage effects across multiple tissues in a mouse model of Down syndrome. *Genome Research, 14*(7), 1258–1267.

Kayaalp, E., Treacy, E., Waters, P. J., Byck, S., Nowacki, P., & Scriver, C. R. (1997). Human phenylalanine hydroxylase mutations and hyperphenylalaninemia phenotypes: A metanalysis of genotype-phenotype correlations. *American Journal of Human Genetics, 61*(6), 1309–1317.

Koolen, D. A., Nillesen, W. M., Versteeg, M. H., Merkx, G. F., Knoers, N. V., Kets, M., et al. (2004). Screening for subtelomeric rearrangements in 210 patients with unexplained mental retardation using multiplex ligation dependent probe amplification (MLPA). *Journal of Medical Genetics, 41*(12), 892–899.

Korenberg, J. R., Chen, X. N., Schipper, R., Sun, Z., Gonsky, R., Gerwehr, S., et al. (1994). Down syndrome phenotypes: the consequences of chromosomal imbalance. *Proceedings of the National Academy of Science of the United States of America, 91*(11), 4997–5001.

Koukoui, S. D., & Chaudhuri, A. (2007). Neuroanatomical, molecular genetic, and behavioral correlates of fragile X syndrome. *Brain Research Reviews, 53*(1), 27–38.

Kozomara, A., & Griffiths-Jones, S. (2011). miRbase: Integrating microRNA annotation and deep-sequencing data. *Nucleic Acids Research, 39*(Database issue), D152–D157.

Kuehn, M. R., Bradley, A., Robertson, E. J., & Evans, M. J. (1987). A potential animal model for Lesch–Nyhan syndrome through introduction of HPRT mutations into mice. *Nature, 326*(6110), 295–298.

Lachiewicz, A. M., & Dawson, D. V. (1994). Behavior problems of young girls Fragile X syndrome: Factor scores on the Conners' parent's questionnaire. *American Journal of Medical Genetics, 51*(4), 364–369.

Lejeune, J., Turpin, B., & Gauthier, M. (1959). Le mongolisme, premier exemple d'aberration autosomique humaine. *Annales de Génétique, 1*(2), 41–49.

Leonard, H., & Wen, X. (2002). The epidemiology of mental retardation: Challenges and opportunities in the new millennium. *Mental Retardation and Developmental Disabilities Research Review, 8*(3), 117–34.

Lightbody, A. A., & Reiss, A. L. (2009). Gene, brain, and behavior relationships in fragile X syndrome: Evidence from neuroimaging studies. *Developmental Disabilities Research Review, 15*(4), 343–352.

Lightbody, A. A., Hall, S. S., & Reiss, A. L. (2006). Chronological age, but not FMRP levels, predicts neuropsychological performance in girls with fragile X syndrome. *American Journal of Medical Genetics. Part B, Neuropsychiatric Genetics, 141B*(5), 468–472.

Lombard, Z., Tiffin, N., Hofmann, O., Bajic, V. B., Hide, W., & Ramsay, M. (2007). Computational selection and prioritization of candidate genes for fetal alcohol syndrome. *BMC Genomics, 8*, 389.

Lyle, R., Gehrig, C., Neergaard-Henrichsen, C., Deutsch, S., & Antonarakis, S. E. (2004). Gene expression from the aneuploid chromosome in a trisomy mouse model of Down syndrome. *Genome Research, 14*(7), 1268–1274.

Manolakos, E., Vetro, A., Kefalas, K., Rapti, S. M., Louizou, E., Garas, A., et al. (2010). The use of array-CGH in a cohort of Greek children with developmental delay. *Molecular Cytogenetics, 3,* 22.

Mason, R. P., & Giorgini, F. (2011). Modeling Huntington disease in yeast: Perspectives and future directions. Prion, 5(4) (Epub ahead of print).

May, P. A., Fiorentino, D., Coriale, G., Kalberg, W. O., Hoyme, H. E., Aragón, A. S., et al. (2011). Prevalence of children with severe fetal alcohol spectrum disorders in communities near Rome, Italy: New estimated rates are higher than previous estimates. *International Journal of Environmental Research and Public Health, 8*(6), 2331–2351.

McDermott, S., Durkin, M. S., Schupf, N., & Stein, Z. A. (2007). Epidemiology and etiology in mental retardation. In J. W. Jacobson, J. A. Mulick, & J. Rojahn (Eds.), *Handbook of intellectual and developmental disabilities* (pp. 3–40). New York: Springer.

McDuffie, A., & Abbeduto, L. (2009). Language disorders in children with mental retardation of genetic origin: Down syndrome, fragile X and Williams syndrome. In R. G. Schwartz (Ed.), *Handbook of child language disorders* (pp. 44–66). New York: Psychology Press.

McKinney, W. T. (1977). Biobehavioral models of depression in monkeys. In E. Usdin & I. Hanin (Eds.), *Animal models in psychiatry and neurology* (pp. 117–126). Oxford: Pergamon Press.

McKusick, V. A. (2007). Mendelian Inheritance in Man and its online version, OMIM. *American Journal of Human Genetics, 80*(4), 588–604.

Menghini, D., Costanzo, F., & Vicari, S. (2011a). Relationship between brain and cognitive processes in persons with Down syndrome. *Behavior Genetics, 41*(3), 381–393.

Menghini, D., Di Paola, M., Federico, F., Vicari, S., Petrosini, L., Caltagirone, C., et al. (2011b). Relationship between brain abnormalities and cognitive profile in Williams syndrome. *Behavior Genetics, 41*(3), 394–402.

Meyer-Lindenberg, A., Mervis, C. B., & Faith Berman, K. (2006). Neural mechanisms in Williams syndrome: A unique window to genetic influences on cognition and behavior. *Nature Reviews Neuroscience, 7*(5), 367–379.

Milner, B., Squire, L. R., & Kandel, E. R. (1998). Cognitive neuroscience and the study of memory. *Neuron, 20*(3), 445–468.

Möller, H. E., Weglage, J., Bick, U., Wiedermann, D., Feldmann, R., & Ullrich, K. (2003). Brain imaging and proton magnetic resonance spectroscopy in patients with phenylketonuria. *Pediatrics, 112*(6 Pt 2), 1580-, 1583.

Morris, C. A., & Mervis, C. B. (2000). Williams syndrome and related disorders. *Annual Review of Genomics and Human Genetics, 1,* 461–484.

Murphy, K. C., Jones, L. A., & Owen, M. J. (1999). High rates of schizophrenia in adults with velo-cardio-facial syndrome. *Archives of General Psychiatry, 56*(10), 940–945.

Musso, M. W., Barker, A. A., Proto, D. A., & Gouvier, W. D. (2012). College students' conceptualizations of deficits involved in mild intellectual disability. *Research in Developmental Disability, 33,* 224–228.

Mwaniki, M. K., Atieno, M., Lawn, J. E., & Newton, C. R. (2012). Long-term neurodevelopmental outcomes after intrauterine and neonatal insults: A systematic review. *Lancet, 379*(9814), 445–452.

Naidu, S., Kaufmann, W. E., Abrams, M. T., Pearlson, G. D.., Lanham, D. C., Fredericksen, K. A., et al. (2001). Neuroimaging studies in Rett syndrome. *Brain Development, 23*(Suppl 1), S62–S71.

Najmabadi, H., Hu, H., Garshasbi, M., Zemojtel, T., Abedini, S. S., Chen, W., et al. (2011). Deep sequencing reveals 50 novel genes for recessive cognitive disorders. *Nature, 478*(7367), 57–63.

Oberlé, I., Rousseau, F., Heitz, D., Kretz, C., Devys, D., Hanauer, A., Boué, J., Bertheas, M. F., & Mandel, J. L. (1991). Instability of a 550-base pair DNA segment and abnormal methylation in Fragile X syndrome. *Science, 252*(5010), 1097–1102.

Okasha, A. (1999). Mental health in the Middle East: An Egyptian perspective. *Clinical Psychology Review, 19*(8), 917–933.

OMIM Online Medelian Inheritance in Man Johns Hopkins University. http://www.ncbi.nlm.nih.gov.gate1.inist.fr/omim.

Osorio, F. G., Navarro, C. L., Cadiñanos, J., López-Mejàa, I. C., Quirós, P. M., Bartoli. C., et al. (2011). Splicing-directed therapy in a new mouse model of human accelerated aging. *Science Translational Medicine, 3*(106), 106ra107.

Participation and Activity Limitation Survey, (P. A. L. S.). (2006). Human resources and skills development Canada. http://www.statcan.gc.ca/. Accessed 6 Mar 2012.

Philip, N., & Bassett, A. (2011). Cognitive, behavioral and psychiatric phenotype in 22q11.2 deletion syndrome. *Behavior Genetics, 41*(3), 403–412.

Popp, M. S., Schulze, B., Granzow, M., Keller, M., Holtgreve–Grez, H., Schoell, B., et al. (2002). Study of 30 patients with unexplained developmental delay and dysmorphic features or congenital abnormalities using conventional cytogenetics and multiplex FISH telomere (M–TEL) integrity assay. *Human Genetics, 111*(1), 31–39.

Prick, B. W., Hop, W. C. J., & Duvekot, J. J. (2012). Maternal phenylketonuria and hyperphenylalaninemia in pregnancy: Pregnancy complications and neonatal sequelae in untreated and treated pregnancies. *The American Journal of Clinical Nutrition, 95*(2), 374–382.

Raux, G., Bumsel, E., Hecketsweiler, B., van Amelsvoort, T., Zinkstok, J., Manouvrier-Hanu, S., et al. (2007). Involvement of hyperprolinemia in cognitive and psychiatric features of the 22q11 deletion syndrome. *Human Molecular Genetics, 16*(1), 83–91.

Reiss, A., L., & Dant, C. C. (2003). The behavioral neurogenetics of fragile X syndrome: Analyzing gene-brain-behavior relationships in child developmental psychopathologies. *Development and Psychopathology, 15*(4), 927–968.

Reymond, A., Marigo, V., Yaylaoglu, M. B., Leoni, A., Ucla, C., Scamuffa, N., et al. (2002). Human chromosome 21 gene expression atlas in the mouse. *Nature, 420*(6915), 582–586.

Robbins, T. W., & Sahakian, B. J. (1979). "Paradoxical" effects of psychomotor stimulant drugs in hyperactive children from the standpoint of behavioral pharmacology. *Neuropharmacology, 18*(12), 931–950.

Ropers, H. H. (2008). Genetics of intellectual disability. *Current Opinion in Genetics & Development, 18*(3), 241–250.

Roubertoux, P. (2004). *Existe–t–il des gènes du comportement.* Paris: Odile Jacob.

Roubertoux, M., & Carlier, M. (2007). From DNA to the mind. *EMBO Reports, 8*(Science & Society, Special Issue), S7–S11.

Roubertoux, P. L., & Carlier, M. (2009). Neurogenetic analysis and cognitive functions in Trisomy 21. In Y. K. Kim (Ed.), *Handbook of behavior genetics* (pp. 175–185). New York: Springer.

Roubertoux, P. L., & Carlier, M. (2011). Good use and misuse of "genetic determinism". *Journal of Physiology-Paris, 105,* 190–194.

Roubertoux, P. L., & Kerdelhué, B. (2006). Trisomy 21: From chromosome to mental retardation. *Behavior Genetics, 36*(3), 346–354.

Rutter, M., Sonuga-Barke, E. J., & Castle, J. (2010). I. Investigating the impact of early institutional deprivation on development: Background and research strategy of the English and Romanian adoptees (ERA) study. *Monographs of the Society for Research in Child Development, 75*(1), 1–20.

Sayed, D., & Abdellatif, M. (2011). MicroRNAs in development and disease. *Physiological Reviews, 91*(3), 827–887.

Saywell, V., Viola, A., Confort-Gouny, S., Le Fur, Y., Villard, L., & Cozzone, P. J. (2006). Brain magnetic resonance study of MECP2 deletion effects on anatomy and metabolism. *Biochemical and Biophysical Research Communications, 340*(3), 776–783.

Schaevitz, L. R., Moriuchi, J. M., Nag, N., Mellot, T. J., & Berger-Sweeney, J. (2010). Cognitive and social functions and growth factors in a mouse model of Rett syndrome. *Physiology & Behavior, 100*(3), 255–263.

Schenk, D., Barbour, R., Dunn, W., Gordon, G., Grajeda, H., Guido, T., et al. (1999). Immunization with amyloid-beta attenuates Alzheimer-disease-like pathology in the PDAPP mouse. *Nature, 400*(6740), 173–717.

Schuurs-Hoeijmakers, J. H., Hehir-Kwa, J. Y., Pfundt, R., van Bon, B. W., de Leeuw, N., Kleefstra, T., et al. (2011). Homozygosity mapping in outbred families with mental retardation. *European Journal of Human Genetics, 19*(5), 597–601.

Seguin, E. (1846). Traitement moral, hygiène et éducation des idiots et des autres enfants arriérés. Paris: Baillère. http://gallica.bnf.fr/Catalogue/noticesInd/FRBNF31346191.htm.

Seltzer, M. M., Barker, E. T., Greenberg, J. S., Hong, J., Coe, C., & Almeida, D. (2011). Differential sensitivity to life stress in FMR1 premutation carrier mothers of children with fragile X syndrome. Health Psychology, Dec 12. (Epub ahead of print)

Sérégaza, Z., Roubertoux, P. L., Jamon, M., & Soumireu-Mourat, B. (2006). Mouse models of cognitive disorders in trisomy 21: A review. *Behavior Genetics, 36*(3), 387–404.

Sharman, R., Sullivan, K., Young, R., & McGill, J. (2010). A preliminary investigation of the role of the phenylalanine: Tyrosine ratio in children with early and continuously treated phenylketonuria: Toward identification of 'safe' levels. *Developmental Neuropsychology, 35*(1), 57–65.

Smith, M. (2006). *Mental retardation and developmental delay. Genetic and epigenetic factors.* Oxford: University Press.

Smith, D. J., Zhu, Y., Zhang, J., Cheng, J. F., & Rubin, E. M. (1995). Construction of a panel of transgenic mice containing a contiguous 2 Mb set of YAC/P1 clones from human chromosome. *Genomics, 27*(3), 425–434.

Smith, D. J., Stevens, M. E., Sudanagunta, S. P., Bronson, R. T., Makhinson, M., Watabe, A. M., et al. (1997). Functional screening of 2 Mb of human chromosome 21q22.2 in transgenic mice implicates minibrain in learning defects associated with Down syndrome. *Nature Genetics, 16*(1), 28–36.

Song, W. J., Sternberg, L. R., Kasten-Sportes, C., Keuren, M. L., Chung, S. H., Slack, A. C., et al. (1996). Isolation of human and murine homologues of the Drosophila minibrain gene: Human homologue maps to 21q22.2 in the Down syndrome "critical region". *Genomics, 38*(3), 331–339.

Stratton, K., Howe, C., & Battaglia, F. (1996). *Fetal alcohol syndrome. Diagnostic, epidemiology, prevention, and treatment.* Washington DC: New York Academy Press.

Strømme, P., Bjørnstad, P. G., & Ramstad, K. (2002). Prevalence estimation of Williams syndrome. *Journal of Child Neurology, 17*(2), 269–271.

Schwartz, C. E., Dean, J., Howard-Peebles, P. N., Bugge, M., Mikkelsen, M., Tommerup, N., et al. (1994). Obstetrical and gynecological complications in fragile X carriers: a multicenter study. *American Journal of Medical Genetics, 51*(4), 400–402.

Tauber, E., Miller-Fleming, L., Mason, R. P., Kwan, W., Clapp, J., Butler, J. N., et al. (2011). Functional gene expression profiling in yeast implicates translational dysfunction in mutant huntingtin toxicity. *The Journal of Biological Chemistry, 286*(1), 410–419.

Taylor, E., & Rogers, J. W. (2005). Practitioner review: Early adversity and developmental disorders. *Journal of Child Psychology and Psychiatry, 46*(5), 451–467.

Tejedor, F., Zhu, X. R., Kaltenbach, E., Ackermann, K., Baumann, A., Canal, I., et al. (1995). Minibrain: a new protein kinase family involved in postembryonic neurogenesis in Drosophila. *Neuron, 14*(2), 287–301.

Tenenbaum, A., Hertz, P., Dor, T., Castiel, Y., Sapir, A., & Wexler, I. D. (2011). Fetal alcohol spectrum disorder in Israel: Increased prevalence in an at-risk population. *The Israel Medical Association Journal: IMAJ, 13*(12), 725–729.

Thomas, M. S. C., Annaz, D., Ansari, D., Scerif, G., Jarrold, C., & Karmiloff-Smith, A. (2009). Using developmental trajectories to understand developmental disorders. *Journal of Speech, Language, and Hearing Research, 52*(2), 336–358.

Thompson, L., & Gillberg, C. (2012). Behavioral problems from perinatal and neonatal insults. *The Lancet, 379*(9814), 392–393.

Tordjman, S., Drapier, D., Bonnot, O., Graignic, R., Fortes, S., Cohen, D., et al. (2007). Animal models relevant to schizophrenia and autism: validity and limitations. *Behavior Genetics, 37*(1), 61–78.

Tordjman, S., Anderson, G. M., Botbol, M., Toutain, A., Sarda, P., Carlier, M., et al. (2012) Autistic disorder in patients with Williams-Beuren syndrome: A reconsideration of the Williams-Beuren syndrome phenotype. *PLoS ONE, 7*(3), e30778. doi:10.1371/journal.pone.0030778.

Turner, G., & Turner, B. (1074). X-linked mental retardation. *Journal of Medical Genetics, 11*, 109–113.

United Nations/Nations Unies (2011). 2009–2011 DemographicyearbookAnnuaire démographique Sixty-first issus/Soixante et unième édition. United Nations/ Nations Unies. Economic & Social Affairs. http://unstats.un.org/unsd/demographic/sconcerns/disability/. Accessed 30 Jan 2012.

Viau, K. S., Wengreen, H. J., Ernst, S. L., Cantor, N. L., Furtado, L. V., & Longo, N. (2011). Correlation of age-specific phenylalanine levels with intellectual outcome in patients with phenylketonuria. *Journal of Inherited Metabolic Disease, 34*(4), 963–971.

Vicari, S. (2006). Motor development and neuropsychological patterns in persons with Down syndrome. *Behavior Genetics, 36*(3), 355–364.

Vissers, L. E., de Ligt, J., Gilissen, C., Janssen, I., Steehouwer, M., de Vries P., et al. (2010). A de novo paradigm for mental retardation. *Nature Genetics, 42*(12), 1109–1112.

Waisbren, S. E., Noel, K., Fahrbach, K., Cella, C., Frame, D., Dorenbaum, A., et al. (2007). Phenylalanine blood levels and clinical outcomes in phenylketonuria: A systematic literature review and meta-analysis. *Molecular Genetics and Metabolism, 92*(1–2), 63–70.

Walter, E., Mazaika, P. K., & Reiss, A. L. (2009). Insights into brain development from neurogenetic syndromes: Evidence from fragile X syndrome, Williams syndrome, Turner syndrome and velo-cardio-facial syndrome. *Neuroscience, 164*(1), 257–271.

Wang, R., Bray, S. M., & Warren, S. (2012). New perspectives on the biology of fragile X syndrome. *Current Opinion in genetics & Development, 22,* 256–263.

Warren, S. F., Brady, N.C. (2007). The role of maternal responsivity in the development of children with intellectual disabilities. *Mental Retardation & Developmental Disabilities Research Reviews, 13*(4), 330–338.

Warren, K. R., & Hewitt, B. G. (2009). Fetal alcohol spectrum disorders: When science, medicine, public policy, and laws collide. *Developmental Disabilities Research Reviews, 15,* 170–175.

Warren, K. R., & Li, T. K. (2005). Genetic polymorphisms: Impact on the risk of fetal alcohol spectrum disorder. *Birth Defects Research Part A Clinical and Molecular Teratology, 73*(4), 195–203.

Warren, S. F., Brady, N., Sterling, A., Fleming, K., & Marquis, J. (2010). Maternal responsivity predicts language development in young children with Fragile X syndrome. *American Journal on Intellectual and Developmental Disabilities, 115*(1), 54–75.

Watanabe, H., Fujiyama, A., Hattori, M., Taylor, T. D., Toyoda, A., Kuroki, Y., et al. (2004). DNA sequence and comparative analysis of chimpanzee chromosome 22. *Nature, 429*(6990), 382–388.

Williams, R. A., Mamotte, C. D., & Burnett, J. R. (2008). Phenylketonuria: an inborn error of phenylalanine metabolism. *The Clinical Biochemist Reviews, 29*(1), 31–41.

WHO (2001). World health assembly resolution. www.who.int/classifications/icf/en/. Accessed 18 April 2007

World Health Organization (WHO). (2002). Towards a common language for functioning, disability and health, ICF. www.designfor21st. org/documents/who_icf_2002.pdf. Accessed 18 April 2007

Wu, C. L., & Melton, D. W. (1993). Production of a model for Lesch–Nyhan syndrome in Hypoxanthine phosphoribosyltransferase-deficient mice. *Nature Genetics, 3*(3), 235–240.

Youngstrom, E. A., Glutting, J. J., & Watkins, M. W. (2003). Stanford–Binet intelligence scale: Fourth Edition (SB4): Evaluating the empirical bases for interpretations. In C. R. Reynolds & R. W. Kamphaus (Eds.), *Handbook of psychological & educational assessment of children* (2nd ed., pp. 217–242). New York: Guilford Press.

Yu, S., Pritchard, M., Kremer, E., Lynch, M., Nancarrow, J., Baker, E., et al. (1991). Fragile X Genotype characterized by an unstable region of DNA. *Science, 252*(5010), 1179–1181.

Part II
Adulthood

Chapter 4
Cognition in Middle Adulthood

William S. Kremen, Caitlin S. Moore, Carol E. Franz, Matthew S. Panizzon and Michael J. Lyons

For purposes of this chapter, we define middle adulthood as the period between the ages of 40 and 60 years. We use the terms middle adulthood, midlife, and middle age interchangeably to represent this period. We first address some important general issues with respect to behavior genetic research on cognition in middle adulthood. Next, we review some of the extant empirical findings, focusing first on general cognitive ability and then on specific abilities. We present a seemingly disproportionate subset of the results from our Vietnam Era Twin Study of Aging (VETSA). The reason is simply that the large majority of behavior genetic studies have focused on younger or older individuals, whereas the VETSA is one of the very few studies with extensive midlife data.

W. S. Kremen (✉) · C. E. Franz · M. S. Panizzon
Department of Psychiatry, University of California, San Diego,
9500 Gilman Drive (MC 0738),
La Jolla, CA 92093-0738 USA
e-mail: wkremen@ucsd.edu

Center of Excellence for Stress and Mental Health,
VA San Diego Healthcare System
La Jolla, 92093-0738 CA, USA

C. E. Franz
e-mail: cfranz@ucsd.edu

M. S. Panizzon
e-mail: mspanizzon@ucsd.edu

C. S. Moore
Department of Psychology, Boston University, 64 Cummington St.,
Boston, MA 02215-2407 USA
e-mail: csmoore@bu.edu

M. J. Lyons
Department of Psychology, Boston University, 64 Cummington Mall,
Room 149, Boston, USA
e-mail: mlyons@bu.edu

D. Finkel, C. A. Reynolds (eds.), *Behavior Genetics of Cognition Across the Lifespan*, Advances in Behavior Genetics 1, DOI 10.1007/978-1-4614-7447-0_4,

4.1 Key Issues

4.1.1 Midlife: An Understudied, but Important Transitional Period for Cognitive Aging

Finch (1991) noted that midlife was an understudied period in gerontological research. About a decade later, Bouchard and McGue (2003) pointed to the "extreme paucity" of genetic studies of general cognitive ability in adult twins. There are now some important behavior genetic studies of cognitive aging in adults, but the emphasis in those studies has been primarily on later life (see Chap. 5 for a review). Indeed, it can be said that behavior genetic research focusing on middle adulthood remains in its infancy. Although we do address some issues of change from early adulthood, we wished to focus primarily on cognitive function between the ages of 40 and 60 years in this chapter. However, only modest numbers of people in this age range have been included in the major twin studies of cognitive aging. For example, middle-aged groups in the Swedish Adoption Twin Study of Aging (SATSA) and Minnesota Twin Study of Adult Development and Aging (MTSADA) combined, comprised about 145 twin pairs with a mean age of about 60 (range 50–64; Finkel et al. 1995a).

When studying older adults, we need a baseline from which to gauge change. One could argue that young adulthood can provide that baseline, but that leaves at least two decades as a "black hole." Midlife provides a more proximal baseline for understanding later life cognitive changes. There is also growing evidence suggesting that treatments for dementia are ineffective because neuronal degeneration has already progressed too far by the time of intervention (Sperling et al. 2011). The National Institute on Aging-Alzheimer's Association workgroups have made mild cognitive impairment (MCI) and earlier preclinical stages of Alzheimer's disease priorities in order to better understand the transition to dementia (Albert et al. 2011; Sperling et al. 2011). Genetic factors are, of course, important determinants of Alzheimer's disease (Gatz et al. 2006; for a review see Chap. 7). Together, these factors argue for more intensive behavior genetic studies of cognitive function during middle adulthood.

4.1.2 Understanding Trajectories of Midlife Cognitive Aging

Middle adulthood is particularly relevant to the central question of when cognitive decline begins. The fact that mean level change for some cognitive functions tends to be small or absent until or just past late midlife (McArdle et al. 1998; Rönnlund et al. 2005; Schaie 2005) may give the impression that cognition in middle adulthood is of little interest in its own right. Longitudinal studies indicate linear declines in episodic memory from age 60, but there is also evidence of declines in processing speed (Finkel et al. 2005; Hertzog et al. 2003; Rönnlund et al. 2005; Schaie 1996) and spatial processing occurring earlier (Finkel et al. 2005). Working memory and executive functions represent cognitive abilities that are very important for effective daily functioning, and they are also among the more age-sensitive cognitive domains (West 1996). As we will highlight throughout this chapter, despite their importance, these two domains have received little attention in behavior genetic studies of cognitive aging in middle-aged or older adults.

Some cross-sectional studies suggest that relatively linear declines in several cognitive abilities take place prior to age 60, beginning in young adulthood (Salthouse 2009). A key methodological issue with regard to these different views has been the idea that cross-sectional studies may overestimate age-related differences due to cohort effects, whereas longitudinal studies may underestimate age-related change due to retest effects. Although it was not a behavior genetic study, the Whitehall II study, which had the largest cohorts of middle-aged and older adults, did find significant cognitive declines over a 10-year period in both longitudinal and cross-sectional analyses of 5-year age cohorts ranging in age from 45 to 70 at baseline (Singh-Manoux et al. 2011). For example, there were longitudinal declines in reasoning, verbal fluency, and episodic memory ranging from 2.9–3.6 % in men and 2.6–4.2 % in women. There were also increases of 0.7–0.8 % in vocabulary. All the changes were statistically significant. Cross-sectional comparisons in Whitehall II suggested declines for men that were similar to their longitudinal findings, but cross-sectional results suggested larger declines than longitudinal data for women. The demonstration of cognitive declines in the youngest age bracket (45–49) of the Whitehall II study argues strongly for cognitive change taking place as early as middle adulthood. The fact that practice effects have been observed even after 5 years (Rönnlund et al. 2005) makes the latter finding all the more striking because there was no adjustment for practice effects.

Adjusting for education had little impact on the longitudinal Whitehall II results, but it did substantially reduce the cross-sectional differences. Adjusting for education is a common approach to account for cohort differences, which presumably reflect, at least in part, educational differences. For example, the average educational attainment of individuals born before 1920 is lower than that for individuals born after 1950. On the other hand, it is worth noting that using educational attainment in this way has some serious limitations as can be seen in our study, the VETSA (Kremen et al. 2006). There were 1,237 middle-aged male twins almost all of whom were between the ages of 51 and 59 (four turned 60 just prior to coming to the laboratory) in wave 1 of the VETSA. We were in the relatively unique situation of having available actual general cognitive ability scores from early adulthood (at an average age of 20 years) for all VETSA participants. The Lothian and Aberdeen Birth Cohort studies (see Chap. 5) similarly had general cognitive data from age 11. Not surprisingly, the majority of VETSA participants had 12 years of formal education at the initial VETSA assessment when they averaged 55 years of age. As an illustration of the limitation of using education in this way, although this subset of VETSA participants all had the same amount of education, there was still substantial variability in their general cognitive ability scores at age 20 with scores ranging from the 10th to the 99th percentile.

The aforementioned studies suggest that without characterizing cognitive function in middle adulthood (i.e., before age 60), key points of transition may be missed. Moreover, non-twin studies show that even with little group mean change, some subgroups still do manifest significant changes. For example, Schaie et al. (2004) found that 15 % of people in their early 40s to early 60s showed significant declines, primarily in executive, attention, and episodic memory functions. These subgroups are perhaps the most likely to be highly informative about risk and protective factors for cognitive aging. Divergent patterns of change after 55 have also been noted for working memory and inductive reasoning (Hertzog et al. 2003).

Another important construct with regard to cognitive trajectories is that of MCI. MCI is now generally considered to be the predementia stage of Alzheimer's disease (Albert et al. 2011) and may, therefore, be considered an important transition phase. Studies of MCI have focused on people who are 65 and older, but we have assessed multiple criteria sets for neuropsychologically defined MCI in the younger VETSA participants. To ensure that MCI classifications truly represented decline from prior levels of function, all definitions were based on scores adjusted for general cognitive ability measured at age 20, on average. To our knowledge, this is the youngest and the only middle-aged sample (adults in their 50s) to be systematically assessed for MCI. Our preliminary findings show that, as in studies of older adults, the prevalence of MCI varies dramatically depending on the definition. The heritability of MCI varied greatly as well. However, we did find that MCI can be identified in people this young. There was also partial validation for the diagnoses in that an index of hippocampal atrophy was greatest in participants with amnestic MCI (Jak et al. 2013). Ongoing follow-up assessments will be the key to determining the optimal definition of MCI. These preliminary findings further highlight the importance of further behavior genetic study of cognition in middle adulthood, and increased emphasis on individual or subgroup differences in level of performance and differences in cognitive trajectories. It could be possible that we were able to detect MCI in the middle-aged VETSA sample because, as suggested by some researchers (Roberts et al. 2012), men are more likely than women to develop MCI. However, in almost all studies cited by Roberts et al. and a review by Panza et al. (2005), there were no sex differences in incidence or prevalence of MCI.

4.1.3 Approaches to Cognitive Aging: Lumping Versus Splitting

As suggested by cross-sectional data (Salthouse 2009), aging-related changes in cognitive trajectories can affect both general ability (*g*) and specific cognitive abilities. Specific cognitive domains that appear to be most affected by aging are processing speed, executive function, and memory. The fact that *g* accounts for 40–50 % of the variance in different cognitive measures (Deary et al. 2004) means a full understanding of cognitive aging will require a greater focus on specific cognitive abilities beyond *g*. In an early approach to this issue, Pedersen et al. (1994) showed that there were significant genetic influences on 12 of 13 cognitive measures that were independent of *g* in the SATSA.

With respect to specific cognitive abilities, behavior genetic studies of cognitive aging—particularly for middle adulthood—have been largely at the level of broad domains such as processing speed, memory, and spatial processing. Components of executive functions and working memory—which is closely related to executive function (Friedman et al. 2006)—are prime candidates because they are mediated by neural systems linked to the prefrontal cortex, the parenchymal region with the largest age-related shrinkage (Raz 2000). However, these functions have received very little attention in behavior genetic studies of middle-aged and older adults. In a nongenetic, longitudinal analysis covering ages 18–95, specific abilities accounted for 61 % of the variance in cognitive change for all ages; 33 % was accounted for by domain-specific change in the four domains that were examined (abstract reasoning, spatial processing, verbal episodic memory, and processing speed), and 28 % was accounted for by test-specific change in the 12 tests on which the domains were based (Tucker-Drob 2011). Because most of the variance is accounted for by specific, rather than global, abilities, we favor an approach that leans even more strongly toward further splitting. That includes extending studies even beyond test-specific variance to examine component processes *within* cognitive tests in order to learn which specific underlying processes may be accounting for age-related changes.

Tests within the same domain involve different cognitive processes, and even a single test always consists of multiple component cognitive processes. Experimental psychology and neuroscience-oriented approaches focus on relatively fine-grained component processes in order to understand how particular cognitive functions are carried out. It is our view that these approaches will be most useful for elucidating the determinants of good or poor performance because one goal of these approaches is to isolate the component processes. For example, spatial processing is a broad cognitive construct that can include several component processes linked to different brain pathways. The dorsal visual stream includes posterior parietal cortex and is responsible for object location and visual control of actions, whereas the ventral stream includes inferior temporal cortex and is responsible for visual perception and object recognition (Goodale and Milner 1992).

Factor analysis has been the major approach to identifying cognitive dimensions in behavior genetic studies of aging, but there are some important limitations to this approach. Components derived from factor analysis are typically still at a relatively broad level. In addition, factor analysis alone—without solid theory and evidence from studies of brain–behavior relationships—is insufficient for determining the subcomponents of cognitive functions or abilities. Suppose, for example, that one factor analyzes digit span, story recall, and a number of visual–spatial tasks. It is likely that digit span and story recall will load on a "memory" factor, but a wealth of neuroscientific evidence indicates that those tests are linked to different neural circuitry underlying different memory systems; digit span, a short-term/working memory test is more strongly associated with prefrontal circuitry, whereas story recall is more strongly associated with the hippocampus (Fuster 1995). Functional neuroimaging also demonstrates that elucidating specific components—even within a single test—is crucial for understanding brain and cognition. Without this level of analysis, it is difficult to know what exactly accounts for good performance or for age-related declines.

In our view, an integrative approach that maps the twin method onto the cognitive neuroscience approach, which aims to isolate component cognitive processes, has powerful advantages. Genetic factors are clearly of importance for explaining individual differences in cognitive abilities (Bouchard and McGue 2003), but the breakdown of component processes at the phenotypic level will not necessarily be the same as it is at the genetic level. However, factor analysis of cognitive measures in behavior genetic studies of middle-aged and older adults are usually performed at the phenotypic level. The resulting phenotypic factors are then typically subjected to biometrical modeling. A somewhat different picture may emerge if a genetic factor analysis is conducted. In this chapter, we show some examples of this approach.

The emphasis in cognitive research has shifted quite a bit since the major twin studies of aging (SATSA and MTSADA) were begun. To their credit, the investigators on these studies have made very important contributions with limited sets of cognitive measures. Moreover, the need for both large samples and extensive measures is a substantial impediment to integrated twin-cognitive neuroscience studies. Greater effort is needed toward developing collaborative twin studies of aging that would include substantial numbers of middle-aged adults with at least a core set of the same cognitive and neuroimaging measures. Besides the issue of sample size, such efforts would require work to determine the best measures to be included.

4.1.4 Importance of Phenotype Definition/Refinement

Simply finding significant genetic influences for particular cognitive functions may seem uninteresting given the general rule of thumb that all human behavioral characteristics are heritable (Turkheimer 2000). However, another rule of thumb, the construct–measurement fallacy states that because a cognitive domain is heritable, it cannot be assumed that any measure within that domain will be heritable (Kremen and Lyons 2011). Almost all studies have found no, or virtually no, heritability for the Wisconsin Card Sorting Test (Kremen et al. 2007a). In the VETSA sample, the first trial of the California Verbal Learning Test (CVLT) was not heritable (see Table 4.1B). Surely, the executive function and memory domains tapped by these measures are influenced by genes as indicated by the heritability of other tests of these abilities, but these particular measures did not consistently manifest individual differences based on genetic factors. Thus, although all cognitive abilities may be heritable, not all measures of those abilities will necessarily show that heritability. We might refer to these as "fallible indicators" (Meehl 1977). So-called missing heritability is acknowledged as a serious problem for genetic association studies (Maher 2008), and inclusion of a cognitive phenotype that is not consistently found to be heritable in twin studies is only likely to compound the problem.

Behavior genetic studies—particularly multivariate twin analyses—are useful tools for refining phenotypes in cognitive aging studies. By separating out component processes with specific genetic influences from the overall score on a cognitive

Table 4.1 Heritability estimates (h^2) for individual cognitive tests assessed during midlife in VET Registry twins

Cognitive test	h^2	95 % CI	Cognitive test	h^2	95 % CI
A. *Earlier VET Registry Study: average age = 47.9 (Range: 41–58)*					
Verbal ability			*Short term memory*		
Word recognition (WRAT) (Kremen et al. 2007b)[a]	0.49	(0.28; 0.74)	Digit span forward (Kremen et al. 2008)[a]	0.22	(0.15; 0.65)
Executive functions			*Working memory*		
Tower of London execution speed (Kremen et al. 2009)[a]	0.32	(0.10; 0.58)	Reading span (Kremen et al. 2007b)[a]	0.49	(0.38; 0.58)
Tower of London attempts to complete (Kremen et al. 2009)[a]	0.29	(0.07; 0.56)	Digit transformation (Kremen et al. 2008)[a]	0.50	(0.41; 0.58)
Wisconsin Card Sorting Perseverations (Kremen et al. 2007a)			Non-spatial delayed alternation task (Kremen et al. 2011a)	0.25	(0.03; 0.45)
Wisconsin Card Sorting Categories (Kremen et al. 2007a)			Spatial delayed alternation task (Kremen et al. 2011a)	0.00	(0.00; 0.25)
B. *VETSA: average age = 55.4 (Range: 51–60)*					
General cognitive ability			*Working memory*		
Age 55 AFQT (Lyons et al. 2009)[a]	0.57	(0.42; 0.74)	Digit span backward	0.41	(0.14; 0.53)
Verbal ability			Spatial span backward	0.35	(0.09; 0.53)
WASI vocabulary	0.42	(0.22; 0.65)	Letter number sequencing	0.38	(0.13; 0.59)
Verbal memory			Letter number sequencing adjusted for digit span forward	0.28	(0.00; 0.42)
Logical memory immediate recall	0.52	(0.30; 0.59)			
Logical memory delayed recall	0.53	(0.34; 0.60)	AX-CPT AX errors (Kremen et al. 2011b)[a]	0.20	(0.01; 0.37)
CVLT-2 trial 1 recall	0.00	(0.00; 0.21)	AX-CPT BX errors (Kremen et al. 2011b)[a]	0.28	(0.03; 0.40)
CVLT-2 trial 2 recall	0.17	(0.00; 0.39)	*Inhibitory control*		
CVLT-2 trial 3 recall	0.15	(0.00; 0.39)	Stroop color/word condition	0.49	(0.25; 0.63)
CVLT-2 trial 4 recall	0.37	(0.17; 0.45)	Stroop color/word interference	0.23	(0.00; 0.39)

Table 4.1 (continued)

Cognitive test	h^2	95 % CI	Cognitive test	h^2	95 % CI
B. *VETSA: average age = 55.4 (Range: 51–60)* (continued)					
CVLT-2 trial 5 recall	0.29	(0.00; 0.43)	AX-CPT AY errors (Kremen et al. 2011b)[a]	0.24	(0.19; 0.39)
CVLT-2 total trials 1–5 (Panizzon et al. 2011)[a]	0.36	(0.10; 0.55)	*Set-shifting*		
CVLT-2 short delay free recall (Panizzon et al. 2011)[a]	0.24	(0.05; 0.48)	D-KEFS Trail Making (condition 4) (Vasilopoulos et al. 2012)[a]	0.62	(0.53; 0.72)
CVLT-2 long delay free recall (Panizzon et al. 2011)[a]	0.35	(0.10; 0.56)	D-KEFS Trail Making (condition 4 adjusted for condition 2)	0.13	(0.00; 0.42)
Visual spatial memory					
Visual reproductions immediate recall	0.38	(0.12; 0.54)	Category switching total correct	0.31	(0.10; 0.40)
Visual reproductions delayed recall	0.42	(0.24; 0.50)	Category switching adjusted for category fluency	0.20	(0.00; 0.29)
Processing speed					
D-KEFS Trail Making (condition 2) (Vasilopoulos et al. 2012)[a]	0.34	(0.26; 0.43)	Category switching accuracy	0.19	(0.00; 0.28)
D-KEFS Trail Making (condition 3) (Vasilopoulos et al. 2012)[a]	0.43	(0.34; 0.53)	*Abstract reasoning*		
Stroop word condition	0.65	(0.54; 0.70)	WASI matrix reasoning	0.43	(0.20; 0.62)
Stroop color condition	0.60	(0.46; 0.66)	*Verbal fluency*		
Short term memory			Letter fluency total correct	0.62	(0.49; 0.68)
Digit span forward	0.51	(0.27; 0.63)	Category fluency total correct	0.54	(0.43; 0.61)
Spatial span forward	0.44	(0.27; 0.52)	*Visual spatial ability*		
			Card/mental rotation	0.46	(0.23; 0.62)
			Hidden figures	0.72	(0.57; 0.77)

VET Vietnam Era Twin, *CI* confidence interval, *VETSA* Vietnam Era Twin Study of Aging, *WASI* Wechsler Abbreviated Scale of Intelligence, *AFQT* (Armed Forces Qualification Test)

[a] Results based on multivariate analyses

test, these analyses essentially reduce the "genetic signal-to-noise ratio." Doing so may increase the likelihood of positive findings in genetic association studies. In addition, the genetic underpinnings of the particular cognitive processes that are most important for cognitive aging may be different from what is observed at the phenotypic level. This integrative behavior genetic and neuroscience-oriented approach has been utilized in twin studies of children or very young adults (e.g., Ando et al. 2001; Luciano et al. 2001), but rarely has it been applied to middle-aged or older adults.

4.2 Empirical Findings on Midlife Cognition

4.2.1 Studies of General Cognitive Ability

In this section, we review findings in the domain of general cognitive ability. Several nongenetic studies have demonstrated considerable stability for general cognitive ability during the period from youth through middle to later adulthood. In their review, Deary et al. (2000) concluded that "the genetic and environmental sources of this remarkable stability of individual differences in human intelligence must be sought." (p. 54). As noted, however, most of the relevant studies of cognitive ability have been based on child and adolescent samples.

4.2.1.1 Heritability of General Cognitive Ability (g)

In the SATSA, Pedersen et al. (2002, 1992) reported that the heritability for a psychometrically derived *g* variable was 0.81. Using hierarchical multiple regression, they found that heritability did not change as a function of age between the ages of 50 and 84 years. In another study of the SATSA sample using a cohort sequential design, Finkel and McGue (1998) observed a significant decrease in the genetic variance for general cognitive ability measured on three occasions separated by 3-year intervals. The younger cohorts in their study ranged from 41 to 68 years and their older cohorts ranged from 62 to 84 years. The heritability of their *g* measure decreased from about 0.80 for the three younger cohorts to about 0.60 for the three older cohorts. The longest longitudinal interval between assessments was 6 years.

Finkel et al. (1995b) compared subsamples from the SATSA and the MTSADA. Using a cross-sectional approach, they divided the samples into three age groups, two of which are relevant to our focus on development from young adulthood to late middle age (younger adults were 27–50 years; middle-aged adults were 50–65 years). Utilizing principal-components factor analysis to derive a *g* factor, they found a heritability of 0.81 for both age groups of the MTSADA and SATSA groups. The data from MTSADA and SATSA suggested no change in the heritability of cognitive ability during the period from young adulthood to late middle age.

Posthuma et al. (2001a) used cross-sectional data from an extended twin design study from the Netherlands Twin Registry (Boomsma 1998). They reported that heritabilities did not differ between cohorts with mean ages of 26.2 and 50.4 years. For the combined sample, heritability was 0.85 for Verbal IQ and 0.69 for Performance IQ. Several other papers utilized samples that overlapped with those in the Posthuma et al. (2001a) paper. Both Brans et al. (2010) and Hulshoff Pol et al. (2006) studied samples of twins and their siblings drawn from a cohort at the University Medical Centre Utrecht (Baaré et al. 2001) and the Netherlands Twin Registry (Boomsma 1998). The mean age in the Hulshoff Pol paper was about 37 years (range not provided). The heritability of verbal IQ was 0.84 and the heritability of Performance IQ was 0.67. Brans et al. (2010) studied twins and siblings (mean age = 29.6 years ± 7.5 years). The heritability of Full Scale IQ was 0.86 (95 % confidence interval [CI]: 0.75; 0.92).

Some studies have investigated samples that might be considered to be on the border between late middle age and old age. For example, Plomin et al. (1994) studied a sample of older twins on two occasions separated by 3 years (mean ages 64 and 67). They observed a heritability of about 0.80 for general cognitive ability at both times. In the VETSA, which is the only longitudinal twin study focused exclusively on midlife, the heritability of general cognitive ability was 0.49 at about age 20 years and 0.57 at about age 55 years (Lyons et al. 2009). This increase in heritability was not statistically significant.

There is something of a consensus that there is an increase in the influence of genetic factors with age, and that the influence of shared environmental factors decreases with age, at least until middle age (McCartney et al. 1990; McGue et al. 1993; Plomin and Spinath 2004). Haworth et al. (2010) analyzed cross-sectional data from a combined sample of 11,000 twin pairs drawn from four countries. Heritability of general cognitive ability (based on different measures in different studies) increased linearly from childhood to late adolescence ($h^2 = 0.41$ at age 9 years; $h^2 = 0.55$ at age 12 years; and $h^2 = 0.66$ at age 17 years). As suggested by Haworth et al., one mechanism that probably influences increasing heritability of cognitive ability is gene–environment (GE) correlation. GE correlation refers to differential exposure to environmental conditions depending on one's genotype. Passive GE correlation (e.g., shared home environments determined by parents' genetic propensities that may be conducive to intellectual development) may be more of a factor in childhood. Active GE correlation (e.g., genetic propensities lead one to seek out individual-specific environments that may be more conducive to intellectual development) may be more of a factor in adulthood.

Vogler (2006) suggested that the heritability of cognitive functioning during adulthood seems to be relatively stable over time with some decline in heritability in older cohorts, and the results of a number of studies are consistent with Vogler's conclusion (Finkel et al. 1995b; Finkel et al. 1998; Lyons et al. 2009; McGue and Christensen 2002; Posthuma et al. 2001a). Reynolds et al. (2005) administered cognitive assessments at ages 50, 60, 70, and 80 years and observed an inverted U-shaped pattern for genetic variance; that is, genetic variance increased somewhat from age 50 to 60 followed by a decrease. Among studies of adults, the limited number of studies that utilized a true longitudinal design, the short time intervals utilized, and the preponderance of subjects older than 65 years preclude drawing strong conclusions about the issue of whether the magnitudes of genetic and environmental influences change over the course of middle adulthood. The VETSA results from age 20 to 55 did indicate just a slight, but nonsignificant increase in genetic variance, but we are unaware of any other longitudinal studies in this age range.

4.2.1.2 Genetic or Environmental Influences on Change in General Cognitive Ability

Several studies have investigated the extent to which genetic factors contribute to stability and change of cognitive ability during adulthood. McGue and Christensen (2002) observed a heritability of 0.76 for the mean score on their cognitive measures. However, they found a heritability of only 0.06 for the linear change in cognitive scores from four testing occasions over a 6-year period. Plomin et al. (1994) found a phenotypic stability of 0.92 over a test/retest interval of 3 years, with genetic factors accounting for almost 90 % of the stability. Reynolds et al.'s (2005) *g* measure demonstrated a nonsignificant heritability of linear change of 0.01, whereas nonshared environmental influences explained 99 % of the variance. The acceleration of cognitive change over time or "change in the change" (the quadratic trend) had a heritability of 0.43 with a contribution from the nonshared environment of 0.57. In the VETSA, stability in general cognitive ability primarily reflected genetic and shared environmental influences; 22.4 % of the correlation between age 20 and 55 performance was due to shared environmental factors, 71.3 % was due to genetic factors, and 6.3 % due to nonshared environments. Largely (98.3 %), changes were due to aspects of the environment, primarily aspects that were not shared by the twins (Lyons et al. 2009).

Longitudinal twin data are required to determine the extent to which the same or different genetic influences are operating during different developmental periods. Several studies have addressed this issue during childhood and adolescence (summarized in Lyons et al. 2009). However, VETSA is unique in addressing this question during the period from young adulthood to late middle age. In VETSA, the genetic correlation for general cognitive ability from early adulthood to late middle age was 1.0, which indicates that the same genes were operating at both times (Lyons et al. 2009).

4.2.1.3 Molecular Genetic Studies

Molecular genetic approaches are increasingly being applied to the investigation of behavioral and psychological traits. Although our focus is primarily on twin studies, we do touch briefly on some molecular genetic studies. Deary and colleagues (2010b) and Sabb et al. (2009) reviewed the published reports of individual genes that influence intelligence or general cognitive ability. Although a number of studies have reported individual genes that contribute to cognitive ability (e.g., Pan et al. 2011; Zhang et al. 2010), Deary et al. concluded that molecular genetic studies have failed so far to produce replicable findings that identify individual genes that influence intelligence. The conclusion of the Sabb et al. review was that the variance in intelligence explained by specific genes that have thus far been identified is only about 5.5 %, leaving most of individual sources of genetic influence unidentified.

A recent report by Davies and colleagues (2011) illustrates an approach that applies molecular genetic methods to assess the total contribution of individual genetic markers to general cognitive ability without specifying particular functional genetic variants presumed to be responsible for the observed associations. This approach, based on the premise that for traits that are likely to be highly polygenic, the influence of many genes, each of small effect, will be missed if stringent significance levels are required for each single nucleotide polymorphism (SNP). This approach had been successfully applied to height data (Yang et al. 2010). Davies et al. (2011) carried out a genome-wide SNP analysis on a sample of 3,511 unrelated adults. They had extensive phenotypic information for cognitive functioning in their participants. They created measures representing crystallized and fluid intelligence. They found that linkage disequilibrium between the examined SNPs and genetic variants that accounted for variation in the cognitive measures explained 40 % of the variation among individuals for crystallized intelligence and 51 % for fluid intelligence. They concluded that their estimates represent a lower bound for the actual heritability of these traits. They also concluded, reflecting an emerging consensus, that the very substantial genetic influence on general cognitive ability in all likelihood reflects the actions of a very large number of genes, each with small effect.

Deary et al. (2012) were the first to apply the same approach to a bivariate analysis in which they examined genetic influences on general cognitive ability measures during childhood and late life. Their conclusion was that some genetic variants influencing *g* in old age are different from those operating during childhood. However, the evidence for their conclusion is ambiguous because the genetic correlation they reported based on the same measure in childhood and old age was not significantly different from 1.0. Therefore, it may well be that the same genes were operating on both occasions as was the case from early adulthood to late middle age in the VETSA data. Resolution of this important issue will require further study.

4.2.2 Studies of Specific Cognitive Abilities

In this section, we review some of the findings about specific cognitive abilities or cognitive domains. Not all cognitive abilities appear to be equally sensitive to aging, and they may not necessarily follow the same pattern as measures of general cognitive ability. We think that it is necessary to elucidate more specific abilities in order to understand genetic and environmental influences on cognitive aging. As noted, there is little behavior genetics research on specific cognitive abilities, especially component processes of specific abilities, in middle adulthood and beyond. As the VETSA is one of the few behavior genetic studies with a detailed cognitive assessment in middle adulthood, we have included a Table (Table 4.1B) of heritability estimates for many of the cognitive measures from that study.

4.2.2.1 Verbal Ability

Cross-sectional twin studies in adulthood have shown moderate-to-high heritability of verbal ability, ranging from 0.52 to 0.85, in middle-aged and older adults (e.g., Finkel et al. 1995b; Pedersen et al. 1992; Posthuma et al. 2001a; Reynolds et al. 2005). Verbal abilities show relatively little change in middle adulthood and old age, and what change there is appears to be largely due to environmental factors (Reynolds et al. 2005). Verbal (letter and category) fluency is a more age-sensitive verbal ability, particularly with regard to risk for Alzheimer's disease. However, measures of verbal fluency are essentially absent from behavior genetic studies of cognition. Letter and category fluency have been found to have moderate-to-high heritability in the VETSA (0.62 and 0.54, respectively; see Table 4.1B).

4.2.2.2 Visual–Spatial Ability

High heritability estimates have typically been found for spatial abilities, ranging from 0.60 to 0.90 in SATSA and MTSADA (Finkel et al. 1995b; Pedersen et al. 1992; Reynolds et al. 2005). In the VETSA, Hidden figures was the most highly heritable individual cognitive test ($h^2=0.72$; see Table 4.1B); however, this test may include a strong executive component. Visual–spatial abilities warrant greater study. They are often more susceptible to aging than most verbal abilities, and there is also evidence for significant genetic influences on change in visual–spatial abilities over time, including a substantial quadratic component (Reynolds et al. 2005).

4.2.2.3 Episodic Memory

The construct of episodic (declarative) memory—involving recollection of facts and events—comprises multiple processes such as attention, encoding, learning, storage, and recall both immediately and after delay. These processes may occur in different modalities as well (e.g., verbal or visual). Genetic and environmental influences on memory in studies of aging have largely examined episodic memory at the broad domain level, so that these component processes and their interrelationships have seldom been examined, particularly in middle-aged adults.

Episodic memory (based on immediate recall) is moderately heritable during adulthood. SATSA participants were divided into young adult, middle-aged (mean age=59 [range 50–64]), and older. There were similar age groups in the MTSADA: young adult, middle-aged (mean age=61 [range 50–64]); and older (Finkel et al. 1995a). Heritabilities for memory in the SATSA middle-age group were 0.50 (Thurstone picture memory [visual]) and 0.40 (Names and faces [visual–verbal]). Heritabilities in the MTSADA were 0.63 for visual–spatial memory (Wechsler visual reproductions) and 0.56 for verbal memory (Wechsler logical memory).

As seen in Table 4.1B, heritabilities of several episodic memory measures were examined in the VETSA and in an earlier study of twins from the Vietnam Era Twin Registry. VETSA participants had a mean age of 55 (range: 51–60), whereas the latter study included 693 participants with a mean age of 48 years (range: 41–58). About 25 % of the VETSA participants also participated in the earlier study. The list of measures in the table is not complete because analyses have not been completed for all measures. At age 55, both verbal and spatial episodic memory abilities appear to be moderately heritable. Most of the more commonly used measures were in the 0.30–0.50 range. However, note that several individual CVLT trials were not significantly heritable, particularly trial 1, which is considered to be more of an auditory attention measure and one on which examinees are often a bit overwhelmed (Delis et al. 2000; See also section 4.2.2.4.).

Only a few studies have included multivariate genetic analyses of different memory abilities. In both the MTSADA and SATSA, heritabilities of memory measures were similar in all age groups so they were combined in multivariate genetic analyses (Finkel et al. 1995a). In both studies, there was common memory factor—which included digit span (a short-term/working memory measure)—with a very high heritability of 0.83. All the MTSADA memory measures showed significant levels of residual genetic variance, i.e., genetic variance that was test-specific. Only one of the SATSA measures (digit span) had significant residual genetic variance. The findings indicated that for logical memories (story recall) and visual reproductions (figure recall) some genetic influences were specific to each of these different modalities. Such findings suggest that significant findings from genome-wide genetic association studies of memory could be obscured by combining verbal and visual–spatial memory tests, or by trying to replicate results across modalities.

Singer et al. (2006) evaluated associations among general visual–spatial memory, spatial working memory, inspection time, reaction time, and general verbal ability in adult female twins. Four visual memory tests were consolidated into a single visual memory factor. General visual memory and spatial working memory had a correlation of 0.42. Heritability estimates were 0.57 for general visual memory and 0.31 for spatial working memory. The genetic correlation between the two was 0.80. Age was significantly negatively associated with both general visual memory ($r=-0.42$) and spatial working memory ($r=-0.43$). The average age in this study was 55 years, but with an age range of 18–76, it is unclear what the results tell us about middle adulthood. Several studies with very wide age ranges have an average age in the midlife range, but caution should be exercised in making strong inferences about middle adulthood based on those samples. On the other hand, the heritability estimate for visual memory in this study of women was very similar to the estimate in the all-male VETSA sample.

Swan et al. (1999) examined genetic and environmental influences on memory components using the CVLT in older adults (mean age=71.8 years; SD=2.9). After factor analyzing CVLT measures at the phenotypic level and then examining genetic and environmental influences on the phenotypic factors, they found a general learning and memory factor. In the VETSA, the genetic architecture of CVLT-II verbal learning, short- and long-delay free recall was assessed in almost 1,200 male twins in their fifties (Panizzon et al. 2011). Learning was defined as the total words recalled across the five learning trials. Because variation in delayed recall impairment is a cardinal symptom of Alzheimer's disease, delayed recall might be influenced by some genetic factors distinct from the other memory component processes. Thus, the individual measures were subjected to a multivariate genetic analysis. A higher-order latent genetic factor with a heritability of 0.36 influenced variation in all three measures; however, there were additional specific genetic influences that accounted for approximately 10% of the overall variance in learning. Both learning and free recall involve storage and retrieval, but acquisition is not involved in free recall because no information is presented. Only the learning trials involve acquisition of presented information. Consequently, these results suggest that the genetic influences specific to the learning trials are genetic influences that are specific to acquisition processes in memory. Thus, in genetic studies, a composite of immediate and delayed recall may be useful, but it is probably best to examine learning measures separately. Ongoing follow-up assessments will be needed to determine if these processes are differentially affected by aging. It is also possible that there could be some different genetic influences on short- and long-delayed recall in later life.

Longitudinal studies of episodic memory that include middle adulthood are rare. There were linear declines in visual–spatial episodic memory (Thurstone's picture memory) over time in SATSA twins such that declines from age 50–60 were similar to declines from age 80–90 (about 4 percentage points per decade; Finkel et al. 2003). Men had greater variability in change than women, but results were mixed with regard to sex differences in the average amount of decline. Latent growth curve analysis showed that the heritability of linear change over time in this memory measure was only 0.06, and the contribution of shared environment was 0.53 (Reynolds et al. 2005). However, the quadratic trend for accelerated decline was highly heritable ($h^2=0.70$). Thus, visual memory declines influenced primarily by environmental factors were observed during late midlife. There was also genetically influenced accelerated change that most likely occurred later in life.

Genetic influences on memory and acceleration of changes in the composite (verbal, visual, digit span) memory measure in the SATSA were evenly split between genetic influences that were shared with, and independent of processing speed (Finkel et al. 2005). Finkel et al. (2007) then found evidence to support the notion that processing speed was a leading indicator of age changes in memory. Much work is still needed with respect to uncovering the specific genetic and environmental mechanisms—including brain changes—that underlie the relationship between processing speed and different components of memory.

In summary, there needs to be more focus on subgroups of people who may experience episodic memory decline during middle adulthood. Behavior genetic studies also need to examine possible leading indicators of change in episodic memory other than processing speed. Finally, further investigation of the genetic and environmental influences on specific component processes is needed. For example, the finding of genetic influences specific to acquisition as opposed to retrieval and recall (Panizzon et al. 2011) calls for further study of which specific memory processes may account for age-related changes.

4.2.2.4 Apolipoprotein E (*APOE*) and Episodic Memory

The *APOE* gene is certainly not the only gene that is important for episodic memory, but it is singled out here because of the importance of the *APOE*-ε4 allele in risk for Alzheimer's disease (Farrer et al. 1997). In nonpathological aging, some of its effects appear to be independent of risk for Alzheimer's disease (Deary et al. 2002). Meta-analyses indicate that the ε4 allele does affect normal cognitive aging, but findings have been mixed as to whether it is associated with memory deficits in middle adulthood or whether the effects appear primarily in later life (Small et al. 2004; Wisdom et al. 2011). Some studies have shown episodic memory deficits in middle-aged ε4 carriers (e.g., under 60 years of age; Flory et al. 2000; Schultz et al. 2008), but others did not find an effect in individuals under 65 (Jorm et al. 2007). A negative finding was also reported in 70-year-old adults (Luciano et al. 2009). Based on the VETSA data (see Table 4.1B), CVLT trial 1 was not heritable. Because it was the measure used in the study of Jorm et al., one would indeed predict their negative finding. Thinning of prefrontal cortex has been associated with the *APOE*-ε4 allele in middle-aged men (Fennema-Notestine et al. 2011), but this effect appeared to precede cognitive differences. Not all ε4 carriers are expected to develop dementia, and it may be that subtle effects in middle age are missed because only a subgroup is affected. Also, although it is important in risk for Alzheimer's disease, *APOE* still accounts for only a small proportion of the variance in Alzheimer's disease.

Not taking other factors into account may also obscure the relatively subtle differences that may be present in nondemented middle-aged individuals. Interactions with other factors, including other genes, may obscure *APOE* effects in middle adulthood. We found verbal memory deficits in ε4 carriers in the first 626 VETSA participants (Schultz et al. 2008), but that finding did not hold up in the full sample of 1,237. However, based on an animal model (Raber et al. 2002), we predicted that protective effects of androgens would result in an *APOE* genotype × testosterone interaction. We did observe such an interaction; smaller hippocampal volumes were observed only in those with both *APOE*-ε4 and low testosterone (Panizzon et al. 2010). The same pattern was found for verbal memory based on Wechsler logical memories. Interestingly, there was no main effect of *APOE* genotype in analyses of *APOE* alone, but the main effect of *APOE* became significant after testosterone was included in the models.

APOE is a cholesterol transporter that can affect brain lipid homeostasis (Poirier 2003). Testosterone is synthesized from cholesterol through a series of steps, and it affects androgen receptor function. The hippocampus is rich in androgen receptors in both men and women and the ε4 allele is associated with reducing the binding of testosterone to androgen receptors (Panizzon et al. 2010; Raber et al. 2002). Thus, having an ε4 allele could lead to reduced efficiency of hippocampal androgen receptors and increased risk for memory impairments. Testosterone declines with age in both men and women beginning in one's thirties, and these processes may have a greater impact in those with greater testosterone declines (Panizzon et al. 2010). It is uncertain whether this pattern exists in women, but testosterone levels have been positively correlated with verbal memory in older women (Wolf and Krischbaum 2001).

4.2.2.5 Short-Term and Working Memory

Working memory represents the ability to temporarily store and process information for future goals (Baddeley 1992). Thus, it comprises processes involving short-term memory (storage) and executive functions. The processes of maintenance and manipulation of information in short-term memory are sometimes referred to as working memory; however, we refer to maintenance alone (e.g., digits forward) as short-term memory and we use the term working memory to refer to processes involving manipulation or processing of information in short-term memory. In contrast to hippocampal-dependent episodic memory, there is a large literature showing that prefrontal cortical regions play a central role in neural systems underlying working memory (Fuster 1995). For these reasons, we think it is best to avoid lumping short-term and working memory measures with episodic memory measures. Even if they go together in a factor analysis, that should not outweigh what is known about brain and memory. Consider the well-known case of H.M. After bilateral medial temporal excision he could not form new (episodic) memories, but his digit span performance remained intact (Corkin 2002). On the other hand, genetic influences that are unique to either episodic or working memory might be found in a genetic factor analysis even if phenotypic factor analysis shows only a single factor.

Short-term/working memory tasks vary substantially across studies and heritability estimates range from 0.00 to 0.65. Digit span, either forward or backward, tends to be moderately heritable in middle adulthood, generally ranging from about 0.40 to 0.65 (Table 4.1A, B and Finkel et al. 1995a; Karlsgodt et al. 2010). For the most part, its heritability appears to be similar in young and middle adulthood (Finkel et al. 1995a). Similar heritabilities have been found for spatial span and letter–number sequencing (Table 4.1B and Karlsgodt et al. 2010), and Posthuma et al. (2003) found a heritability of 0.65 for a composite arithmetic and letter–number sequencing measure. In contrast, variants of classic spatial and nonspatial delayed response or delayed alternation tasks appear to have little or no heritability; estimates ranged from 0.00 to 0.25 (Table 4.1A and Karlsgodt et al. 2010; Kremen et al. 2011a).

As already noted, given close links between working memory and executive functions, there is a strong need to elucidate executive and other components of working memory beyond individual tests. Almost all behavior genetic studies of this kind have been conducted in children or very young adults (e.g., Ando et al. 2001; Friedman et al. 2008). Such studies are needed in middle and later adulthood because non-twin studies indicate that these functions are among the more susceptible to age effects, there is substantial age-related prefrontal shrinkage relative to other parenchymal regions, and these functions are important for successful daily functioning.

A study of young adults (average age 19.9 years) decomposed the genetic structure of spatial and verbal memory in relation to general verbal and visual ability using rotation-arrow and verification-word tasks (Ando et al. 2001). These tasks could be subdivided into spatial or verbal storage, and spatial or verbal executive efficiency functions. Heritabilities were 0.43 for the verbal executive and 0.49 for the spatial executive components. The best fitting model indicated multiple sources of genetic influences on the four functions including a common factor, two modality-specific factors (verbal and spatial), and a storage-specific factor that explained a significant amount of the variance (11–43 %). Genetic influences specific to the verbal and visual working memory storage components were also identified.

As described earlier (section 4.2.2.3), the spatial working memory measure of Singer et al. (2006) had a heritability of 0.31. Its correlation with general visual memory was 0.42, and 0.16 with general verbal ability; it had little or no correlation with processing speed (e.g., r with inspection time=0.00). The observed associations were largely due to shared genetic influences, but they also clearly indicate genetic influences that were specific to spatial working memory. Again, as noted earlier, given the age range of 18–76, it is difficult to know how to interpret these findings with respect to middle adulthood.

A series of investigations into working memory were conducted in Vietnam Era Twin Registry twins (average age 48 years; range 41–58). The approach taken was not to focus simply on measures that might be associated with working memory, but to include the simpler abilities that comprise more basic cognitive components of a more complex working memory test. Kremen et al. (2007b) examined overlap between short-term memory (digits forward), reading ability (word recognition), and verbal working memory (reading span). Overall heritabilities were 0.27 for digits forward, 0.51 for reading, and 0.52 for reading span. A common latent phenotype explained all the genetic influences on reading and digits forward, but there were specific genetic influences only on reading span. The interpretation was that the specific genetic influences must reflect the executive component of reading span, i.e., the dual processing required for simultaneously reading aloud and holding some element in memory.

Verbal working memory was assessed in the same sample using a digit transformation task in which participants heard four digits and after a pause were instructed to add 3 or 4 to each digit. Heritabilities of add-3 and add-4 were roughly double the heritability of storage only (digits forward). The additional executive function demands of the add-3 or add-4 tasks (i.e., both storage and manipulation) appeared to increase the variance of individual genetic differences from 25 % for digits forward to 48 % and 53 % for the digit transformation scores. A two-factor model suggested the possibility of a second set of genetic factors specifically influencing the executive (manipulation) component. These results suggested that new genetic influences might come into play if demand continues to increase beyond a certain threshold, a threshold that may change with task difficulty and with age. Together, these studies suggest that, in part, there are genetic influences that are specific to the executive components of working memory, and those are likely to be the most age-sensitive processes.

In summary, measures of working memory have only partial genetic overlap with episodic memory, processing speed, or general cognitive ability, suggesting the presence of some distinct underlying influences. Separating executive components from other abilities that affect performance on working memory tasks also appears to be important. There were no genetic influences on change over time in digit span performance in the SATSA (Reynolds et al. 2005), but it is important to remember that different working memory measures tap different processes. To our knowledge, there are no other studies that examine the role of genetic and environmental influences on age-related change or age differences in working memory during adulthood despite the fact that performance on most working memory tasks decreases with age. A few functional MRI twin studies in young adults have shown heritable activation in prefrontal and other brain regions using Sternberg task and n-back working memory tasks (Blokland et al. 2008; Koten et al. 2009). In their extended pedigree study with a very wide age range, Karlsgodt et al. (2010) found a genetic correlation of 0.59 between spatial delayed response and integrity of the superior longitudinal fasciculus, which connects frontal and parietal cortices. These studies further support the value of more extensive examination of working memory and prefrontal cortex in behavior genetics studies of cognitive aging. The VETSA project has begun work along these lines in middle-age adults (see section 4.2.2.7).

4.2.2.6 Executive Function

Executive functions refer to cognitive control processes that help to integrate more simple functions by regulating thinking and behavior. They include planning, organization, shifting mental set (cognitive flexibility), and inhibitory control (resistance to interfering, task-irrelevant stimuli). Working memory is closely linked to many executive functions. For example, one well-known approach includes working memory among three major executive functions: (1) mental set shifting, (2) updating and monitoring working memory representations, and (3) inhibition of prepotent responses (Miyake et al. 2000). Thus, some tests discussed in this section include working memory components and the decision on whether to include them in this or the working memory section (4.2.2.5) is necessarily somewhat arbitrary. One such test is the Wisconsin Card Sorting Test, a measure of executive function that requires set shifting, but also has a strong working memory updating component. Review of a handful of studies estimating heritability of this test showed that it was almost always at or near zero (Chou et al. 2009; Kremen et al. 2007a). Although it is a classic executive function test, twin studies have thus shown that it is unlikely to be useful for genetic association studies of executive function.

Kremen et al. (2009) examined the genetic architecture of the Tower of London test, a measure of planning and problem-solving in VET Registry twins (mean age = 48 years; Table 4.1A). Interestingly, even though phenotypic analyses revealed only one general factor, the best fitting genetic model indicated two correlated genetic factors: speed and efficiency. Genetic influences accounted for 38 % of the variance in the six Tower of London measures.

A major theory of the cause of age-related cognitive declines focuses on reduced efficiency of the executive function of inhibitory control or response inhibition

(Hasher and Zacks 1988). Barkley (1997) distinguished three interrelated processes believed to constitute response inhibition: (1) inhibiting a prepotent response, (2) stopping an ongoing response, and (3) interference control. The Stroop color-word task has a condition called "interfernce" but it primarily involves inhibition of a prepotent response. Johnson et al. (2003) found that genetic influences accounted for 54% of the variance in color-word scores, and 35% of variance in the Stroop interference condition. Color-word scores were correlated −0.33 with age in their sample, which ranged in age from under 30 to over 70. Correlations of interference with age were not reported, but color-word and interference were highly correlated. Interestingly, color-word performance tended to increase with age until about 60 and then decline, suggesting that late middle adulthood may be a key transition period. In the mostly younger sample (mean age=24 years) of Taylor (2007), heritability was 0.57 for color-word performance but only 0.17 for interference. In the VETSA, heritability of the color-word performance was 0.49, and heritability of the interference score was 0.23. The latter two heritability estimates for Stroop interference did not reach statistical significance, the cross-sectional results from all these studies suggest that heritability of Stroop color-word and interference control is consistent from young to late adulthood. However, the age-related performance differences reported by Johnson et al. suggest that ongoing follow-up of the VETSA sample may capture a key transition period. The extent to which expected declines after age 60 may be due to genetic or environmental influences remains to be seen.

Tests of executive function always measure the complex executive processes along with the more simple processes that underlie them. Performance on these processes is also likely to be genetically correlated. Thus, adjusting the more complex function for the more simple ones to isolate the executive component may run the risk of removing too much of the genetic variance. It appears that this is what happened in the case of Stroop interference and adjusted set-shifting measures in the VETSA (see Table 4.1B). One strategy for addressing this problem may be to utilize multivariate twin analysis. Vasilopoulos et al. (2012) used this approach to examine the Delis-Kaplan Executive Function System Trail Making test. Conditions included: visual search ($h^2=0.35$), number sequencing ($h^2=0.34$), letter sequencing ($h^2=0.43$), and letter–number switching ($h^2=0.62$). There was a common genetic factor, and the heritabilities of number and letter sequencing conditions (processing speed and sequencing) were each completely explained by the common genetic factor. However, a significant 21% of the genetic variance in set shifting (switching condition) was accounted for by measure-specific genetic influences that were independent of the common factor. As seen in Table 4.1B, adjusting the Trails set-shifting condition for the nonswitching conditions yielded a lower and nonsignificant heritability, but the multivariate analysis strongly suggested genetic influences specific to set-shifting ability. Like previous phenotypic factor analytic studies, a phenotypic factor analysis in this study resulted in only one factor.

Neuropsychologists have long considered set shifting to be a key executive function and the inability to show a separate factor has been somewhat puzzling. There was only one latent factor in the twin analysis but the twin method was still able to show specific genetic influences on set shifting, consistent with its being a different cognitive function. For genetic studies of cognitive aging, it will be important to differentiate the speed and executive components of this test as these may manifest differential change with age.

4.2.2.7 Context Processing

Context processing, which addresses working memory and executive function, has received considerable attention in the phenotypic literature. It refers to internally represented, task-relevant information being used to influence planning and behavior (Braver et al. 2005). Although it is probably unfamiliar to most behavior genetics researchers, it is a mechanism that can account for several age-related cognitive changes. Reduced prefrontal dopamine availability with aging is associated with response slowing and signal-to-noise reductions resulting in internal representations (context maintenance) that are more susceptible to decay over time (working memory) and to the effects of task-irrelevant inputs (inhibitory control; Braver et al. 2005; Kremen et al. 2011b). Since the development of the context-processing model in the mid-1990s, a key point has been that there was a single mechanism underlying working memory and cognitive control components of context processing.

The AX-CPT (continuous performance test) has been frequently used to assess context processing. Participants are presented letters, one at a time, on a screen and they must press a target button for an X, but only when it is immediately preceded by an A. By making most of the cue-probe pairs AX trials like the one described, a prepotent response tendency is created toward incorrect X probes (referred to as BX trials, i.e., X preceded by a non-A cue) and toward false alarms when A cues are followed by a non-X probe (referred to as AY trials). If context processing is intact, BX trials will require little inhibitory control because representation of the non-A cue will be well maintained. Older adults, who have context processing deficits, have poorer BX performance than young adults do. However, if context maintenance and response preparation are strong, AY trials will require greater inhibitory control because the A cue provides a strong expectation of a target probe. Less efficient context maintenance in older adults reduces the strength of this expectation, and hence, reduces the need for inhibitory control. Consistent with this framework, older adults make fewer errors and have faster response times on AY trials compared with young adults (e.g., Braver et al. 2005). This pattern represents a very rare instance of faster responding in older adults, thus indicating that processing speed alone cannot account for age-related cognitive declines. In sum, AX and BX trials emphasize proactive cognitive control (using context cues to prepare responses), whereas AY trials emphasize reactive control (adjusting responses after seeing the probe).

The first twin study of context processing was conducted in the VETSA sample (ages 51–59; Kremen et al. 2011b). Heritability of signal detection (an overall index of the ability to differentiate target from nontarget) was 0.40, and about one-quarter of the variance in each component was accounted for by genetic influences. A single common genetic factor accounted for accuracy on AX, BX, and AY trials with specific genetic influences only for AY trials. There were significant genetic correlations between general cognitive ability and proactive control (AX and BX performance), but not reactive control. The presence of independent genetic influences underlying reactive control processes indicates that, in contrast to phenotypic studies, there must be more than one underlying mechanism.

There is some cross-sectional, phenotypic evidence that older adults tend to rely on reactive control more than young adults (Braver et al. 2009), suggesting that there should be a shift with age in the balance of proactive and reactive control. Given these findings, elucidating the genetic factors underlying proactive and reactive control processes is likely to be an important component of understanding cognitive aging. Age-related changes could be due to the fact that proactive control requires more metabolic/neuronal resources (Braver et al. 2007). It has also been argued that reduced efficiency of inhibitory control underlies many aging-related cognitive deficits (Hasher and Zacks 1988), and reduced use of proactive control makes one more susceptible to proactive interference. Longitudinal studies will be needed to determine whether and when a shift from proactive to reactive control takes place.

Reaction time was unrelated to chronological age in the narrow age range of the VETSA sample. Interestingly, however, participants with slower reaction time on noncontext processing trials were performing more like older adults. That is, slower reaction time on noncontext processing (BY) trials (typical of older adults) was associated with fewer AY errors and more BX errors. Although this pattern did not generalize to other reaction time measures, it could be a marker for larger or earlier cognitive declines as the VETSA participants age.

4.2.2.8 Processing Speed

Processing speed is a key cognitive variable in studies of aging given evidence (mainly cross-sectional) that age-related processing speed declines account for many other cognitive declines (Salthouse 1985). The concept of processing speed may be viewed as relatively simple. Everyone knows what is meant by processing speed, but its measurement is not necessarily straightforward. It is conceptualized as a simple or basic process (a cognitive primitive) underlying more complex cognitive abilities, but it is very difficult to avoid confounding by other cognitive abilities in tests of processing speed. For example, digit symbol—a classic processing speed measure—also involves elements of visual scanning and working memory. Deary et al. (2010a) reported phenotypic correlations among five widely used processing speed measures in older adults. Not counting correlations between variants of the same test (e.g., simple and choice reaction time), the median correlation was only 0.28. Their results highlight an important feature, namely, that different processing speed measures may actually measure fairly different phenomena. Similar variability may be reflected in the heritability estimates.

Most processing speed measures—including components of the trail making test, inspection time, and reaction time measures—have heritabilities generally in the 0.30–0.40 range (Finkel and McGue 2007; Posthuma et al. 2001a; Singer et al. 2006; Vasilopoulos et al. 2012). Several of those estimates were based on samples with very wide age ranges. Heritability of digit symbol/symbol digit appears to be somewhat higher, in the 0.60–0.70 range in the MTSADA and SATSA (Finkel et al. 1995b). All these may underestimate the genetic influences on "true" processing speed based on the findings of Posthuma et al. (2001b). Utilizing electroencephalography (EEG) recordings to assess speed of encoding and accessing of information, they found a heritability of 0.83 in a middle-aged cohort.

Some cross-sectional studies have shown a significant though modest relationship between processing speed and general cognitive ability in late middle-age and older adulthood, and this association is almost entirely mediated by genetic influences (e.g., Finkel and Pedersen 2000; Posthuma et al. 2001a). In longitudinal biometric dual-change score models, genetic variance for processing speed was a leading determinant of variation in changes for spatial and memory, but not verbal, abilities across time in the SATSA (Finkel et al. 2009). Luciano et al. (2005) examined causal genetic models between processing speed (inspection time) and IQ in a cross-sectional study. Rather than a causal relationship in either direction, their evidence suggested that the covariation between intelligence and processing speed was explained by pleiotropy (i.e., some of the same genes influencing both phenotypes). Based on their bivariate model for processing speed and Performance IQ, for example, we estimated a genetic correlation of approximately 0.55. Although this does indicate pleiotropy, it indicates substantial nonshared genetic influences as well. In sum, processing speed can be a leading indicator, but the results suggest that it may be neither a necessary nor a sufficient condition.

4.3 Summary and Future Directions

Middle adulthood remains an understudied period in phenotypic and behavior genetic studies of cognitive aging. Further study focused specifically on middle adulthood is needed for understanding cognitive changes during midlife itself and for understanding midlife cognition as a "gateway" to cognition in old age. The available data suggest that there are some cognitive changes during midlife itself. However, because these changes tend to be small on average, a greater focus on subgroups of individuals who are manifesting earlier declines will be important. One dilemma is that in order to have adequate power to draw conclusions about a subgroup, larger sample sizes will be needed. Most studies, to date, include a relatively wide age range, but not enough middle-adult participants to examine subgroup or individual differences, particularly within-individual differences in trajectories over time.

Behavior genetic studies of cognitive aging in both middle and older adulthood have also had somewhat limited cognitive test batteries. It will be important for test selection to be guided by the goal of elucidating more specific cognitive component processes that may drive cognitive aging, i.e., a greater emphasis on splitting rather than lumping. Behavior genetic studies can be particularly enlightening in this regard. As shown in some of our examples, behavior genetic analysis of a well-chosen set of cognitive measures may reveal underlying influences that are different from what can be observed at just the phenotypic level. This approach not only helps to elucidate cognitive processes, it also serves to refine phenotypes for other studies including genetic association studies. Although we have emphasized splitting, behavior genetic analyses may also indicate that combining measures into a more general phenotype is sometimes optimal. That would be the case, for example, for the short- and long-delay free recall phenotypes in our study of the CVLT (section 4.2.2.3). The same genes influenced both phenotypes, and pleiotropic effects like this do justify combining measures. However, without this sort of decomposition of cognitive components, it is not possible to know the most appropriate way to combine or differentiate measures.

Measures of general cognitive ability can be advantageous given their greater reliability over specific ability measures and greater ease of administration. General ability measures also tend to have higher heritabilities than specific ability measures, which may mean greater statistical power for behavioral genetic analyses. However, as suggested by Tucker-Drob's (2011) results, they are disadvantageous in that the variance in cognitive change is mostly accounted for by specific domains and tests. Moreover, as can be seen from cognitive neuroscience and neuroimaging studies, it seems fairly clear that if the ultimate goal is to understand the mechanisms underlying cognitive aging, it will be necessary to study specific cognitive abilities and even subprocesses underlying specific abilities.

As we have noted, the emphasis on elucidating component processes is consistent with cognitive science and cognitive neuroscience approaches. Simply examining an overall score on a test might miss a key underlying subprocess that may be driving cognitive changes with age. When we refer to mapping the twin method onto these approaches, we simply mean examining these subprocesses in behavior genetic studies rather than only at the phenotypic level. With respect to cognitive neuroscience, it would mean the behavior genetic examination of brain–behavior relationships (e.g., with neuroimaging; for a review, see Chap. 8). Feasibility is clearly an important issue when it comes to neuroimaging twin studies, but recent studies have been able to obtain relatively large twin samples in neuroimaging studies (VETSA, Older Adult Twin Study, NIMH, Australian Twin Sample). Collaborative efforts are extremely important so that sample sizes can be increased. Methods have been in place for some time now for combining magnetic resonance imaging (MRI) samples across sites and platforms. The Enhanced Neuroimaging through Meta-Analysis (ENIGMA) consortium, for example, includes about 10,000 subjects with MRI and genotyping (including some twins). It may actually be easier to combine samples with structural MRI data than to combine samples with neurocognitive data because neurocognitive studies often use different tests even though they may be assessing the same cognitive domains.

A clear limitation is that much of the behavior genetic data on cognition in middle adulthood comes from the VETSA, which has only male participants. One of the better (albeit non-behavior-genetic) studies for examining sex differences may be the Whitehall II study (Singh-Manoux et al. 2011) because it is a very large study with data on middle-aged adults ($n=7{,}390$). They found no sex differences (overlapping confidence intervals) in percent change over 10 years in their five different cognitive measures. Nevertheless, factors such as different probabilities of various health events, different risks for depression, differences in life expectancy, or differences in age-related hormonal changes may play a role in cognitive function in middle age and beyond.

Another important limitation of behavior genetic studies of middle and older adults is the lack of working memory and executive function tests. Although processing speed is most often studied as a leading indicator, working memory and executive functions (including context processing) are also strong candidates for genetically mediated leading indicators (also suggested by candidate gene studies). In a longitudinal phenotypic study, Hultsch et al. (1998) found that working memory was a stronger predictor of later episodic memory than processing speed in older adults. Given the importance of these functions for daily living and the evidence of age-related changes in prefrontal cortex (which plays a key role in mediating these functions), substantially increased emphasis on these functions along with a greater focus on underlying component cognitive processes is warranted in longitudinal behavior genetic studies of middle-aged and older adults.

Acknowledgments Funding for this work was provided by grants from the National Institute on Aging (R01 AG018386, R01 AG018384, R01 AG022381, R01 AG037985, and R01 AG022982). This material is also the result of work supported in part with resources of the Center of Excellence for Stress and Mental Health at the VA San Diego Healthcare System. The Cooperative Studies Program of the Office of Research and Development of the US Department of Veterans Affairs has provided financial support for the development and maintenance of the Vietnam Era Twin (VET) Registry. Numerous organizations have provided invaluable assistance in the conduct of this study, including: Department of Defense; National Personnel Records Center, National Archives and Records Administration; the Internal Revenue Service; National Opinion Research Center; National Research Council, National Academy of Sciences; the Institute for Survey Research, Temple University. Most importantly, the authors gratefully acknowledge the continued cooperation and participation of the members of the VET Registry and their families. Without their contribution, this research would not have been possible.

References

Albert, M. S., DeKosky, S. T., Dickson, D., Dubois, B., Feldman, H. H., Fox, N. C., et al. (2011). The diagnosis of mild cognitive impairment due to Alzheimer's disease: Recommendations from the National Institute on Aging-Alzheimer's Association workgroups on diagnostic guidelines for Alzheimer's disease. *Alzheimer's & Dementia, 7*, 270–279.

Ando, J., Ono, Y., Wright, M. J. (2001). Genetic structure of spatial and verbal working memory. *Behavior Genetics, 31,* 615–624.

Baaré, W. F. C., Hulshoff Pol, H. F., Boomsma, D. I., Posthuma, D., de Geus, E. J. C., Schnack, H. G., et al. (2001). Quantitative genetic modeling of variation in human brain morphology. *Cerebral Cortex, 11,* 816–824.

Baddeley, A. D. (1992). Working memory. *Science, 255,* 556–559.

Barkley, R. A. (1997). Behavioral inhibition, sustained attention, and executive functions: Constructing a unifying theory of ADHD. *Psychological Bulletin, 121,* 65–94.

Blokland, G. A., McMahon, K. L., Hoffman, J., Zhu, G., Meredith, M., Martin, N. G., et al. (2008). Quantifying the heritability of task-related brain activation and performance during the N-back working memory task: A twin fMRI study. *Biological Psychology, 79,* 70–79.

Boomsma, D. I. (1998). Twin registers in Europe: An overview. *Twin Research, 1,* 34–51.

Bouchard, T. J., Jr., McGue, M. (2003). Genetic and environmental influences on human psychological differences. *Journal of Neurobiology, 54,* 4–45.

Brans, R. G., Kahn, R. S., Schnack, H. G., van Baal, G. C., Posthuma, D., van Haren, N. E., et al. (2010). Brain plasticity and intellectual ability are influenced by shared genes. *Journal of Neuroscience, 30,* 5519–5524.

Braver, T. S., Satpute, A. B., Rush, B. K., Racine, C. A., Barch, D. M. (2005). Context processing and context maintenance in healthy aging and early stage dementia of the Alzheimer's type. *Psychology and Aging, 20,* 33–46.

Braver, T. S., Gray, J. R., Burgess, G. C. (2007).Explaining the many varieties of working memory variation: Dual mechanisms of cognitive control. In A. R. A. Conway, C. Jarrold, M. J. Kane, A. Miyake, J. N. Towse (Eds.), *Variation in working memory* (pp. 76–108). New York: Oxford University Press.

Braver, T. S., Paxton, J. L., Locke, H. S., Barch, D. M. (2009). *Flexible neural mechanisms of cognitive control within human prefrontal cortex.* Proceedings of the National Academy of Sciences USA, 106, 7351–7356.

Chou, L. N., Kuo, P. H., Lin, C. C., Chen, W. J. (2010). Genetic and environmental influences on the Wisconsin Card Sorting Test performance in healthy adolescents: A twin/sibling study. *Behavior Genetics, 40*(1), 22–30.

Corkin, S. (2002). What's new with the amnesic patient H.M.? *Nature Reviews. Neuroscience, 3,* 153–160.

Davies, G., Tenesa, A., Payton, A., Yang, J., Harris, S. E., Liewald, D., et al. (2011). Genome-wide association studies establish that human intelligence is highly heritable and polygenic. *Molecular Psychiatry, 16,* 996–1005.

Deary, I. J., Whalley, L. J., Lemmon, H., Crawford, J. R., Starr, J. M. (2000). The stability of individual differences in mental ability from childhood to old age: Follow-up of the 1932 Scottish mental survey. *Intelligence, 28,* 49–55.

Deary, I. J., Whiteman, M. C., Pattie, A., Starr, J. M., Hayward, C., Wright, A. F., et al. (2002). Cognitive change and the *APOE*4 allele. *Nature, 418,* 932.

Deary, I. J., Wright, A. F., Harris, S. E., Whalley, L. J., Starr, J. M. (2004). Searching for genetic influences on normal cognitive ageing. *Trends in Cognitive Sciences, 8,* 178–184.

Deary, I. J., Johnson, W., Starr, J. M. (2010a). Are processing speed tasks biomarkers of cognitive aging? *Psychology and Aging, 25,* 219–228.

Deary, I. J., Penke, L., Johnson, W. (2010b). The neuroscience of human intelligence differences. *Nature Reviews. Neuroscience, 11,* 201–211.

Deary, I. J., Yang, J., Davies, G., Harris, S. E., Tenesa, A., Liewald, D., et al. (2012). Genetic contributions to stability and change in intelligence from childhood to old age. *Nature, 482*(7384), 212–215.

Delis, D. C., Kramer, J. H., Kaplan, E., Ober, B. A. (2000). *California Verbal Learning Test.* San Antonio: Psychological Corporation.

Farrer, L. A., Cupples, L. A., Haines, J. L., Hyman, B., Kukull, W. A., Mayeux, R., et al. (1997). Effects of age, sex, and ethnicity on the association between apolipoprotein E genotype and Alzheimer disease. A meta-analysis. APOE and Alzheimer Disease Meta Analysis Consortium. *Journal of the American Medical Association, 278,* 1349–1356.

Fennema-Notestine, C., Panizzon, M. S., Thompson, W. R., Chen, C. H., Eyler, L. T., Fischl, B., et al. (2011). Presence of ApoE epsilon4 allele associated with thinner frontal cortex in middle age. *Journal of Alzheimer's Disease, 26*(Suppl 3), 49–60.

Finch, C. E. (1991). Middle-age: An evolving frontier in gerontology. *Neurobiology of Aging, 12,* 1–2.

Finkel, D., McGue, M. (1998). Age differences in the nature and origin of individual differences in memory: A behavior genetic analysis. *International Journal of Aging and Human Development, 47,* 217–239.

Finkel, D., McGue, M. (2007). Genetic and environmental influences on intraindividual variability in reaction time. *Experimental Aging Research, 33,* 13–35.

Finkel, D., Pedersen, N. L. (2000). Contribution of age, genes, and environment to the relationship between perceptual speed and cognitive ability. *Psychology and Aging, 15,* 56–64.

Finkel, D., Pedersen, N. L., McGue, M. (1995a). Genetic influences on memory performance in adulthood: Comparison of Minnesota and Swedish twin data. *Psychology and Aging, 10,* 437–446.

Finkel, D., Pedersen, N. L., McGue, M., McClearn, G. E. (1995b). Heritability of cognitive abilities in adult twins: Comparison of Minnesota and Swedish data. *Behavior Genetics, 25,* 421–431.

Finkel, D., Pedersen, N. L., Plomin, R., McClearn, G. E. (1998). Longitudinal and cross-sectional twin data on cognitive abilities in adulthood: The Swedish Adoption/Twin Study of Aging. *Developmental Psychology, 34,* 1400–1413.

Finkel, D., Reynolds, C. A., McArdle, J. J., Gatz, M., Pedersen, N. L. (2003). Latent growth curve analyses of accelerating decline in cognitive abilities in late adulthood. *Developmental Psychology, 39,* 535–550.

Finkel, D., Reynolds, C. A., McArdle, J. J., Pedersen, N. L. (2005). The longitudinal relationship between processing speed and cognitive ability: Genetic and environmental influences. *Behavior Genetics, 35,* 535–549.

Finkel, D., Reynolds, C. A., McArdle, J. J., Pedersen, N. L. (2007). Age changes in processing speed as a leading indicator of cognitive aging. *Psychology and Aging, 22,* 558–568.

Finkel, D., Reynolds, C. A., McArdle, J. J., Hamagami, F., Pedersen, N. L. (2009). Genetic variance in processing speed drives variation in aging of spatial and memory abilities. *Developmental Psychology, 45,* 820–834.

Flory, J. D., Manuck, S. B., Ferrell, R. E., Ryan, C. M., Muldoon, M. F. (2000). Memory performance and the apolipoprotein E polymorphism in a community sample of middle-aged adults. *American Journal of Medical Genetics (Neuropsychiatric Genetics), 96,* 707–711.

Friedman, N. P., Miyake, A., Corley, R. P., Young, S. E., Defries, J. C., Hewitt, J. K. (2006). Not all executive functions are related to intelligence. *Psychological Science, 17,* 172–179.

Friedman, N. P., Miyake, A., Young, S. E., Defries, J. C., Corley, R. P., Hewitt, J. K. (2008). Individual differences in executive functions are almost entirely genetic in origin. *Journal of Experimental Psychology General, 137,* 201–225.

Fuster, J. M. (1995). *Memory in the cerebral cortex*. Cambridge: MIT Press.

Gatz, M., Reynolds, C. A., Fratiglioni, L., Johansson, B., Mortimer, J. A., Berg, S., et al. (2006). Role of genes and environments for explaining Alzheimer disease. *Archives of General Psychiatry, 63,* 168–174.

Goodale, M. A., Milner, A. D. (1992). Separate visual pathways for perception and action. *Trends in Neurosciences, 15,* 20–25.

Hasher, L., Zacks, R. T. (1988).Working memory, comprehension, and aging: A review and new view. In G.H. Bower (Ed.), *The psychology of learning and motivation* (pp. 193–225). New York: Academic Press.

Haworth, C. M., Wright, M. J., Luciano, M., Martin, N. G., de Geus, E. J., van Beijsterveldt, C. E., et al. (2010). The heritability of general cognitive ability increases linearly from childhood to young adulthood. *Molecular Psychiatry, 15,* 1112–1120.

Hertzog, C., Dixon, R. A., Hultsch, D. F., MacDonald, S. W. (2003). Latent change models of adult cognition: Are changes in processing speed and working memory associated with changes in episodic memory? *Psychology and Aging, 18,* 755–769.

Hulshoff Pol, H. E., Schnack, H. G., Posthuma, D., Mandl, R. C., Baare, W. F., van Oel, C., et al. (2006). Genetic contributions to human brain morphology and intelligence. *Journal of Neuroscience, 26,* 10235–10242.

Hultsch, D. F., Hertzog, C., Dixon, R. A., Small, B. J. (1998). *Memory change in the aged.* Cambridge: Cambridge University Press.

Jak, A. J., Spoon, K. M., Panizzon, M. S., Franz, C. E., Thompson, W. K., Jacobson, K. C., et al. (2013). Hippocampal atrophy varies by neuropsychological definition of mild cognitive impairment in men in their 50s. Paper to be presented at the annual meeting of the International Neuropsychological Society, Waikoloa, HI.

Johnson, W., Bouchard, T. J., Jr., Segal, N. L., Keyes, M., Samuels, J. (2003). The Stroop Color-Word Test: Genetic and environmental influences; reading ability, mental ability, and personality correlates. *Journal of Educational Psychology, 95,* 58–65.

Jorm, A. F., Mather, K. A., Butterworth, P., Anstey, K. J., Christensen, H., Easteal, S. (2007). APOE genotype and cognitive functioning in a large age-stratified population sample. *Neuropsychology, 21,* 1–8.

Karlsgodt, K. H., Kochunov, P., Winkler, A. M., Laird, A. R., Almasy, L., Duggirala, R., et al. (2010). A multimodal assessment of the genetic control over working memory. *Journal of Neuroscience, 30,* 8197–8202.

Koten, J. W., Jr., Wood, G., Hagoort, P., Goebel, R., Propping, P., Willmes, K., et al. (2009). Genetic contribution to variation in cognitive function: An FMRI study in twins. *Science, 323,* 1737–1740.

Kremen, W. S., Thompson-Brenner, H., Leung, Y. J., Grant, M. D., Franz, C. E., Eisen, S. A., et al. (2006). Genes, environment, and time: The Vietnam Era Twin Study of Aging (VETSA). *Twin Research and Human Genetics, 9,* 1009–1022.

Kremen, W. S., Eisen, S. A., Tsuang, M. T., Lyons, M. J. (2007a). Is the Wisconsin Card Sorting Test a useful neurocognitive endophenotype? *American Journal of Medical Genetics (Neuropsychiatric Genet), 144B,* 403–406.

Kremen, W. S., Jacobsen, K. C., Xian, H., Eisen, S. A., Eaves, L. J., Tsuang, M. T., et al. (2007b). Genetics of verbal working memory processes: A twin study of middle-aged men. *Neuropsychology, 21,* 569–580.

Kremen, W. S., Xian, H., Jacobson, K. C., Eaves, L. J., Franz, C. E., Panizzon, M. S. (2008). Storage and executive components of working memory: Integrating cognitive psychology and behavior genetics in the study of aging. *Journal of Gerontology: Psychological Sciences, 63B,* 84–91.

Kremen, W. S., Jacobson, K. C., Panizzon, M. S., Xian, H., Eaves, L. J., Eisen, S. A. (2009). Factor structure of planning and problem-solving: A behavioral genetic analysis of the tower of London task in middle-aged twins. *Behavior Genetics, 39,* 133–144.

Kremen, W. S., Lyons, M. J. (2011).Behavior genetics of aging. In K. W. Schaie, S.L. Willis (Eds.), *Handbook of the psychology of aging* (pp. 93–107). San Diego: Elsevier.

Kremen, W. S., Mai, T., Panizzon, M. S., Franz, C. E., Blankfeld, H. M., Xian, H., et al. (2011a). A twin study of spatial and non-spatial delayed response performance in middle age. *Brain and Cognition, 76,* 43–51.

Kremen, W. S., Panizzon, M. S., Xian, H., Barch, D. M., Franz, C. E., Grant, M. D., et al. (2011b). Genetic architecture of context processing in late middle age: More than one underlying mechanism. *Psychology and Aging, 26,* 852–863.

Luciano, M., Wright, M. J., Smith, G. A., Geffen, G. M., Geffen, L. B., Martin, N. G. (2001). Genetic covariance among measures of information processing speed, working memory, and IQ. *Behavior Genetics, 31,* 581–592.

Luciano, M., Posthuma, D., Wright, M. J., de Geus, E. J., Smith, G. A., Geffen, G. M., et al. (2005). Perceptual speed does not cause intelligence, and intelligence does not cause perceptual speed. *Biological Psychology, 70,* 1–8.

Luciano, M., Gow, A. J., Taylor, M. D., Hayward, C., Harris, S. E., Campbell, H., et al. (2009). Apolipoprotein E is not related to memory abilities at 70 years of age. *Behavior Genetics, 39,* 6–14.

Lyons, M. J., York, T. P., Franz, C. E., Grant, M. D., Eaves, L. J., Jacobson, K. C., et al. (2009). Genes determine stability and the environment determines change in cognitive ability during 35 years of adulthood. *Psychological Science, 20,* 1146–1152.

Maher, B. (2008). Personal genomes: The case of the missing heritability. *Nature, 456,* 18–21.

McArdle, J. J., Prescott, C., Hamagami, F., Horn, J. L. (1998). A contemporary method for development-genetic analyses of age changes in intellectual abilities. *Developmental Neuropsychology, 14,* 69–114.

McCartney, K., Harris, M., Bernieri, F. (1990). Growing up and growing apart: A developmental meta-analysis of twin studies. *Psychological Bulletin, 107,* 226–237.

McGue, M., Bouchard, T. J., Iacono, W. G., Lykken, D. T. (1993).Behavior genetics of cognitive abilities: A life-span perspective. In R. Plomin, G. E. McClearn (Eds.), *Nature, nurture and psychology* (pp. 59–76). Washington, DC: American Psychological Association.

McGue, M., Christensen, K. (2002). The heritability of level and rate-of-change in cognitive functioning in Danish twins aged 70 years and older. *Experimental Aging Research, 28,* 435–451.

Meehl, P. E. (1977). *MAXCOV-HITMAX: A taxonomic search method for loose genetic syndromes. In Psychodiagnosis: Selected Papers* (pp. 200–224). New York: Norton.

Miyake, A., Friedman, N. P., Emerson, M. J., Witzki, A. H., Howerter, A., Wager, T. D. (2000). The unity and diversity of executive functions and their contributions to complex "Frontal Lobe" tasks: A latent variable analysis. *Cognitive Psychology, 41,* 49–100.

Pan, Y., Wang, K. S., Aragam, N. (2011). NTM and NR3C2 polymorphisms influencing intelligence: Family-based association studies. *Progress in Neuro-Psychopharmacology & Biological Psychiatry, 35,* 154–160.

Panizzon, M. S., Hauger, R., Dale, A. M., Eaves, L. J., Eyler, L. T., Fischl, B., et al. (2010). Testosterone modifies the effect of APOE genotype on hippocampal volume in middle-aged men. *Neurology, 75,* 874–880.

Panizzon, M. S., Lyons, M. J., Jacobson, K. C., Franz, C. E., Grant, M. D., Eisen, S. A., et al. (2011). Genetic architecture of learning and delayed recall: A twin study of episodic memory. *Neuropsychology, 25,* 488–498.

Panza, F., D'Introno, A., Colacicco, A. M., Capurso, C., Del Parigi, A., Caselli, R. J., et al. (2005). Current epidemiology of mild cognitive impairment and other predementia syndromes. *American Journal of Geriatric Psychiatry, 13,* 633–644.

Pedersen, N. L., Plomin, R., Nesselroade, J. R., McClearn, G. E. (1992). A quantitative genetic analysis of cognitive abilities during the second half of the life span. *Psychological Science, 3,* 346–353.

Pedersen, N. L., Plomin, R., McClearn, G. E. (1994). Is there G beyond *g*?(Is there genetic influence on specific cognitive abilities independent of genetic influence on general cognitive ability?). *Intelligence, 18,* 133–143.

Pedersen, N. L., Lichtenstein, P., Svedberg, P. (2002). The Swedish Twin Registry in the third millennium. *Twin Research, 5,* 427–432.

Plomin, R., Pedersen, N. L., Lichtenstein, P., McClearn, G. E. (1994). Variability and stability in cognitive abilities are largely genetic later in life. *Behavior Genetics, 24,* 207–215.

Plomin, R., Spinath, F. M. (2004). Intelligence: genetics, genes, and genomics. *Journal of Personality and Social Psychology, 86,* 112–129.

Poirier, J. (2003). Apolipoprotein E and cholesterol metabolism in the pathogenesis and treatment of Alzheimer's disease. *Trends in Molecular Medicine, 9,* 94–101.

Posthuma, D., de Geus, E. J., Boomsma, D. I. (2001a). Perceptual speed and IQ are associated through common genetic factors. *Behavior Genetics, 31,* 593–602.

Posthuma, D., Neale, M. C., Boomsma, D. I., de Geus, E. J. C. (2001b). Are smarter brains running faster? Heritability of alpha peak frequency, IQ, and their interrelation. *Behavior Genetics, 31,* 567–579.

Posthuma, D., Baare, W. F., Hulshoff Pol, H. E., Kahn, R. S., Boomsma, D. I., De Geus, E. J. (2003). Genetic correlations between brain volumes and the WAIS-III dimensions of verbal comprehension, working memory, perceptual organization, and processing speed. *Twin Research, 6,* 131–139.

Raber, J., Bongers, G., LeFevour, A., Buttini, M., Mucke, L. (2002). Androgens protect against apolipoprotein E4-induced cognitive deficits. *Journal of Neuroscience, 22,* 5204–5209.

Raz, N. (2000). Aging of the brain and its impact on cognitive performance: Integration of structural and functional findings. In F. I. M. Craik, T. A. Salthouse (Eds.), *The handbook of aging and cognition* (pp. 1–90). Hillsdale: Erlbaum.

Reynolds, C. A., Finkel, D., McArdle, J. J., Gatz, M., Berg, S., Pedersen, N. L. (2005). Quantitative genetic analysis of latent growth curve models of cognitive abilities in adulthood. *Developmental Psychology, 41,* 3–16.

Roberts, R. O., Geda, Y. E., Knopman, D. S., Cha, R. H., Pankratz, V. S., Boeve, B. F., et al. (2012). The incidence of MCI differs by subtype and is higher in men: The Mayo Clinic Study of Aging. *Neurology, 78,* 342–351.

Rönnlund, M., Nyberg, L., Backman, L., Nilsson, L. G. (2005). Stability, growth, and decline in adult life span development of declarative memory: Cross-sectional and longitudinal data from a population-based study. *Psychology and Aging, 20,* 3–18.

Sabb, F. W., Burggren, A. C., Higier, R. G., Fox, J., He, J., Parker, D. S., et al. (2009). Challenges in phenotype definition in the whole-genome era: Multivariate models of memory and intelligence. *Neuroscience, 164,* 88–107.

Salthouse, T. A. (1985).Speed of behavior and its implications for cognition. In J. E. Birren, K. W. Schaie (Eds.), *Handbook of the psychology of aging* (pp. 400–426). New York: Van Nostrand Reinhold.

Salthouse, T. A. (2009). When does age-related cognitive decline begin? *Neurobiology of Aging, 30,* 507–514.

Schaie, K. W. (1996). *Intellectual development in adulthood: The Seattle Longitudinal Study.* Cambridge: Cambridge University Press.

Schaie, K. W. (2005). *Developmental influences on adult intelligence: The Seattle Longitudinal Study.* New York: Oxford University Press.

Schaie, K. W., Willis, S. L., Caskie, G. I. (2004). The Seattle Longitudinal Study: Relationship between personality and cognition. *Aging, Neuropsychology, and Cognition, 11,* 304–324.

Schultz, M. R., Lyons, M. J., Franz, C. E., Grant, M. D., Boake, C., Jacobson, K. C., et al. (2008). Apolipoprotein E genotype and memory in the sixth decade of life. *Neurology, 70,* 1771–1777.

Singer, J. J., Macgregor, A. J., Cherkas, L. F., Spector, T. D. (2006). Genetic influences on cognitive function using the Cambridge Neuropsychological Test Automated Battery. *Intelligence, 34,* 421–428.

Singh-Manoux, A., Kivimaki, M., Glymour, M. M., Elbaz, A., Berr, C., Ebmeier, K. P., et al. (2011). Timing of onset of cognitive decline: Results from Whitehall II prospective cohort study. British Medical Journal, 343, d7622.

Small, B. J., Rosnick, C. B., Fratiglioni, L., Backman, L. (2004). Apolipoprotein E and cognitive performance: A meta-analysis. *Psychology and Aging, 19,* 592–600.

Sperling, R. A., Aisen, P. S., Beckett, L. A., Bennett, D. A., Craft, S., Fagan, A. M., et al. (2011). Toward defining the preclinical stages of Alzheimer's disease: recommendations from the National Institute on Aging-Alzheimer's Association workgroups on diagnostic guidelines for Alzheimer's disease. *Alzheimer's & Dementia, 7,* 280–292.

Swan, G. E., Reed, T., Jack, L. M., Miller, B. L., Markee, T., Wolf, P. A., et al. (1999). Differential genetic influence for components of memory in aging adult twins. *Archives of Neurology, 56,* 1127–1132.

Taylor, J. (2007). Heritability of Wisconsin Card Sorting Test (WCST) and Stroop Color-Word Test performance in normal individuals: Implications for the search for endophenotypes. *Twin Research and Human Genetics, 10,* 829–834.

Tucker-Drob, E. M. (2011). Global and domain-specific changes in cognition throughout adulthood. *Developmental Psychology, 47,* 331–343.

Turkheimer, E. (2000). Three laws of behavior genetics and what they mean. *Current Directions in Psychological Science, 9,* 160–164.

Vasilopoulos, T., Franz, C. E., Panizzon, M. S., Xian, H., Grant, M. D., Lyons, M. J., et al. (2012). Genetic architecture of the Delis-Kaplan Executive Function System Trail Making Test: Evidence for distinct genetic influences on executive function. *Neuropsychology, 26,* 238–250.

Vogler, G. P. (2006).Behavior genetics and aging. In J. E. Birren, K. W. Schaie (Eds.), *Handbook of the psychology of aging* (pp. 41–55). San Diego: Elsevier.

West, R. L. (1996). An application of prefrontal cortex function theory to cognitive aging. *Psychological Bulletin, 120,* 272–292.

Wisdom, N. M., Callahan, J. L., Hawkins, K. A. (2011). The effects of apolipoprotein E on non-impaired cognitive functioning: A meta-analysis. *Neurobiology of Aging, 32,* 63–74.

Wolf, O. T., Krischbaum, C. (2001). Endogenous estradiol and testosterone levels are associated with cognitive performance in older women and men. *Hormones and Behavior, 41,* 259–266.

Yang, J., Benyamin, B., McEvoy, B. P., Gordon, S., Henders, A. K., Nyholt, D. R., et al. (2010). Common SNPs explain a large proportion of the heritability for human height. *Nature Genetics, 42,* 565–569.

Zhang, J. P., Burdick, K. E., Lencz, T., Malhotra, A. K. (2010). Meta-analysis of genetic variation in DTNBP1 and general cognitive ability. *Biological Psychiatry, 68,* 1126–1133.

Chapter 5
Normative Cognitive Aging

Wendy Johnson, Matt McGue and Ian J. Deary

Over the past 20 years or so, characterization of the nature of age differences and changes with age in cognitive function has improved dramatically. In general, average cognitive function declines across much of the adult life-span, and this decline has come to be understood as normative, though the rates of decline and the ages at which they commence vary across different domains of function. This decline takes place in the context of much larger variation among individuals of any given age and the rates of decline show individual differences as well. Although characterization of these normative patterns is now quite good, understanding of what drives the changes is much more limited. Some researchers posit that inherent neurobiological processes are of primary importance, while others focus on psychosocial factors. Behavior genetic approaches to investigating possible explanations offer unique opportunities to distinguish among these kinds of possibilities and to explore their interplay. In this chapter, we review studies that have used a variety of different behavior genetic analytical approaches to investigate some of the thorniest questions facing cognitive aging, but we also highlight areas ripe for future behavior genetic approaches.

We do not directly address dementia (see Chap. 7) or mild cognitive impairment (see Chap. 4). It may, nonetheless, have impacted many of the studies we discuss. Dementia becomes very common at older ages, reaching rates of 25–30% for Alzheimer's disease alone in those over age 85 (e.g., Blennow et al. 2006). This implies that rates of all-cause dementia are considerably higher. Although most studies of cognitive aging screen for dementia, the methods used focus on thresholds of

W. Johnson (✉) · I. J. Deary
Department of Psychology, University of Edinburgh, 7 George Square, Edinburgh, EH8 9JZ, UK
e-mail: Wendy.johnson@ed.ac.uk

M. McGue
Department of Psychology, University of Minnesota,
75 E. River Road, Minneapolis, MN 55455, USA
e-mail: mcgue001@umn.edu

I. J. Deary
e-mail: i.deary@ed.ac.uk

D. Finkel, C. A. Reynolds (eds.), *Behavior Genetics of Cognition Across the Lifespan*,
Advances in Behavior Genetics 1, DOI 10.1007/978-1-4614-7447-0_5,

cognitive impairment and many are quite insensitive to declines in function in people who have had particularly high function in midlife. Moreover, decline develops over a period as long as a decade, so that many who do not yet qualify for diagnosis may still show incipient symptoms. Thus many aging samples likely contain participants in early and undiagnosed stages of dementia. Because dementia shows genetic influences, this could have its own biasing effects on estimates of genetic influences on normative cognitive aging.

Cognitive function is not a unitary construct. Molecular genetic studies sometimes address general cognitive ability, especially general reasoning ability. This is warranted because much research shows that the common factor that may be derived from diverse cognitive tests declines with age. Sometimes studies also or instead address more specific cognitive domains, such as aspects of memory or processing speed for which decline with age is especially sharp. Some of these functions appear to be affected by age, even after accounting for age effects on general cognitive ability. Some studies examine even more specific cognitive tests and tasks and components. Because of differences in the normative patterns of aging across these different aspects of cognitive function, replication or generalization of specific estimates of magnitudes of influences or specific effects should not necessarily be expected.

In this chapter, we review quantitative genetic studies that have taken both cross-sectional and longitudinal approaches, as well as studies that have examined the extent to which different aspects of cognitive function and variables with which it is associated show common genetic influences. We then turn to behavior genetic contributions to special topics in cognitive aging including intra-individual variability and terminal decline, the problems of sample selectivity, and gene–environment correlation. Following these topics involving aggregate genetic contributions to individual differences, we consider molecular genetic approaches to identifying individual genes involved in cognitive aging.

5.1 Quantitative Genetic Approaches

Quantitative genetic approaches involve the analysis of twin and family data in order to identify, quantify, and characterize the factors that contribute to individual differences in a trait (i.e., phenotypic variability). At the initial stages of inquiry, the focus is often on quantifying the independent contributions of three major factors: additive genetic influences (commonly termed "A"), shared environmental influences that act to make people who live together similar (termed "C"), and nonshared environmental influences, including measurement error, that act to make people different (termed "E"). As research on a specific phenotype advances, there is increasing emphasis on the exploration of models of gene–environment interplay. Quantitative genetic research on late-life cognitive function is generally at the initial stage of decomposing phenotypic variance into underlying biometric components and has relied almost exclusively on the analysis of monozygotic (MZ) and dizygotic (DZ) twin similarity. These twin studies have sought to address three major questions. First, in cross-sectional research, what are the contributions of genetic and environmental factors to cognitive function at various ages? Second, in longitudinal research, what are the genetic and environmental contributions to both stability and change in cognitive function? Third, in multivariate research, what are the factors that underlie the genetic and environmental components of variance?

5.1.1 Cross-Sectional Twin Research

5.1.1.1 General Cognitive Ability

One of the most robust findings in the behavioral genetics literature is that genetic factors contribute to individual differences for most behavioral traits (Turkheimer 2000). Late-life cognitive function does not provide an exception to this general rule. Table 5.1 summarizes major cross-sectional twin studies in this area (Lee et al. 2010). Several features of the information summarized in the table are worthy of comment. First, as compared to many areas within behavioral genetics, there are few twin samples and the sizes of these samples are small relative to twin samples for many other behavioral traits. Even the samples derived from the large Scandinavian registries are modest when compared to behavioral genetic research on other phenotypes. This is a reflection of the challenges (e.g., mortality, frailty, emigration) associated with ascertaining and assessing large representative twin samples in late life. Second, despite the modest number and sizes of the relevant twin studies, a consistent pattern of findings is evident. In late life, MZ twins are consistently more similar in general cognitive ability than DZ twins, resulting in heritability estimates that are moderate to large (i.e., 50–80 %), and comparable to those from younger adult samples. Moreover, the estimates of the proportion of variance attributable to shared environmental factors have been consistently low. Indeed, in only one study was it estimated to be anything other than zero.

The failure to observe evidence for shared environmental influences on late-life cognitive function is perhaps to be expected. The magnitudes of shared environmental influences on a broad array of behavioral phenotypes drop off markedly during the transition from late adolescence to early adulthood (Bergen et al. 2007), the period when twins are likely to establish separate residences. Little, if any, shared environmental influence on the cognitive function of older twins, many of whom have not lived together for 50 years or more, may simply be consistent with this more general observation, though it has implications for theories positing that cognitive function is strongly shaped by early-life environmental circumstances. Of theoretical interest is also whether heritability estimates for late-life cognitive function differ from estimates at other adult ages. It is well known that the heritability of general cognitive ability increases from childhood through early adulthood (Haworth et al. 2010). It is less clear whether heritability changes from early through late adulthood. Finkel and Reynolds (2010) plotted estimates of the heritability of general cognitive ability from cross-sectional twin studies as a function of sample age. As every study did not report heritability estimates based on the same biometric

Table 5.1 Major cross-sectional twin studies of late-life general cognitive ability

Citation	Study	Assessment	Age	MZ r (No. of pairs)	DZ r (No. of pairs)	Biometric estimates		
						a^2	c^2	e^2
(Pedersen et al. 1992)	SATSA	First principal component	Mean=65.6	0.79 (108)[a]	0.27 (167)[a]	0.81	0.00	0.19
(Finkel et al. 1995)	MTSADA	First principal component	All≥65	0.86 (18)	0.47 (15)	0.81	0.00[b]	0.19
(McClearn et al. 1997)	OCTO Twin	First principal component	All≥80	0.78 (52)	0.38 (65)	0.62	0.11	0.27
(McGue and Christensen 2001)	LSADT	Cognitive composite	All≥75	0.51 (152)	0.26 (213)	0.54	0.00	0.46
(Giubilei et al. 2008)	ITR	Raven matrices	Mean=67.6	0.60 (35)	0.33 (58)	0.56	0.00[b]	0.44
(Lee et al. 2012)	OATS	First principal component	All≥65	0.74 (117)	0.26 (98)	0.74	0.00	0.26

In cases with multiple relevant publications from same study, only the largest cross-sectional study is included here

SATSA Swedish Adoption/Twin Study of Aging, *MTSADA* Minnesota Twin Study of Adult Development and Aging, *LSADT* Longitudinal Study of Aging Danish Twins, *ITR* Italian Twin Registry, *OATS* Older Australian Twin Study, *NR* Twin correlations not reported for the latent factor

[a] Reared-together and reared-apart twin pairs combined for SATSA

[b] Estimates reported only for AE model with c^2 fixed at 0

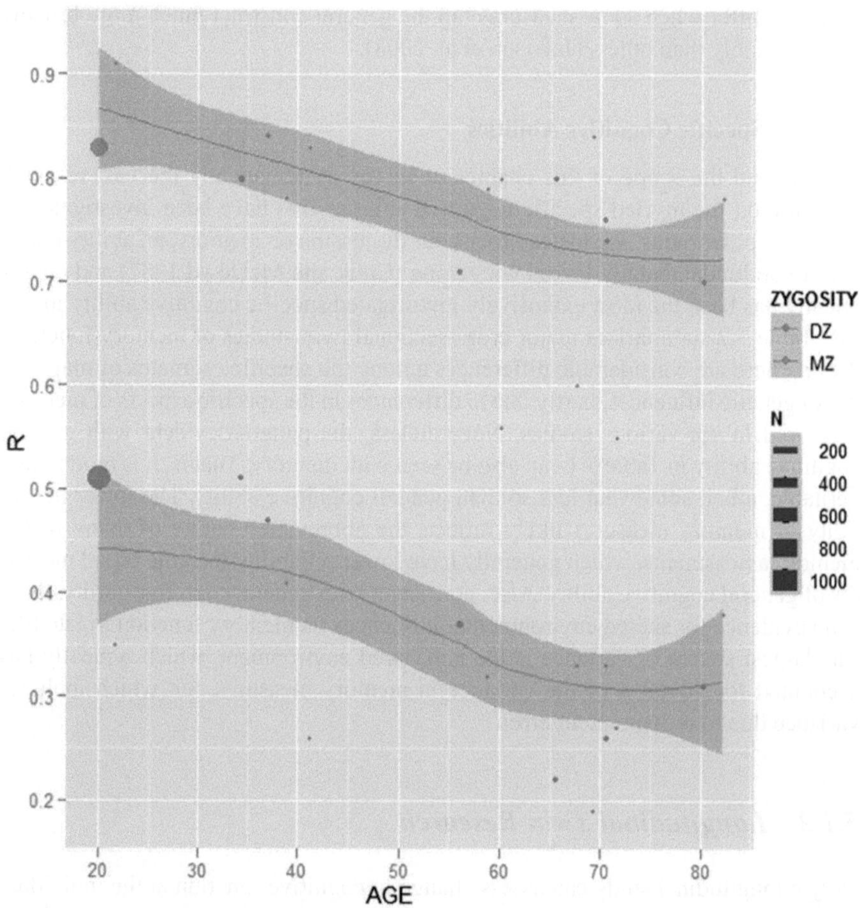

Fig. 5.1 Association between age and twin correlation for general cognitive ability in adulthood. Each *dot* represents a twin correlation reported in a single study. The size of the *dot* indicates the size of the twin sample

model (i.e., some reported heritability for the AE model and others used the ACE model), we plot the MZ and DZ twin correlations for general cognitive ability reported in cross-sectional studies of adult twins as a function of age, in Fig. 5.1. The figure clearly shows that both the MZ and DZ correlations tend to decrease with age, with MZ correlations decreasing at a slightly more rapid rate than DZ correlations. This pattern is consistent with the conclusion drawn by Finkel and Reynolds (2010), that late-life reductions in the heritability of general cognitive ability reflect an increasing importance of nonshared environmental factors. Despite this general consistency, there is considerable variation in the sizes of the twin correlations. Some of this can be attributed to differences in the measures of general cognitive ability used. Although different measures tend to be well correlated (Johnson et al. 2004), they can differ when some measures tap the general construct much more broadly and/or reliably than others (Johnson et al. 2008).

5.1.1.2 Specific Cognitive Abilities

It is beyond the scope of this chapter to survey systematically the vast research literature on the myriad specific cognitive abilities that have been investigated in studies of older twins. We instead focus our discussion on memory, an ability that is seen to be fundamentally linked with aging (Craik and McDowd 1987) and consequently has been the most extensively investigated specific cognitive ability in late life. Table 5.2 summarizes major cross-sectional twin studies of memory function. Again, there are considerable differences among the specific estimates of magnitudes of genetic influences, partly due to differences in the specific aspects of memory measured in the various studies. Nevertheless, the patterns evident with general cognitive ability in Table 5.1 can also be seen with memory. That is, it is moderately heritable, albeit somewhat less so than general cognitive ability. The lower heritability of memory measures likely reflects the abbreviated nature of many of the memory assessments, which generally have lower reliability than the broad measures of general cognitive ability. Also, consistent with general cognitive ability, there is no evidence for shared environmental influences on memory function in late life. The largest source of variance is the nonshared environment, which typically has accounted for 50–60% of the variance in memory measures, but which includes variance due to measurement error.

5.1.2 Longitudinal Twin Research

Only a longitudinal study can assess change in cognitive function at the individual level, and thus allow for an investigation of the factors that contribute to individual cognitive aging. Undertaking a longitudinal study in a late-life sample is, however, challenging, much more so than cross-sectional research. To be informative, longitudinal studies should ideally assess large samples and include multiple follow-up assessments, requiring time spans of many years. The costs of longitudinal research, in both funding and researcher time, are considerable. These challenges are further compounded by sample loss to follow-up due to illness or mortality, which can greatly diminish their size, and in twin samples by the need to recruit most participants in pairs. Because of the logistical challenges associated with undertaking longitudinal twin research in late life, the number of relevant data sources for studies of genetic influences on general cognitive ability is limited. Those available, however, have been quite extensively utilized, and two have been maintained over many assessments.

Most longitudinal twin studies of aging have been analyzed by fitting latent growth curve models (Lee et al. 2010; Neale and McArdle 2000). Briefly, latent growth curve analysis involves using the individual sequences of observed

Table 5.2 Major cross-sectional twin studies of memory in late-life

Citation	Study	Age	Assessment	MZ r (No. of pairs)	DZ r (No. of pairs)	Biometric estimates		
						a^2	c^2	e^2
(Pedersen et al. 1992)	SATSA	Mean = 65.6	Digit span forward	0.34	0.21	0.37	0.00	0.63
			Digit span backward	0.44	0.17	0.44	0.00	0.56
			Names and faces immediate	0.24	0.21	0.32	0.00	0.68
			Names and faces delay	0.40	0.20	0.42	0.00	0.58
			Picture memory	0.36	0.16	0.37	0.00	0.63
				(108–113)[a]	(178–188)[a]			
(Finkel and McGue 1993)	MTSADA	Mean = 67.1	Text recall immediate	0.61	0.22	0.57	0.00	0.43
			Text recall delay	0.64	0.07	0.56	0.00	0.44
			Figure memory	0.66	0.18	0.64	0.00	0.36
				(87–89)	(62–63)			
(McClearn et al. 1997)	OCTO Twin	All ≥ 80	Composite of digit span and picture memory	0.45 (63)	0.26 (82)	0.52	0.00	0.48
(McGue and Christensen 2001)	LSADT	All ≥ 75	Word recall	0.42	0.29	0.43	0.03	0.53
			Digit span	0.45	0.25	0.26	0.09	0.65
				(46–50)	(57–64)			

Table 5.2 (continued)

Citation	Study	Age	Assessment	MZ r (No. of pairs)	DZ r (No. of pairs)	Biometric estimates		
						a^2	c^2	e^2
(Swan et al. 1999)	NHLBI Twin Study	Mean=71.8	Verbal learning and memory	0.56	0.12	0.56	0.00	0.44
			Response discrimination	0.28	0.06	0.24	0.00	0.46
			Learning strategy	0.36	0.38	0.00	0.27	0.73
			Recognition	0.01	0.11	0.00	0.00	1.00
				(*N*=94)	(*N*=89)			
(Giubilei et al. 2008)	ITR	Mean=67.6	Story recall	0.59 (35)	0.25 (58)	0.54	0.00	0.46

In cases with multiple relevant publications from same study, only the largest cross-sectional study is included here
SATSA Swedish Adoption/Twin Study of Aging, *MTSADA* Minnesota Twin Study of Adult Development and Aging, *LSADT* Longitudinal Study of Aging Danish Twins, *ITR* Italian Twin Registry[a] Reared-together and reared-apart twin pairs combined for SATSA

phenotypes to estimate the components of longitudinal curves. In studies of cognitive aging, these curves might more accurately be called decay rather than growth curves, although we retain the standard nomenclature. At a minimum, two components are estimated. The first is the initial value or intercept, which in effect captures individual differences that are stable over the multiple assessments. Because the intercept is estimated as a latent variable and moreover is a function of the multiple phenotypic observations, the impact of measurement error is minimized. The second component is the slope, or rate of linear change across time. The slope is arguably the component of greatest interest, as it captures how individuals are changing on average across the observation period. In some cases, nonlinear models of change are investigated by estimating a quadratic component, which reflects acceleration/deceleration in the rate of change. Alternatively, a nonlinear model might involve estimating a change point, after which the rate of linear change is different from the original. Reliable estimates of these additional components, however, require relatively large numbers of longitudinal observations (Bryk and Raudenbush 2002). There have thus been only a few attempts in the behavioral genetics literature to characterize individual differences in nonlinear components of change.

The most extensive longitudinal twin study of cognitive aging is the Swedish Adoption/Twin Study of Aging (SATSA). SATSA began in 1984 with a sample of 303 reared-together and reared-apart twins aged 50 years and older. It includes up to six waves of assessments spanning nearly 20 years (Finkel and Pedersen 2004). Plomin, Pedersen, Lichtenstein, and McClearn (Plomin et al. 1994) provided the first longitudinal analysis of SATSA cognitive data. However, their analysis was restricted to just the first two waves of cognitive assessment, which were separated by only 3 years. As a consequence, rather than focus on cognitive change, which was minimal over this time span, they focused on cognitive stability, which was substantial as reflected by a longitudinal correlation of 0.92. They estimated the heritability of general cognitive ability to be 80 % and concluded that 90 % of the stability of cognitive function across the two time points could be ascribed to genetic influences.

Since this initial publication, the SATSA sample has been assessed cognitively an additional two times, bringing the maximal retest interval to 13 years. Reynolds et al. (2005) provided the most comprehensive and up-to-date longitudinal analysis of the SATSA cognitive data. We focus here on findings for their measure of general cognitive ability, the first principal component of a battery of ten tests of specific cognitive ability. In their growth curve analysis of up to four assessments on a sample of 362 pairs of twins, they concluded that the intercept was highly heritable (91 %), but that the rate of linear change was not (heritability estimate 1 %). However, they did report a significant heritable effect on the quadratic component (43 %). The finding of genetic influences on the quadratic but not the linear component is somewhat counterintuitive, especially because it was observed in the context of overall decreasing twin similarity with age. A possible face-value interpretation is that genetic influences are important to the large variance in stable individual differences, but the variance in cognitive change (primarily decline) that takes place in "early old age" is much smaller by comparison so that there is little power to identify its sources. There is greater variance, of which some is genetic, in the rate and timing of acceleration in decline in "late old age" that may be associated with overtly declining health.

The second major longitudinal twin study of cognitive aging is the Longitudinal Study of Aging Danish Twins (LSADT). LSADT utilizes a cohort-sequential design. It was begun in 1995 and includes up to six waves of in-person assessment spanning 10 years (Christensen et al. 1999). A total of 1,112 same-sex twin pairs of known zygosity aged 70 years and older have participated in LSADT. McGue and Christensen (2007) provides the most comprehensive and up-to-date longitudinal analysis of the LSADT cognitive data. We focus here on their analysis of LSADT's measure of general cognitive ability, which is a composite of five brief individual cognitive measures of fluency, forward and backward digit span, and immediate and delayed word recall, which means that it is somewhat limited than SATSA's as it emphasizes on several aspects of memory. The heritability of the intercept (39%) was significant but more moderate than that reported in SATSA. There was also a significant shared environmental effect on the intercept (30%), unlike in SATSA. However, in agreement with SATSA, they reported nonsignificant estimates for both the genetic (18%) and shared environmental (2%) contributions to the slope. They did not fit a quadratic component, which might explain the larger, though nonsignificant, estimate of genetic influence on the slope.

Longitudinal twin studies of cognitive aging are in general agreement. Although the heritabilities of cognitive abilities broadly construed at any point in adulthood are significant and at least moderate (with estimates generally at least 50%), change in cognitive performance appears to be predominantly due to nonshared environmental factors. Despite the consistency of findings, several factors caution against drawing strong conclusions about the absence of genetic influences on cognitive change. First, retest intervals have generally fallen in the 4–10-year range, which may be too limited a time period to allow for reliable assessment of individual change. Second, late-life cognitive assessment can be confounded by the effects of impending death (Bosworth and Schaie 1999), which could attenuate twin similarity for cognitive ability when twins are not concordant for time at death (Johansson et al. 2004). Third, practice effects, which have been shown to exist and to vary in older samples even when retest intervals are long (Rabbitt et al. 2004; Singer et al. 2003) probably need further consideration than they have received to date. Finally, and perhaps most significantly, change has typically been assessed linearly. Yet the two studies that investigated nonlinear models of change (McArdle and Plassman 2009; Reynolds et al. 2005), did find evidence for genetic influences on these higher-order moments of cognitive change.

5.1.3 Multivariate Twin Research

Multivariate methods have been used to explore the nature of genetic effects on late-life cognitive function. A reasonable guiding hypothesis is that genetic factors influence cognitive function because they influence the brain structures and processes upon which higher-level cognitive function depends (Fjell and Walhovd 2010). For example, the speed with which individuals process information is thought to reflect the integrity of underlying neural systems that support higher-level cognitive function (Kennedy and Raz 2009). Processing speed is typically measured using psychometric tests such as digit symbol coding, experimental tests such as processing speed, or psychophysical tests such as inspection time (Deary 2000). Processing speed shows marked decreases with age. It is also moderately to highly heritable. In an early report from SATSA, the estimated heritabilities for measures of speed ranged from 51 to 64 % (Pedersen et al. 1992). Similarly, in a recent publication based on the Older Australian Twin Study, the estimated heritability for five speed measures ranged from 0.35 to 0.62 (Lee et al. 2012). Of interest, the lowest heritability estimate in this study was for Choice Reaction Time, which another twin study had also reported to have low heritability (Finkel and McGue 2007).

Some consider processing speed to be simply another domain of cognitive function, both with respect to the hierarchical structure of cognitive abilities and within cognitive aging. From this perspective, processing speed has its own variance like other cognitive ability tests, as well as variance shared with all other cognitive ability tests, as captured by its loading on the general cognitive factor (Carroll 1993; Salthouse 2004). Consistent with this, one bivariate behavior genetic study found common genetic variance between processing speed and general intelligence, but neither appeared to be causal to the other (Luciano et al. 2005). Others hypothesize that the observed declines in speed underlie declines in other cognitive functions, especially so-called fluid abilities (Finkel et al. 2007; Salthouse 1996). Indeed, a meta-analysis concluded that speed can account statistically for a large share of the variance in a broad array of cognitive measures (Verhaeghen and Salthouse 1997), and the few studies that have appeared have been consistent in indicating common genetic variance between measures of speed and other aspects of cognitive function (Finkel et al. 2005; Lee et al. 2012). More directly corroborating the hypothesis that declines in speed underlie cognitive aging, a SATSA study found that that the genetic contribution to processing speed appeared to drive age-related changes in memory and spatial but not verbal ability (Finkel et al. 2009). Processing speed may, of course, not be the only leading indicator of more general cognitive aging, though it is the only one to have been examined to date in behavior genetic studies.

5.2 Special Topics in Cognitive Aging

5.2.1 Intraindividual Variability in Cognitive Abilities

Many cognitive tasks are designed to include a series of items that tap the same basic skills, knowledge, or perceptual or manipulative capacities at different levels of difficulty. Items on these tasks are usually scored as correct or incorrect, and a single score consisting of the correct number is generated. With scores of this type, the ideal is that, if a person were to carry out the same set of items again, s/he would generate the same score. Of course, this ideal is never attained. Differences in scores with repeated assessments over even short time periods always occur. Some of these are practice effects; some might reflect state differences such as recent caffeine consumption or relative fatigue; and some are outright error of measurement. Despite this, differences in scores over extended time spans are considered to reflect change in true score (plus the other sources of change as relevant). For tasks such as this, it is most common to examine means for a study group overall, and variance in scores reflects interindividual differences. Tasks that are intended to assess fundamental cognitive processes, however, are generally designed differently. The idea in designing such tasks is that any complex cognitive task must require execution of several simpler cognitive processes. Identification and measurement of such processes would facilitate understanding of individual differences in performance on the more complex cognitive tasks to which they contribute.

The kinds of very simple tasks used to reflect fundamental cognitive processes take very little time to carry out and almost everyone can do them. For example, one common task measuring reaction time requires the participant to push a button when s/he sees a symbol flash onto a screen. That's all. The measure taken is the time from presentation of the symbol flash on the screen to the participant's button press. As this task (like most tasks of fundamental processes) is so simple, it is possible to get participants to do it many times without fatigue. Doing so reveals variance in response times both across and within individual participants. Reliability of estimates of variance across individuals can be increased dramatically by using the average within-person variance for participants across many task trials. This is typically done, and such averages generally show substantial correlations with age as well as with performance on more complex cognitive tasks.

However, the intraindividual variation also tends to show stability both over time (Hultsch et al. 2000; Rabbitt et al. 2001; Ram et al. 2005) and across tasks of fundamental cognitive processes and more complex cognitive tasks (Fuentes et al. 2001; Hultsch et al. 2000; Li et al. 2001). Moreover, people who tend to perform not very well on cognitive tests of all kinds tend to show greater variability on tests of fundamental cognitive processes (Li et al. 2001; Li et al. 2001). This is true both when cognitive ability has been low throughout life, and when pre-existing cognitive ability has been impaired by dementia (e.g., Hultsch et al. 2000), schizophrenia (e.g., Winterer and Weinberger 2004), brain injury (e.g., Stuss et al. 1994), or even just normal aging (e.g., Deary and Der 2005; West et al. 2002). Moreover, increases in variability have been linked to decreases in performance in the same individuals over time (MacDonald et al. 2003). Greater performance variability on fundamental cognitive tasks also appears to be associated with concurrent lower performance on more complex cognitive tasks independent of the association between mean level performance on fundamental cognitive tasks and performance on more complex cognitive tasks (e.g., Li et al. 2004). In addition, increases in variability on fundamental cognitive tasks over time have been linked to decreases over the same period in performance on more complex cognitive tasks more strongly than vice versa (e.g., Bielak et al. 2010; Lovden et al. 2007). Taken together, these observations suggest that variation around the average within-person performance level is systemic and thus potentially meaningful, that greater variability in performance on fundamental cognitive tasks may be related to impairments in central nervous system function that also impair performance on more complex cognitive tasks, and that this performance serves as a sort of leading indicator, or canary-in-the-coal-mine warning, of coming general cognitive decline.

If so, variability in performance on these tasks should show genetic influences, as do mean levels of performance, as well as most other psychological measures that show stability over periods of days or weeks. To our knowledge, this has been investigated twice, once in a small younger sample, and once in a larger older sample. Vernon (1989) administered eight reaction time tasks to a sample of 102 twin pairs ranging in age from 15 to 37 years. The tasks generated 11 measures of intra-individual variability, with heritability estimates ranging from 41 to 98 %. Finkel and McGue (2007) carried out a much more extensive examination. They used a sample of 738 participants including 316 twin pairs, ranging in age from 27 to 95 years, with median 62 years. The participants completed simple and four-choice reaction time tasks, though the number of trials administered for each task (15) was much smaller than is customary for such tasks, limiting the reliability of the means and standard deviations. In both tasks, Finkel and McGue (2007) estimated genetic influences on mean and intraindividual standard deviation separately for decision and movement times, under the presumption that decision time directly reflects central nervous system function, and movement time motor processes that are more peripheral to cognitive function.

Univariate estimates of genetic and environmental influences indicated that genetic influences accounted for 20–35 % of the variance in movement and decision time means and movement time standard deviation, but effectively none of the variance in decision time standard deviation. For the latter, shared environmental influences accounted for 13 % of the variance; these influences accounted for 0–7 % of the variance in the other measures. Age accounted for 3–12 % of the variance in all the measures. Multivariate analyses found genetic and age influences common to the four measures, even decision time standard deviation. Some nonshared environmental influences were common to all but decision time standard deviation, but all measures except movement time standard deviation also showed nonshared environmental influences unique to each measure. There was little consistency in the results of these two studies, and both suffered from substantial limitations that likely contributed to this lack of consistency. For the Vernon (1989) study, small sample size was a primary limitation, and its age range rendered it irrelevant to cognition in old age. Its tasks and the kind of analysis carried out also differed. The analysis in the Finkel and McGue (2007) study was considerably more sophisticated, but the wide range and strongly negative skew of the sample's age distribution likely introduced sources of variance that undermine the relevance of its results to cognition in old age. Moreover, the number of trials, which would now be considered rather small, undoubtedly contributed to error variance. Clearly, given the gathering evidence that intraindividual variability in fundamental cognitive task performance is an early indicator of cognitive decline, additional behavior genetic studies in this area are warranted.

5.2.2 Terminal Decline

The concept of terminal decline has attracted considerable research attention in the area of cognitive aging research. It emerged from clinical observations, and generates interest because it offers hope of practical applications to cope with the social burden presented by cognitive declines in aging populations. It is burdened by measurement problems that make even confirming its existence difficult, and has not been well studied from a behavior genetic perspective. Still, its importance as a current topic of investigation in the field of cognitive aging implies that anyone interested in the behavior genetics of cognition should be familiar with it and how behavior genetics might contribute to our understanding of it.

The concept of terminal decline, or drop, grew out of observations that cross-sectional analyses of age differences in cognitive function suggested much sharper declines in function with age than longitudinal studies and that study participation appeared to be biased towards higher-performing individuals in better physical health. The idea of terminal decline is that, beginning some period before death, cognitive functions begin to decline very sharply. This idea has great appeal because, if the period and rate of terminal decline could be identified, aging individuals and clinicians could have forewarning of impending death. Early researchers on the topic in the 1960s (Jarvik and Falek 1963; Kleemeier 1962; Lieberman 1966; Riegel and Riegel 1972) speculated that all observed differences in average cognitive function with age might be attributed to sharp declines in the performance of those who did not survive the next few years after test administration, with survivors remaining stable until they too reached their last few years of life. That is, in cross-sectional samples, decreasing average cognitive function with age may result from increasing proportions of these samples being in this period of terminal decline at older ages.

The subsequent 50 years have seen development of longitudinal databases and new statistical techniques that make it possible to track individual changes over time much more closely than was possible then. These developments have generally indicated that the idea of complete stability until some sharp decline shortly before death is too optimistic, but they generally support the idea that cognitive decline is steeper in later old age than in earlier old age. This makes for a rather blurry image of terminal decline. Unfortunately, this is at least partly because these research developments have also soundly confirmed the omnipresence of two measurement problems: a tendency for both longitudinal and cross-sectional samples of older adults to be increasingly biased with increasing participant age, to varying degrees in different samples, towards healthier and better-performing individuals (e.g., Rabbitt et al. 2008), and the need for measurement to continue until most of the sample population has died. Because of the difficulties these create in assessing and summarizing patterns of intraindividual change accurately, questions of rates of normative change, the length of some period of terminal decline and even its existence remain hotly debated (e.g., Batterham et al. 2011; Gerstorf et al. 2011; MacDonald et al. 2011; Piccinin et al. 2011; Rabbitt et al. 2011). The debate is fueled by focus on change-point analytical methods that are based on an assumption that cognitive decline can be best modeled as linear, with one constant slope pertaining prior to the beginning of the terminal period, and another pertaining afterwards (e.g., Sliwinski et al. 2006; Wilson et al. 2003), rather than, for example, gradually accelerating with age such as might be described by a quadratic function. That is, the methods most commonly used to measure the length of the terminal period and the rates of change before and after it rely on the assumption that the phenomenon of terminal drop is real.

To our knowledge, only one study has attempted to investigate how genetic influences may be involved in declines in cognitive function when linked directly to time to death. Johansson et al. (2004) first used latent growth models to observe that time to death predicted rate of change in several different aspects of cognitive function in a sample of twins over age 80 at study inception, considering the twins as individuals. They went on to examine the patterns of intraclass twin correlations for initial levels and rates of change in the different aspects of cognitive function. For levels, these showed the typical pattern of large correlations in MZ twin pairs and smaller but still substantial correlations in DZ twins, indicating substantial genetic influences. For rate of change, the correlations were generally small in absolute magnitude and many were negative for MZ twins. For DZ twins, many were negative, one even strongly so, and those that were positive were generally small. There was no meaningful evidence of genetic influence on rates of change. Johansson et al. (2004) also examined the individual assessment correlations separately in MZ and DZ twin pairs both of whom participated in all four of the assessments for which they had data, comparing them to those for twin pairs who were intact at only three or two assessments or the first assessment. Results were very mixed, but there was a small tendency for the correlations to be lower at the last assessment for which the pairs were intact, suggesting that they were becoming less similar in the period before at least one of them died. This would be consistent with the existence of some form of terminal decline to which genetic influences did not contribute, especially since there was no evidence of differences in this (very tentative) pattern between MZ and DZ twins.

Clearly, more research is needed on the topic of terminal decline, or perhaps more generally on the topic of the shapes of the typical trajectories of decline in different cognitive functions in old age. Most research on this topic is driven by empirical observations rather than theory, and it may be helpful to develop clearer theoretical rationales for one form of decline or another so that models of alternative hypothesized processes can be pitted against each other (Platt 1964). For example, it is reasonable to postulate that some aspect of cognitive decline accelerates once at a point some time before death that is the same for all or most individuals, as assumed in implementing latent change models. But it is just as reasonable to posit that this decline accelerates at some point, say at age 70, that is similar for all, regardless of when they will die, and also reasonable to postulate that decline is continuous in old age, but accelerates. In this latter case, there is no fixed "change point" but rather decline that is much faster for people close to death than for people further from death, whatever their specific ages. This is more consistent with studies that have implemented models with linear and quadratic terms. Most difficult to evaluate of all, it is also just as reasonable to postulate that decline accelerates just once sometime before death, but that the timing of this acceleration depends on the cause of death so that it varies from individual to individual, depending on cause of death (Rabbitt et al. 2011), and/or perhaps some other factors. It is interesting that, as discussed earlier, several SATSA studies to date that have modeled decline using quadratic functions have suggested that genetic influences are more apparent on the term representing quadratic than on the term representing linear change. Is this real? Does this generalize or is it unique to SATSA? Is it specific to certain aspects of cognitive function and not to others? It should be observed in additional samples before we draw any conclusions. Complicating things still further, Pedersen et al. (2003) demonstrated that failure to model terminal decline can inflate the apparent heritability of linear change in SATSA, though they did not consider nonlinear (quadratic) change. But if the observation that accelerating (quadratic) change is more heritable than linear change replicates, does it imply that accelerating decline is a better model than terminal decline? Does it imply that genetic variation contributes primarily to factors related to mortality and not to cognitive aging prior to inception of mortality-related deterioration? At this point, there is simply not enough evidence to form even a tentative conclusion.

5.2.3 Selection Effects and Gene–Environment Correlation

Most samples in studies of cognitive aging, whether cross-sectional or longitudinal, tend to show higher average cognitive function and socioeconomic status compared to the comparably aged population at large. There are two reasons for this. First, study samples in general tend to show somewhat higher average cognitive function and socioeconomic status than the otherwise-comparable population because these factors contribute to understanding the importance and relevance of research (e.g., Nishiwaki et al. 2005); that is, they tend to be somewhat select. Second, the resulting sample range restriction or selectivity is exaggerated in aging samples because cognitive functions and mortality tend to be positively related (24% reduction in hazard rate for each standard deviation increase in intelligence; Calvin et al. 2011), and older participants who are close to death and thus in poor physical health and potentially suffering terminal decline are less likely to participate in research studies regardless of their original levels of cognitive function and socioeconomic status. This has the effect that, within cross-sectional study samples and initial samples in longitudinal studies that have large age ranges, older participants tend to have had higher midlife socioeconomic status and general cognitive function, often represented by tests of crystallized knowledge such as word-reading accuracy, than younger participants, thus leading to underestimates of the extent of cognitive decline with age (Rabbitt et al. 2008).

Longitudinal studies with narrow age ranges can avoid these problems. But, since the primary reason for attrition from such studies is often death or disability, the samples still become increasingly selected for high midlife cognitive function (Lachman et al. 1982) no matter how measured. Thus, even narrow age-range longitudinal studies can underestimate the extent of normative cognitive decline. To the extent that genetic influences on cognitive ability vary with levels of socioeconomic status, estimates of genetic influences on all aspects of the processes involved may be affected. Offsetting this, however, is the possibility that some study participants were in early stages of undiagnosed dementia.

Sample selectivity brings with it other challenges in understanding cognitive aging, some of which twin samples intended for behavior genetic analyses are especially well positioned to address. There is high interest in the "use it or lose it" hypothesis, or the idea that maintenance of intellectual, or even physical, activity in old age may slow the rate of cognitive decline, and substantial evidence at least for an association between greater activity and slower decline (Schooler and Mulatu 2001). Establishing that such activity is actually causal in reducing the rate of cognitive decline is not, however, straightforward. There are two basic reasons for this. First, the causal influences may flow in the opposite direction. That is, people who are suffering cognitive decline may withdraw from activities because they have become too difficult. Longitudinal samples, whether of twins or not, are the best means to address this possibility, though in practice it is difficult to sequence exposure and outcome measures in order to resolve it conclusively (e.g., Hoffman et al. 2011). Second, other variables may create the association through confounding. Confounding takes place when some third variable(s): (1) actually causes the outcome, (2) is correlated with the exposure, and (3) is not affected by the exposure (McNamee 2003). Although almost any kind of variable could act as a confounder, one of the most likely possibilities, given the pervasive presence of genetic influences on behavioral traits is that the genetic and environmental influences that contribute to motivation towards and enjoyment of engagement in activities may also contribute to preservation of good cognitive function, thus creating what behavior geneticists consider gene–environment correlation. That is, individuals may actively select, consciously or unconsciously, environments that reinforce the genetically influenced characteristics that originally led them to seek those environments (see Chap. 6 for additional discussion). Two sets of behavior genetic models can be of particular help in addressing this possibility.

The first is the co-twin control model. Because MZ twins share a common genotype and, generally, early rearing environment, one twin within a pair can provide control for genetic and familial environmental background for the other. Thus, if, in twin pairs where one is exposed to some environment and the other is not, the exposed twins have an outcome that the nonexposed twins do not, this provides unusually strong evidence that the environmental exposure is actually causative. Control is weaker when discordant DZ pairs are compared because they are less genetically similar, but DZ pairs can still provide important information, and many studies have included them because discordant MZ pairs are rare for many kinds of environments, rendering sample sizes small. There are always some qualifications to this, of course. Twins may not be completely representative of the more general population, cause may actually flow in the opposite direction unless some longitudinal control is in place, or some unmeasured third variable may confound the association through nonshared environmental influences (McGue et al. 2010). There may be inaccuracies even when results appear to refute the causal inference, due to lack of reliability of the measure of difference between co-twins in the outcome. Despite all this, the model provides one of the best tests of confounding by gene–environment correlation.

The co-twin control model has been applied in several studies involving cognitive function in old age. For example, Potter et al. (2006) investigated the association between occupational complexity and cognitive function in a large sample of US male veterans who were on average in their mid-60s at time of baseline assessment. Although the intellectual complexity of the jobs these men held before retirement was significantly associated with their cognitive function, this association did not hold up within MZ twin pairs who were discordant for job complexity. This suggests that rather than reflecting solely an environmental influence, the association of job complexity with cognitive function likely arises at least partly because intellectually demanding jobs are filled by individuals who are cognitively able (Finkel et al. 2009).

McGue and Christensen (2007) had somewhat better luck in demonstrating potentially causal effects. They examined differences in cognitive, primarily memory function in aging Danish MZ twins discordant for level of social activity. They observed that, within MZ pairs, the twin with the greater amount of social activity also showed better cognitive function at any assessed point in time, but there were no differences between the twins in rate of decline in function over time. The effect of social activity in discordant MZ twins was smaller, however, than the effect in the overall sample, indicating that gene–environment correlation was also important in understanding the association. This kind of result, where evidence for both directly causal effects and gene–environment correlation is present, is probably most typical of adequately powered studies investigating many different phenotypes. In the cognitive aging literature, different studies have not yet measured either phenotypes or environments in similar enough ways to draw overall conclusions.

The second behavior genetic model that is useful in evaluating the possibility that gene–environment correlation confounds apparent risk–outcome associations is Purcell's (2002) model of gene–environment interaction in the presence of gene–environment correlation. Although this model has received criticism (Rathouz et al. 2008), it is useful in many situations. Its primary advantage is that it makes it possible to get some sense of the processes underlying and linking gene–environment interaction and correlation (Johnson 2007, 2011). This is because it reveals both when genetically and environmentally influenced variance differs with level of environmental exposure, and when and to what degree the genetic and environmental influences on environmental exposures and outcome phenotypes are linked. For example, Johnson et al. (2009) used this model to explore the associations among educational attainment and primarily memory-related cognitive and physical function in Danish twins aged 70 and over. General biological aging, chronic illnesses that affect both physical and cognitive function such as diabetes, and high lifetime-stable cognitive ability that facilitates lifestyle choices and health habits have been offered as (not mutually exclusive) possible explanations for the widely observed link between physical and cognitive function in old age, with education generally assumed to be a protective factor.

Study results were complex, but likely indicative of the sorts of intertwined processes we should expect to be involved in cognitive aging. Physical function did not moderate genetic or environmental influences on cognitive function, though both their genetic and environmental influences were substantively linked. This suggested that physical deterioration did not precede or cause deterioration in cognitive function, but that, instead, they declined together for some of the same reasons. Cognitive function, however, did moderate genetic and both shared and nonshared environmental influences on physical function, with greater variance from all sources associated with lower cognitive function. This, in conjunction with the basic association between cognitive and physical function, suggested that lifetime-stable cognitive ability supported the development of lifestyle factors that maintained both physical and cognitive function, especially because the pattern of genetic and nonshared environmental correlations suggested that the lifestyle factors acted to minimize expression of genetic vulnerabilities. There was no evidence that educational attainment provided resources to minimize or prevent the sorts of chronic illnesses that affect physical function because it did not moderate physical function. It did, however, moderate variance in cognitive function, suggesting that education acted in ways similar to lifetime-stable cognitive function in facilitating lifestyles that helped to maintain health. Results of this kind are at best suggestive of processes, however, and this area of research badly needs additional methods that can more rigorously distinguish among the kinds of possibilities this study addressed.

5.3 Molecular Genetic Approaches

The principal molecular genetic approaches to studying cognitive abilities in old age are candidate gene and genome-wide association studies (GWAS). In candidate gene association studies, associations between particular genes and traits are investigated, while in GWAS hundreds of thousands or even a million genetic markers throughout the genome are scanned for association. To date, there have been few GWAS studies of cognitive aging, and those that have been carried out require replication and have offered very little with respect to mechanistic pathways that might be associated with differential cognitive aging. There have, however, been several genetic studies of cognitive aging that have gone far beyond candidate gene and GWAS studies.

5.3.1 Candidate Gene Studies

In candidate gene studies, researchers consider whether variants in specific individual genes might be associated with people's differences in cognitive aging. To carry out such studies, some choice must be made of which genes might hold variants that could be associated with differences in cognitive aging from the 20,000+ protein-coding genes in the human genome. Typically, to date, single nucleotide polymorphisms (SNPs) have been chosen, because these can be tested easily and the minor allele will be possessed by reasonable numbers of subjects in most samples. Harris and Deary (2011), Deary et al. (2009), and Payton (2009) have recently reviewed these studies. As with most other phenotypes, they have provided few replicable results, with the exception of small effects from the gene for apolipoprotein E (*APOE*). Payton found total agreement for none of the 50 genes that had been studied with respect to normal cognitive aging in the 14-year period between 1995 and 2009, concluding that the field is "largely bereft of consensus and adequate research design.... Sadly however, if the question were to be asked 'after 14 years of cognitive research what genes can we conclusively say are responsible for the variation in general cognition or its decline with age in healthy individuals?' the answer would have to be 'none'" (p. 465). Problems he identified in many studies included poor and varying assessments of the cognitive phenotype, especially those studies using the Mini-Mental State Exam due to its commonly observed ceiling effects; the possibility of sex-specific effects; poor sample sizes; population stratification; and failure to adjust for vascular risk factors that are known to be associated with dementia and cognitive decline. He also addressed the failure to consider either gene–environment or gene–gene interaction, citing the example of how variation in the *FADS2* gene interaction with breastfeeding appeared to affect children's intelligence (Caspi et al. 2007) and the example of how the brain-derived neurotrophic factor (*BDNF*) and *REST* gene variants interacted in their association with general intelligence in a group of older people without dementia (Miyajima et al. 2008).

Obvious candidate genes for normative cognitive aging are those that have been associated with Alzheimer's disease because of its long period of development. There are three genes that show mutations that are strongly associated with early onset of this disease: amyloid precursor protein (*APP*), and presenilin 1 and 2 (*PS1*, *PS2*; see Hamilton et al. 2011), but these account for only a very small percentage of Alzheimer's cases. The much more common form of Alzheimer's that may confound studies of normative cognitive aging has an older age of onset. The best known and replicated genetic risk for this form of the disease is possession of the epsilon 4 allele of the gene for *APOE* (Corder et al. 2003). Close to this gene on human chromosome 19 is the gene for translocase of the outer mitochondrial membrane 40 homolog (*TOMM40*), and variation in this gene, too, is associated with late-onset Alzheimer's-type dementia (Roses et al. 2010). Large-scale GWAS studies of Alzheimer's disease have also found replicated associations between the disease and genetic variation in the following genes: *BIN1*, *CLU*, *CR1*, *PICALM*, and the genetic region *BLOC1S3/EXOC3L2/MARK4* (Hamilton et al. 2011; Seshadri et al. 2010).

These genes were examined for associations with verbal declarative memory, abstract reasoning, and executive function in the Lothian Birth Cohorts of 1921 (mean age 79) and 1936 (mean age 70; Hamilton et al. 2011). The tests—involving 158 SNPs—were done without adjusting for childhood IQ score to examine cognition in old age, and with this adjustment to examine cognitive aging, and with and without adjustment for *APOE* e4 status. After adjusting for multiple testing, no single SNP was associated with any cognitive ability. However, one haplotype from *TRAPPC6A* was associated with abstract reasoning in those lacking an *APOE* e4 allele. Also suggested, but with less strong evidence, was an interaction between *APP* and *BIN1* in affecting verbal declarative memory in older people who carried the *APOE* e4 allele.

Given the robust association of the *APOE* e4 allele with Alzheimer's disease, some have suggested that this allele may be associated with better cognition at younger ages. This would be an example of antagonistic pleiotropy (Williams 1957), or effects of one gene on more than one trait, at least one of which is advantageous and one disadvantageous. A meta-analysis of 20 studies that compared general cognitive function in *APOE* e4 carriers and noncarriers in children, adolescents and young adults, however, found no significant differences (Ihle et al. 2012). This null finding casts doubt on the antagonistic pleiotropy hypothesis, at least with respect to general cognitive function (Tuminello and Han 2011), though samples in many of the studies meta-analyzed were small, limiting ability to detect small effects. On the other hand, it is now clear that possession of the e4 allele of *APOE* is associated not just with Alzheimer's disease but with lower cognitive function in old age more generally. Wisdom et al. (2011) carried out a meta-analysis including 40,942 nondemented adults in 77 studies. They found that e4 carriers scored more poorly on tests of episodic memory (often tests of verbal declarative memory; $d=-0.14, p<0.01$), executive function ($d=-0.06, p<0.05$), perceptual speed ($d=-0.07, p<0.05$), and general cognitive ability ($d=-0.05, p<0.05$). The detriment in episodic memory and general cognitive ability associated with the e4 allele increased with age, consistent with observations of increasing genetic variance in memory with age (Reynolds et al. 2005). There were no consistently significant differences in attention, primary memory, verbal ability, or visuospatial skill, though patterns were similar and there were fewer studies testing these domains. For example, the effect size (d) for primary memory was -0.11, but was not significant owing to smaller sample size. There was variability in the tests used to test the same-named domain, and that many of the studies testing what was termed general cognitive ability used the minimum mean square error (MMSE), with its marked ceiling effect. In the Lothian Birth Cohort 1921, *APOE* e4 carriers scored significantly lower than noncarriers on a well-validated test of general intelligence at age 79 years, despite the two groups' showing no significant difference at age 11 (Deary et al. 2002). When cognitive aging was studied in the same individuals from ages 79–83 and 87, e4 carriers showed more deterioration in verbal declarative memory and abstract reasoning, but there was no significant difference in executive function (Schiepers et al. 2012). Those individuals with a longer allelic variant of *TOMM40*—which is linked with *APOE*—showed similar results. These small effects could have resulted from presence in the sample of preclinical or undiagnosed cases of Alzheimer's disease.

Beyond genes that have been associated with Alzheimer's disease, three genes stand out as having been extensively studied in relation to cognition, including in older people. Interest in these genes derives substantially because some see the cognitive decrements seen in psychiatric disorders, especially schizophrenia, as integral to those disorders and suspect that genetic susceptibility to the disorder may affect general cognition even if an individual escapes the disorder itself (e.g., Autry and Monteggia 2012). One is the gene for *BDNF*, which has a common functional polymorphism (Val66Met). This variant has been linked to memory function in humans and other species. A review of this genetic variant's association with cognitive abilities, including memory phenotypes, found that results to date were inconsistent but that, "the general consensus from the numerous studies has been that in healthy white populations, when challenged with various cognitive or motor learning behavioral tasks, humans with one or more copies of the BDNF Met allele have altered performance suggestive of a decrease in plasticity [ability to retain new information]" (Dincheva et al. 2012, p. 36). Another is the gene for catechol-*O*-methyl transferase, which has a functional polymorphism (Val158Met). The Met allele leads to lower levels of dopamine and to degradation of other neurotransmitters, especially in the frontal cortex. This gene has been extensively studied in people with schizophrenia as well as in healthy subjects, and has been associated with cognitive functions involving prefrontal cortex: executive function, working memory, fluid-type intelligence, and attention (Dickinson and Elevag 2009). A meta-analysis concluded, there was evidence that people who were homozygous for the Met allele might score higher on general IQ-type tests (Barnett et al. 2008). Finally, variants in the gene for dystrobrevin-binding protein 1 (*DTNBP1*) were originally but inconsistently associated with schizophrenia. A meta-analysis of nine SNPs in this gene, across 10 cohorts (total $N=7{,}592$), found that, overall, minor allele carriers had lower general cognitive ability or IQ-type scores (Zhang et al. 2010). The subjects' ages ranged from young to older adults, so this does not refer specifically to cognitive aging. It also requires replication.

Many other individual genes have been discussed by Payton (2009), Reinvang et al. (2010), and Harris and Deary (2011), though none has produced consistent findings. Beyond individual genes, which, given their action, have emerged as candidate genes for cognitive aging, there are studies of groups of many genes which are associated with given functions. For example, there have been studies of cognitive aging with respect to genes that are associated with oxidative stress (Harris et al. 2007) and longevity (Lopez et al. 2011). From the former, the prion protein gene (*PRNP*) emerged as being possibly associated with cognitive aging. From the latter, the genesynaptojanin-2 (*SYNJ2*) emerged as being possibly linked with cognitive abilities. Telomere length is related to cognitive stress. Telomeres are nucleoprotein complexes at the ends of chromosomes, and they tend to be shorter in the presence of oxidative stress. Telomere length is thought to act as a biomarker of successful aging. However, a large study on age-homogeneous individuals found no association between telomere length at age 70 and cognitive change since childhood and other cognitive and physical phenotypes (Harris et al. 2010). Others have found, in a sample of younger adult women, telomere length to be associated with level of cognitive ability (Valdes et al. 2010).

5.3.2 Genome-Wide Association Studies

The currently available main methodological alternative to testing specific genetic variants for association with cognitive functions in old age is to record each participant's alleles at a large number of SNPs placed throughout the genome and to test for associations between any of these markers and cognitive abilities. This GWAS approach typically uses hundreds of thousands of SNPs. Moreover, based on assumed-known haplotype patterns, studies often use the measured alleles at these markers to impute the alleles at additional genetic loci, often increasing the number of associations considered to be well over a million. Therefore, type 1 statistical errors present a large problem, though potential inaccuracies in the imputation process should not be disregarded. Because of the large probability of type 1 errors, and because it has become clear that, for almost all complex quantitative traits, individual SNP effects are very small, these studies demand large sample sizes. The p value taken to be genome-wide significant in such studies is $< 10^{-8}$, and replication is expected to be sought in independent cohorts prior to initial publication.

To date, there is only one published GWAS study of general cognitive abilities in old age (Davies et al. 2011). This measured approximately 500,000 SNPs in over 3,500 older people (from late-middle age to 79 years old). They came from five cohorts in Scotland and England, the so-called CAGES consortium: Cognitive Aging Genetics in England and Scotland. The cognitive phenotypes tested were fluid- and crystallized-type intelligence. For crystallized intelligence, the National Adult Reading test was used in the Scottish Cohorts and the Mill Hill Vocabulary test in the English cohorts. Fluid intelligence was based on a principal components analysis of diverse tests in the Scottish Cohorts, and from a combination of Alice Heim 4 and Cattell Culture Fair tests in the English cohorts. This points out an additional difficulty with GWAS studies: in forming consortia of studies to increase sample size, it is common to combine cognitive test scores reflecting constructs that are only superficially similar, thus blurring the measurement of the intended phenotype and offsetting the increased power to detect effects provided by the combined samples.

In the CAGES consortium study, there was no genome-wide significant SNP for fluid or crystallized intelligence. Considering all SNPs located within single genes rather than individual SNPs, one gene was significantly associated with fluid intelligence: forming-binding protein 1-like (*FNBP1L*). This did not replicate in a Norwegian sample, ranging in age from 18 to 78 years. The next analysis in this study used the so-called Genome-wide Complex Trait Analysis (GCTA) method (Visscher et al. 2010; Yang et al. 2010). This uses all ~500,000 measured SNPs simultaneously in a model that creates an association matrix and allows estimation of the correlation between the phenotype and the extent of genetic similarity in the sample, consisting of conventionally unrelated individuals. Therefore, for the first time based on DNA testing, estimates were provided for the narrow-sense (purely additive) heritability of fluid (0.51, s.e. = 0.11, $p = 1.2 \times 10^{-7}$) and crystallized (0.40, s.e. = 0.11, $p = 5.7 \times 10^{-5}$) intelligence in older age. Further analysis using this method found that there was a tendency for longer chromosomes to explain more cognitive ability variance. Finally, the study attempted to predict intelligence in each cohort by using the genetic information from all autonomic SNPs in the others. The correlations had means of 0.110 and 0.081 for fluid and crystallized intelligence, respectively, and were 0.076 and 0.092, respectively, in a separate Norwegian sample that had not been used in the GWAS. Therefore, this study suggests that a substantial proportion of the variance in cognitive ability in older ages is accounted for by genetic variants in linkage disequilibrium with common SNPs.

The GCTA method applied to the GWAS data in the study by Davies at al. (2011) was extended to study cognitive aging and lifetime cognitive stability in the three Scottish cohorts (total $N=1{,}940$) of the CAGES consortium (Deary et al. 2012), in a demonstration of the method's potential as well as its extensive data requirements and limitations. All three had taken the same general cognitive ability test—the Moray House Test No. 12—at age 11. They also took various cognitive tests in old age: age 65 for the Aberdeen Birth Cohort 1936, age 70 for the Lothian Birth Cohort 1936, and age 79 for the Lothian Birth Cohort 1921. The GCTA method was used to estimate the genetic contribution to general fluid intelligence in old age after adjusting for childhood intelligence, thus providing an estimate for the proportion of genetic influence on lifetime cognitive change. This was 0.24, though with a large standard error of 0.20, meaning that it was far from significant. In this study, the Lothian cohorts had also taken the same Moray House Test in childhood and old age. Using this test score in both childhood and old age, the ~500,000 SNPs accounted for 7.4% (s.e. = 0.24) of variation in the residual change score. Clearly, these estimates carry little meaning as they were not significant and their confidence intervals contained both 0 and 1. The analysis method requires much larger sample sizes and likely more direct measures of change from peak cognition in adulthood.

The first GWAS study of cognitive aging, based on repeated measures of 17 tests on almost 750 subjects in the Religious Orders Study, at least age 75 at enrollment, found that *APOE* was significantly associated with cognitive change in old age (De Jager et al. 2012). Replication was conducted in three cohorts providing over 2,000 additional subjects. Replicated, too, was an SNP that affected the expression of the genes *PDE7A* and *MTFR1*, which are, respectively, involved in inflammation and oxidative stress.

Because processing speed is considered by some to be fundamental to cognitive function in general and cognitive aging in particular, it has been the subject of its own GWAS (Luciano et al. 2011). The cohorts involved in the GWAS of processing speed were mostly in older age, including the Lothian Birth Cohorts of 1921 (age 82) and 1936 (age 70), and the Helsinki Birth Cohort Study (age 64). The Brisbane study was younger, at 16 years. The total *N* for the study was almost 4,000 subjects. The four cohorts included were remarkable for having experimental (reaction time) and psychophysical (inspection time) measures of processing speed, and not just psychometric tests. Processing speed in each case was assessed using a factor analysis-derived general factor of processing speed from multiple tests. There were no genome-wide significant associations. There were some suggestively significant results ($p<10^{-5}$), some plausible candidate genes (e.g., *TRIB3*). Biological pathways analysis, which examines whether SNPs that have suggestive significance in a GWAS analysis are over-represented in biological pathways of interest, suggested association with the gene processes of cell junction, focal adhesion, receptor binding, and cellular metabolic processes. Several of the GWAS-identified genes apparently involved in these processes were also considered relevant to Alzheimer disease mechanisms.

Integrity of the myelin-sheathed brain white matter might provide one mechanism through which processing speed could be involved in cognitive aging, as these myelin sheaths allow faster neural transmission. Using diffusion-tensor magnetic resonance brain imaging on a relatively large subsample of the Lothian Birth Cohort 1936 ($N=535$), there was a general factor of brain white matter integrity across many major tracts (Penke et al. 2010). Furthermore, it was also shown that this general factor was associated significantly with processing speed: older people with brain white matter of higher integrity had faster processing speed, as assessed using a general factor based on reaction and inspection time measures. This was the basis for a GWAS of brain white matter integrity, the phenotype being the same general factor of white matter integrity (Lopez et al 2012). There were no genome-wide significant associations. There was suggestive significance for *ADAMTS18*, which has roles in tumor suppression and hemostasis, and *LOC388630*, whose function was unknown. Biological pathways analysis found over-representation of genes related to cell adhesion and neural transmission pathways.

5.3.3 The Future and Other Molecular Genetic Approaches

The general absence of replicable associations and the very small effect sizes of most of those that were replicable in the first reports of GWAS studies from complex quantitative traits led to realism about the likely effect sizes of individual genetic variants. There was concern about the "missing heritability" for such traits, because the heritability accounted for by common SNPs was so far below that which had been estimated by behavior genetic studies using twin and adoption studies. The development and application of the GCTA method—in which heritability is estimated by fitting all SNPs, typically hundreds of thousands, simultaneously—has revised this. It seems that there is less missing heritability, but the new problem is that, for many quantitative traits, there will be very large numbers of very small genetic contributions. This makes mechanistic studies very difficult, in their traditional form, and calls into question the very premise of underlying clearly identifiable causal mechanisms, as usually implicitly defined. The GWAS studies of cognitive abilities that will appear in the near future will be larger, in the tens of thousands, to try to find some individual, replicable contributions. The CHARGE consortium will soon report large GWAS studies of memory, processing speed, executive function, and general cognitive ability, mostly in older people. The COGENT consortium will also report a GWAS on general cognitive ability of mostly older people. The CAGES consortium will report a GWAS on age-related cognitive change.

These are just larger GWAS studies, of the same design as Davies et al. (2011). Future studies are also likely to be designed to consider gene-by-environment and gene-by-gene interactions at the GWAS level. There will be studies of additional types of genetic variation, beyond SNPs, such as copy number variations. One such small study has already appeared (Yeo et al. 2011). It found that people who had more rare genetic deletions had lower intelligence. There will be studies that use SNP arrays with increasingly large numbers of SNPs, including greater numbers of SNPs imputed from information based on whole-genome sequencing. There will soon be studies based on whole-genome sequencing and sequencing of only protein-coding regions throughout the genome, which will look for genetic associations with rare and even private mutations (i.e., those unique to particular families).

There will be studies relating intelligence and cognitive aging to individual differences in DNA methylation and other forms of gene expression. DNA methylation changes with age as well as with environmental experiences (including in important brain areas: Hernandez et al. 2011). To the extent that the age changes are regular, individual differences in DNA methylation can provide, in part, records of environmental effects on the gene expression, which in turn can affect phenotypes and their successful aging, including cognition (Feil and Fraga 2012). DNA methylation and gene expression can both be examined at the genome-wide level—on arrays with hundreds of thousands of markers. Both will take the study of intelligence and cognitive aging and genetics to more mechanistic levels. However, both come with a problem that does not affect SNP testing: tissue specificity. DNA methylation and other forms of control of gene expression vary across tissues, and even within the brain there is region-by-region variation in gene expression. It remains to be discovered how much overall individual differences in expression are common across tissues, how much these differences relate to cognitive abilities and cognitive aging, and thus how much it will be necessary to study brain tissues to understand associations between gene expression and cognition. Gene expression studies certainly promise more by way of understanding mechanisms because they capture actual gene function, not just presence of polymorphisms (Geschwind and Konopka 2009). Another emerging avenue of investigation is the output of genetic expression in the form of protein concentrations. A small pilot study of the urinary proteome and general intelligence has already appeared. It indicated some proteins that might have roles in cognition (Lopez et al. 2011). Noncoding RNAs and their regulatory networks (Qureshi and Mehler 2011) are also emerging as potentially relevant to cognitive aging, and to aging-related phenotypes and processes such as the brain's plasticity and its response to stress.

5.4 Conclusions

Genetic studies of normal cognitive aging have made clear that genetic influences continue to be involved in late-life cognitive function. It is far less clear, however, to what degree they are involved in the declines in function with age that have come to be considered normative. And, despite huge technological advances in probing the human genome, we remain far from understanding how they are involved or which particular genes contribute, with the striking exception of *APOE*. The open questions from the genetic perspective, however, closely parallel those in all approaches to the study of cognitive aging. As populations throughout the world continue to "gray," growing older due to declining birth rates as well as increased longevity, meeting the challenge of understanding cognitive aging is of tremendous social importance. In many ways, genetic studies are well-positioned to offer important insights into the processes involved. They face the same measurement and sample selection difficulties as the rest of the field, but generally to no worse degree. And they afford unique opportunities to disentangle some of the causal knots faced by other approaches. We are excited by these challenges and opportunities and look forward to future progress in the field.

References

Autry, A. E., & Monteggia, L. M. (2012). Brain-derived neurotrophic factor and neuropsychiatric disorders. *Pharmacoloogical Reviews, 64,* 238–258.

Barnett, J. H., Scoriels, L., & Munafo, M. R. (2008). Meta-analysis of the cognitive effects of the catechol-O-methyltransferase gene Val158/108Met polymorphism. *Biological Psychiatry, 64,* 137–144.

Batterham, P. J., Mackinnon, A. J., & Christensen, H. (2011). The effect of education on the onset and rate of terminal decline. *Psychology and Aging, 26,* 339–350.

Bergen, S. E., Gardner, C. O., & Kendler, K. S. (2007). Age-related changes in heritability of behavioral phenotypes over adolescence and young adulthood: A meta-analysis. *Twin Research and Human Genetics, 10*(3), 423–433.

Bielak, A. A., Hultsch, D. F., Strauss, E., MacDonald, S. W., & Hunter, M. A. (2010). Intraindividual variability in reaction time predicts cognitive outcomes 5 years later. *Neuropsychology, 24,* 731–741.

Blennow, K., de Leon, M. J., & Zetterberg, H. (2006). Alzheimer's disease. *Lancet, 368,* 387–403.

Bosworth, H. B., & Schaie, K. W. (1999). Survival effects in cognitive function, cognitive style, and sociodemographic variables in the Seattle Longitudinal Study. *Experimental Aging Research, 25*(2), 121–139.

Bryk, A. S., & Raudenbush, S. W. (2002). *Hierarchical linear models: Applications and data analysis methods (2nd ed.)*. Thousand Oaks: Sage.

Calvin, C. M., Deary, I. J., Fenton, C., Roberts, B. A., Der, G., Leckenby, N., & Batty, G. D. (2011). Intelligence in youth and all-case mortality: Systematic review with meta-analysis. *International Journal of Epidemiology, 40,* 626–644.

Carmelli, D., DeCarli, C., Swan, G. E., Jack, L. M., Reed, T., Wolf, P. A., & Miller, B. L. (1998). Evidence for genetic variance in white matter hypersensitivity volume in normal elderly male twins. *Stroke, 29*(6), 1177–1181.

Carroll, J. B. (1993). *Human cognitive abilities: A survey of factor analytic studies*. Cambridge: Cambridge University Press.

Caspi, A., Williams, B., Kim-Cohen, J., Craig, I. W., Milne, B. J., Poulton, R., et al. (2007). Moderation of breastfeeding effects on the IQ by genetic variation in fatty acid metabolism. *Proceedings of the National Academy of Sciences of the United States of America, 104,* 18860–18865.

Christensen, K., Holm, N. V., McGue, M., Corder, L., & Vaupel, J. W. (1999). A Danish population-based twin study on general health in the elderly. *Journal of Aging and Health, 11,* 49–64.

Corder, E. H., Saunders, A. M., Strittmatter, W. J., Schmechel, D. E., Gaskell, P. C., Small, G. W., et al. (2003). Gene dose of apolipoprotein E type 4 allele and the risk of Alzheimer's disease in late onset. *Science, 261,* 921–923.

Craik, F. I. M., & McDowd, J. M. (1987).Age differences in recall and recognition. *Journal of Experimental Psychology-Learning Memory and Cognition, 13*(3), 474–479. doi:10.1037//0278-7393.13.3.474.

Davies, G., Tenesa, A., Payton, A., Yang, J., Harris, S. E., Liewald, D., et al. (2011). Genome-wide association studies establish that human intelligence is highly heritable and polygenic. *Molecular Psychiatry, 16,* 996–1005.

De Jager, P. L., Shulman, J. M., Chibnik, L. B., Keenan, B. T., Raj, T., Wilson, R. S., et al. (2012). A genome-wide scan for common variants affecting the rate of age-related cognitive decline. *Neurobiology of Aging*. doi:10.1016/j.neurobiolaging.2011.09.033.

Deary, I. J. (2000). *Looking down on human intelligence*. Oxford: Oxford University Press.

Deary, I. J., & Der, G. (2005). Reaction time, age, and cognitive ability: Longitudinal findings from age 16 to 63 years in representative population samples. *Aging, Neuropsychology, and Cognition, 12,* 187–215.

Deary, I. J., Whiteman, M. C., Pattie, A., Starr, J. M., Hayward, C., Wright, A. F., Carothers, A., & Whalley, L. J. (2002). Cognitive change and the *APOE* e4 allele. *Nature, 418,* 932.

Deary, I. J., Yang, J., Davies, G., Harries, S. E., Tenesa, A., et al. (2012). Genetic contributions to stability and change in intelligence from childhood to old age. *Nature, 482,* 212–215.

Deary, I. J., Johnson, W., & Houlihan, L. (2009). Genetic foundations of human intelligence. *Human Genetics, 126,* 613–624.

Dickinson, D., & Elevag, B. (2009). Genes, cognition and brain through a COMT lens. *Neuroscience, 164,* 72–87.

Dincheva, I., Glatt, C. E., & Lee, F. S. (2012). Impact of the BDNF Val66Met polymorphism on cognition: implications for behavioral genetics. *The Neuroscientist*. doi:10.1177/1073858411431646.

Feil, R., & Fraga, M. F. (2012). Epigenetics and the environment: emerging patterns and implications. *Nature Reviews Genetics, 13,* 97–109.

Finkel, D., & McGue, M. (1993). The origins of individual differences in memory among the elderly: A behavior genetic. *Psychology and Aging, 8,* 527–537.

Finkel, D., & McGue, M. (2007). Genetic and environmental influences on intraindividual variability in reaction time. *Experimental Aging Research, 33*(1), 13–35.

Finkel, D., & Pedersen, N. (2004). Processing speed and longitudinal trajectories of change for cognitive abilities: the Swedish Adoption/Twin Study of Aging. *Aging Neuropsychology and Cognition, 11,* 325–345.

Finkel, D., & Reynolds, C. A. (2010). Behavioral Genetic Investigations of Cognitive Aging. In Y. K. Kim (Ed.), *Handbook of Behavior Genetics* (pp. 101–112). New York: Springer.

Finkel, D., Pedersen, N. L., McGue, M., & McClearn, G. E. (1995). Heritability of cognitive abilities in adult twins: Comparison of Minnesota and Swedish data. *Behavior Genetics, 25,* 421–431.

Finkel, D., Reynolds, C. A., McArdle, J. J., & Pedersen, N. (2005). The longitudinal relationship between processing speed and cognitive ability: Genetic and environmental influences. *Behavior Genetics, 35,* 535–549.

Finkel, D., McArdle, J. J., Reynolds, C. A., & Pedersen, N. L. (2007). Age changes in processing speed as a leading indicator of cognitive aging. *Psychology and Aging, 22*(3), 558–568.

Finkel, D., Andel, R., Gatz, M., & Pedersen, N. L. (2009). The role of occupational complexity in trajectories of cognitive aging before and after retirement. *Psychology and Aging, 24,* 563–573.

Finkel, D., Reynolds, C. A., McArdle, J. J., Hamagami, F., & Pedersen, N. L. (2009). Genetic variance in processing speed drives variation in aging of spatial and memory abilities. *Developmental Psychology, 45,* 820–834.

Fjell, A. M., & Walhovd, K. B. (2010). Structural brain changes in aging: Courses, causes and cognitive consequences. *Reviews in the Neurosciences, 21*(3), 187–221.

Fuentes, K., Hunter, M. A., Strauss, E., & Hultsch, D. F. (2001). Intraindividual variability in cognitive performance in persons with chronic fatigue syndrome. *Clinical Neuropsychologist, 15,* 210–227.

Gatz, M., Fratiglioni, L., Johansson, B., Berg, S., Mortimer, J. A., Reynolds, C. A., Fiske, A., Pedersen, N. L. (2005). Complete ascertainment of dementia in the Swedish Twin Registry: The HARMONY study. *Neurobiology of Aging, 26,* 439–447.

Gerstorf, D., Ram, N., Hoppmann, C., Willis, S. L., & Schaie, K. W. (2011). Cohort differences in cognitive aging and terminal decline in the Seattle Longitudinal Study. *Developmental Psychology, 47,* 1026–1041.

Geschwind, D. H., & Konopka, G. (2009). Neuroscience in the era of functional genomics and systems biology. *Nature, 461,* 908–915.

Giubilei, F., Medda, E., Fagnani, C., Bianchi, V., De Carolis, A., Salvetti, M., Stazi, M. A. (2008). Heritability of neurocognitive functioning in the elderly: Evidence from an Italian twin study. *Age and Aging, 37*(6), 640–646. doi:10.1093/aging/afn132.

Hamilton, G., Harris, S. E., Davies, G., Liewald, D. C., Tenesa, A., Starr, J. M., Porteous, D., & Deary, I. J. (2011). Alzheimer's disease genes are associated with measures of cognitive aging in the Lothian Birth Cohorts of 1921 and 1936. *International Journal of Alzheimer's Disease, 2011,* 505984.

Harris, S. E., & Deary, I. J. (2011). The genetics of cognitive ability and cognitive aging in healthy older people. *Trends in Cognitive Sciences, 15,* 388–394.

Harris, S. E., Fox, H., Wright, A. F., Hayward, C., Starr, J. M., Whalley, L. J., & Deary, I. J. (2007). A genetic association analysis of cognitive ability and cognitive aging using 325 markers for 109 genes associated with oxidative stress or cognition. *BMC Genetics, 8,* 43.

Harris, S. E., Martin-Ruiz, C., von Zglinicki, T., Starr, J. M., & Deary, I. J. (2010). Telomere length and aging biomarkers in 70 year-olds: The Lothian Birth Cohort 1936. *Neurobiology of Aging.* doi:10.1016/j.neurobiolaging.2010.11.013.

Haworth, C. M. A., Wright, M. J., Luciano, M., Martin, N. G., de Geus, E. J. C., van Beijsterveldt, C. E. M., Plomin, R. (2010).The heritability of general cognitive ability increases linearly from childhood to young adulthood. *Molecular Psychiatry, 15*(11), 1112–1120. doi: 10.1038/mp.2009.55.

Hernandez, D. G., Nalls, M. A., Gibbs, J. R., Arepalli, S., van der Brug, M., Chong, S., et al. (2011). Distinct DNA methylation changes highly correlated with chronological age in the human brain. *Human Molecular Genetics, 20,* 1164–1172.

Hoffman, L., Hofer, S. H., & Sliwinski, M. J. (2011). On the confounds among retest gains and age-cohort differences in the estimation of within-person change in longitudinal studies: A simulation study. *Psychology and Aging, 26,* 778–791.

Hultsch, D. F., MacDonald, S. W., Hunter, M. A., Levy-Benchton, J., & Strauss, E. (2000). Intraindividual variability in cognitive performance in older adults: Comparison of adults with mild dementia, adults with arthritis, and healthy adults. *Neuropsychology, 14,* 588–598.

Ihle, A., Bunce, D., & Kliegel, M. (2012). *APOE* e4 and cognitive function in early life: a meta-analysis. *Neuropsychology.* doi:10.1037/a0026769.

Jarvik, L. F., & Falek, A. (1963). Intellectual stability and survival in the aged. *Journal of Gerontology, 18,* 173–176.

Johansson, B., Hofer, S. M., Allaire, J. C., Maldonado-Molina, M. M., Piccinin, A. M., Berg, S., et al. (2004). Change in cognitive capabilities in the oldest old: The effects of proximity to death in genetically related individuals over a 6-year period. *Psychology and Aging, 19,* 145–156.

Johnson, W. (2007). Genetic and environmental influences on behavior: Capturing all the interplay. *Psychological Review, 114,* 423–440.

Johnson, W. (2011). What do genes have to do with cognition? In S. Kreitler (Ed.), *Cognition and motivation.* New York: Cambridge University Press.

Johnson, W., Bouchard, T. J., Krueger, R. F., McGue, M., & Gottesman, I. I. (2004). Just one *g*: Consistent results from three test batteries. *Intelligence, 32,* 95–107.

Johnson, W., te Nijenhuis, J., & Bouchard, T. J. (2008). Still just one g: Consistent results from five test batteries. *Intelligence, 36,* 81–95.

Johnson, W., Deary, I. J., McGue, M., & Christensen, K. (2009). Genetic and environmental transactions linking cognitive ability, physical fitness, and education in late life. *Psychology and Aging, 24,* 48–62.

Kennedy, K. M., & Raz, N. (2009). Aging white matter and cognition: Differential effects of regional variations in diffusion properties on memory, executive functions, and speed. *Neuropsychologia, 47*(3), 916–927.

Kleemeier, R. W. (1962). Intellectual changes in the senium. *Proceedings of the American Statistical Association, 1,* 290–295.

Lachman, R., Lachman, J. L., & Taylor, D. W. (1982). Reallocation of mental resources over the productive life-span: Assumptions and task analyses. In F. I. Craik, & S. Trehub (Eds.), *Aging and cognitive process* (pp. 304–350). New York: Plenum Press.

Lee, T., Henry, J. D., Trollor, J. N., & Sachdev, P. S. (2010). Genetic influences on cognitive functions in the elderly: A selective review of twin studies. *Brain Research Reviews, 64,* 1–13.

Lee, T., Mosing, M. A., Henry, J. D., Trollor, J. N., Lammel, A., Ames, D., Sachdev, P. S. (2012). Genetic influences on five measures of processing speed and their covariation with general cognitive ability in the elderly: The Older Australian Twins Study. *Behavior Genetics, 42*(1), 96–106. doi:10.1007/s10519-011-9474-1.

Li, S.-C., Aggen, S. H., Nesselroade, J. R., & Baltes, P. B. (2001). Short-term fluctuations in elderly people's sensorimotor functioning predict text and spatial memory performance: the Macarthur successful aging studies. *Gerontology, 47,* 100–116.

Li, S.-C., Lindenberger, U., & Sikstrom, S. (2001). Aging cognition: From neuromodulation to representation. *Trends in Cognitive Sciences, 5,* 479–486.

Li, S.-C., Lindenberger, U., Hommel, B., Aschersleben, G., Prinz, W., & Baltes, P. B. (2004). Transformations in the couplings among intellectual abilities and constituent cognitive processes across the life span. *Psychological Science, 15,* 155–163.

Lieberman, M. A. (1966). Observations on death and dying. *Journal of Gerontology, 6,* 70–72.

Lopez, L. M., Harris, S. E., Luciano, L., Liewald, D., Davies, G., Gow, A. J., et al. (2011). Evolutionary conserved longevity genes and human cognitive abilities in elderly cohorts. *European Journal of Human Genetics*. doi:10.1038/ejhg.2011.201.

Lopez, L. M., Mullen, W., Zurbig, P., Harris, S. E., Gow, A. J., Starr, J. M., et al. (2011). A pilot study of urinary peptides as biomarkers for intelligence in old age. *Intelligence, 39,* 46–53.

Lopez, L. M., Bastin, M. E., Munoz Maniega, S., Penke, L., Davies, G., Christoforou, A., et al. (2012). A genome-wide search for genetic influences and biological pathways related to the brain's white matter integrity. *Neurobiology of Aging, 33*(8), 1847.

Lovden, M., Li, S.-C., Shing, Y. L., & Lindenberger, U. (2007). Within-person trial-to-trial variability precedes and predicts cognitive decline in old and very old age: Longitudinal data from the Berlin Aging Study. *Neuropsychologia, 45,* 2827–2838.

Luciano, M., Hansell, N., Lahti, J., Davies, G., Medland, S. E., Raikkonen, K., et al. (2011). Whole genome association scan for genetic polymorphisms influencing information processing speed. *Biological Psychology, 86,* 193–202.

Luciano, M., Posthuma, D., Wright, M. J., de Geus, E. J., Smith, G. A., Geffen, G. M., et al. (2005). Perceptual speed does not cause intelligence, and intelligence does not cause perceptual speed. *Biological Psychology, 70,* 1–8.

MacDonald, S. W., Hultsch, D. F., & Dixon, R. A. (2003). Performance variability is related to change in cognition: Evidence from the Victoria Longitudinal Study. *Psychology and Aging, 18,* 510–523.

MacDonald, S. W., Hultsch, D. F., & Dixon, R. A. (2011). Aging and the shape of cognitive change before death: Terminal decline or terminal drop? *Journals of Gerontology. Series B, Psychological Sciences and Social Sciences, 66,* 292–301.

McArdle, J. J., & Plassman, B. J. (2009). A biometric latent curve analysis of memory decline in older men of the NAS-NRC Twin Registry. *Behavior Genetics, 39,* 472–495

McClearn, G. E., Johansson, B., Berg, S., Pedersen, N. L., Ahern, F., Petrill, S. A., & Plomin, R. (1997). Substantial genetic influence on cognitive abilities in twins 80 or more years old. *Science, 276,* 1560–1563.

McGue, M., & Christensen, K. (2001). The heritability of cognitive functioning in very old adults: Evidence from Danish twins aged 75 years and older. *Psychology and Aging, 16,* 272–280.

McGue, M., & Christensen, K. (2007). Social activity and healthy aging: A study of aging Danish twins. *Twin Research and Human Genetics, 10*(2), 255–265.

McGue, M., Osler, M., & Christensen, K. (2010). Causal inference and observational research: The utility of twins. *Perspectives on Psychological Science, 5*(5), 546–556.

McNamee, R. (2003). Confounding and confounders. *Occupational and Environmental Medicine, 60,* 227–234.

Miyajima, F., Quinn, J. P., Horan, M., Pickles, A., Ollier, W. E., Pendleton, N., et al. (2008). Additive effect of BDNF and REST polymorphisms is associated with improved general cognitive ability. *Genes, Brain and Behavior, 7,* 714–719.

Neale, M. C., & McArdle, J. J. (2000). Structured latent growth curves for twin data. *Twin Research, 3,* 165–177.

Need, A., Attix, D. K., McElvoy, J. M., Cirulli, E. T., Linney, K. L., Hunt, P., et al. (2009). A genome-wide study of common SNPs and CNVs in cognitive performance in the CANTAB. *Human Molecular Genetics, 18,* 4650–4661.

Nishiwaki, Y., Clark, H., Morton, S. M., & Leon, D. A. (2005). Early life factors, childhood cognition and postal questionnaire response rate in middle age: The Aberdeen Children of the 1950s study. *BMC Medical Research Methodology, 5,* 16.

Payton, A. (2009). The impact of genetic research on our understanding of normal cognitive aging: 1995 to 2009. *Neuropsychological Review, 19,* 451–477.

Pedersen, N. L., Plomin, R., Nesselroade, J. R., & McClearn, G. E. (1992). A quantitative genetic analysis of cognitive abilities during the second half of the life span. *Psychological Science, 3*(6), 346–353.

Pedersen, N. L., Ripatti, S., Berg, S., Reynolds, C., Hofer, S. M., Finkel, D., Gatz, M., & Palmgren, J. (2003). The influence of mortality on twin models of change: Addressing missingness through multiple imputation. *Behavior Genetics, 33,* 161–169.

Penke, L., Munoz Maniega, S., Murray, C., Gow, A. J., Valdes Hernandez, M. C., Clayden, J. D., et al. (2010). A general factor of brain white matter integrity predicts information processing speed in healthy older people. *Journal of Neuroscience, 30,* 7559, 7674.

Piccinin, A. M., Muniz, G., Matthews, F. E., & Johansson, B. (2011). Terminal Decline from within- and between-person perspectives, accounting for incident dementia. *Journals of Gerontology. Series B, Psychological Sciences and Social Sciences, 66,* 391–401.

Platt, J. R. (1964). Strong inference: Certain systematic methods of scientific thinking may produce much more rapid progress that others. *Science, 146,* 347–353.

Plomin, R., Pedersen, N., Lichtenstein, P., & McClearn, G. E. (1994). Variability and stability in cognitive abilities are largely genetic later in life. *Behavior Genetics, 24,* 207–215.

Potter, G. G., Plassman, B. L., Helms, M. J., Foster, S. M., & Edwards, N. W. (2006). Occupational characteristics and cognitive performance among elderly male twins. *Neurology, 67*(8), 1377–1382.

Purcell, S. (2002). Variance component models for gene-environment interaction in twin analysis. *Twin Research, 5,* 554–571.

Qureshi, I. A., & Mehler, M. F. (2011). Non-coding RNA networks underlying cognitive disorders across the life-span. *Trends in Molecular Medicine, 17,* 337–346.

Rabbitt, P. M., Osman, P., Moore, B., & Stollery, B. (2001). There are stable individual differences in performance variability, both from moment to moment and from day to day. *The Quarterly Journal of Experimental Psychology A, Human Experimental Psychology, 4,* 981–1003.

Rabbitt, P., Diggle, P., Holland, F., & McInnes, L. (2004). Practice and drop-out effects during a 17-year longitudinal study of cognitive aging. *Journals of Gerontology. Series B, Psychological Sciences and Social Sciences, 59*(2), 84–97.

Rabbitt, P., Lunn, M., & Wong, D. (2008). Death, dropout, and longitudinal measurements of cognitive change in old age. *Journals of Gerontology. Series B, Psychological Sciences and Social Sciences, 63,* 271–278.

Rabbitt, P., Lunn, M., Pendleton, N., & Yardafagar, G. (2011). Terminal pathologies affect rates of decline to different extents and age accelerates the effects of terminal pathology on cognitive decline. *Journals of Gerontology. Series B, Psychological Sciences and Social Sciences, 66,* 325–334.

Ram, N., Rabbitt, P., Stollery, B., & Nesselroade, J. R. (2005). Cognitive performance inconsistency: Intraindividual change and variability. *Psychology and Aging, 20,* 981–1003.

Rathouz, P. J., Van Hulle, C. A., Rodgers, J. L., Waldman, I. D., & Lahey, B. B. (2008). Specification, testing, and interpretation of gene-by-measured-environment interaction models in the presence of gene-environment correlation. *Behavior Genetics, 38,* 301–315.

Reinvang, I., Deary, I. J., Fjell, A. M., Steen, V. M., Espeseth, T., & Parasuraman, R. (2010). Neurogenetic effects on cognition in aging brains: a window of opportunity for intervention? *Frontiers in Aging Neuroscience, 2,* 143.

Reynolds, C. A., Finkel, D., McArdle, J. J., Gatz, M., Berg, S., & Pedersen, N. L. (2005). Quantitative genetic analysis of latent growth curve models of cognitive abilities in adulthood. *Developmental Psychology, 41*(1), 3–16.

Riegel, K. F., & Riegel, R. M. (1972). Development, drop, and death. *Developmental Psychology, 6,* 306–319.

Roses, A. D. (2010). An inherited variable poly-T repeat genotype in TOMM40 in Alzheimer disease. *Archives of Neurology, 67,* 536–541.

Salthouse, T. A. (1996). The processing speed theory of adult age differences in cognition. *Psychological Review, 103,* 403–428.

Salthouse, T. A. (2004). Localizing age-related individual differences in a hierarchical structure. *Intelligence, 32,* 541–561.

Salthouse, T. A. (2006). Mental exercise and mental aging evaluating the validity of the "use it or lose it" hypothesis. *Perspectives on Psychological Science, 1*(1), 68–87.

Schiepers, O. J. G., Harris, S. E., Gow, A. J., Pattie, A., Brett, C. E., Starr, J. M., & Deary, I. J. (2012). APOE E4 status predicts age-related cognitive decline in the ninth decade: longitudinal follow-up of the Lothian Birth Cohort 1921. *Molecular Psychiatry, 17,* 315–324.

Schooler, C., & Mulatu, M. S. (2001). The reciprocal effects of leisure time activities and intellectual functioning in older people: A longitudinal analysis. *Psychology and Aging, 16*(3), 466–482.

Seshadri, S., Fitzpatrick, A. L., Ikram, A., DeStephano, A. L., Gudnason, V., Boada, M., et al. (2010). Genome-wide analysis of genetic loci associated with Alzheimer disease. *Journal of the American Medical Association, 303,* 1832–1840.

Singer, T., Lindenberger, U., & Baltes, P. B. (2003). Plasticity of memory for new learning in very old age: A story of major loss? *Psychology and Aging, 18*(2), 306–317.

Sliwinski, M. J., Stawski, R. S., Hall, C. B., Katz, M., Verghese, J., & Lipton, R. (2006). Distinguishing preterminal and terminal cognitive decline. *European Psychologist, 11,* 172–181.

Stuss, D. T., Pogue, J., Buckle, L., & Bondar, J. (1994). Characterization of stability of performance in patients with traumatic brain injury: Variability and consistency on reaction time tests. *Neuropsychology, 8,* 316–324.

Swan, G. E., & Carmelli, D. (2002). Evidence for genetic mediation of executive control: A study of aging male twins. *Journals of Gerontology. Series B, Psychological Sciences and Social Sciences, 57*(2), 133–143.

Swan, G. E., Reed, T., Jack, L. M., Miller, B. L., Markee, T., Wolf, P. A., & DeCarli, C. (1999). Differential genetic influence for components of memory in aging adult twins. *Archives of Neurology, 56,* 1127–1132.

Tuminello, E. R., & Han, S. D. (2011). The apolipoprotein e antagonistic pleiotropy hypothesis: review and recommendations. *International Journal of Alzheimer's Disease, 2011,* 72616.

Turkheimer, E. (2000).Three laws of behavior genetics and what they mean. *Current Directions in Psychological Science, 9*(5), 160–164. doi:10.1111/1467-8721.00084.

Valdes, A. M., Deary, I. J., Aviv, A., Gardner, J., Kimura, M., Lu, X., Spector, T. D., & Cherkas, L. F. (2010). Leukocyte telomere length is associated with cognitive performance in healthy women. *Neurobiology of Aging, 31,* 986–992.

Verhaeghen, P., & Salthouse, T. A. (1997). Meta-analyses of age-cognition relations in adulthood: Estimates of linear and nonlinear age effects and structural models. *Psychological Bulletin, 122*(3), 231–249.

Vernon, P. A. (1989). The heritability of measures of speed of information processing. *Personality and Individual Differences, 10,* 573–576.

Visscher, P. M., Yang, J., & Goddard, M. E. (2010). A commentary on "Common SNPs explain a large proportion of the heritability for human height" by Yang et al. *Twin Research and Human Genetics, 13,* 517–524.

West, R., Murphy, K. J., Armilio, M. L., Craik, F. I., & Stuss, D. T. (2002). Lapses of intention and performance variability reveal age-related increases in fluctuations of executive control. *Brain and Cognition, 49,* 402–419.

Williams, G. C. (1957). Pleiotropy, natural selection, and the evolution of senescence. *Evolution, 11,* 398–411.

Wilson, R. S., Beckett, L. A., Bienias, J. L., Evans, D. A., & Bennett, D. A. (2003). Terminal decline in cognitive function. *Neurology, 60,* 1782–1787.

Winterer, G., & Weinberger, D. R. (2004). Genes, dopamine, and cortical signal-to-noise ratio in schizophrenia. *Trends in Neuroscience, 27,* 683–690.

Wisdom, N. M., Callahan, J. L., & Hawkins, K. A. (2011). The effects of apolipoprotein E on non-impaired cognitive functioning: a meta-analysis. *Neurobiology of Aging, 32,* 63–74.

Yang, J., Benyamin, B., McEvoy, B. P., Gordon, S., Henders, A. K., Nyholt, D. R., et al. (2010). Common SNPs explain a large proportion of the heritability for human height. *Nature Genetics, 42,* 565–569.

Yeo, R. A., Gangestad, S. W., Liu, J. Y., Calhoun, V. D., & Hutchison, K. E. (2011). Rare copy number deletions predict individual variation in intelligence. *PLoS One, 6,* e16339. doi:10.137/jounal.pone.0016339.

Zhang, J.-P., Burdick, K. E., Lencz, T., & Malhotra, A. K. (2010). Meta-analysis of genetic variation in DTNBP1 and general cognitive ability. *Biological Psychiatry, 68,* 1126–1133.

Chapter 6
Gene by Environment Interplay in Cognitive Aging

Chandra A. Reynolds, Deborah Finkel and Catalina Zavala

Successful aging is defined, in part, by maintenance of cognitive functioning (Rowe and Kahn 1997) and researchers and entrepreneurs alike are eager to uncover the secrets to slowing or delaying cognitive aging (Buetnner 2008; Fozard et al. 2000). Changes in the relative contribution of environmental factors to cognitive functioning over the course of adulthood suggest that revisiting the concept of Gene–Environment (GE) interplay in middle and late adulthood may increase our understanding of the processes of cognitive aging and provide fertile ground for the development of intervention strategies. Shared environmental influences have a significant impact on individual differences in intelligence in childhood; however, the proportion of variance explained by shared environment drops to negligible levels as early as young adulthood (Plomin et al. 2008). Phenotypic and biometrical studies of cognitive aging provided some early hints that GE interplay may be important to normative cognitive aging. First, variance in cognitive performance tends to increase over the life course for memory, speed of processing, and other fluid abilities but less so for crystallized abilities (Christensen 2001; Morse 1993). Second, twin and adoption studies have both indicated that although heritability increases from childhood into adulthood, increasing from approximately 40 % to a peak of 80 % for general cognitive ability in late adulthood, this increase is followed by a downturn in old-old age to 40–60 % heritability (Finkel and Reynolds 2009; Reynolds 2008a; Reynolds 2008b). These patterns initially result from increasing genetic variance, until about age 65, but are subsequently explained by increasing *nonshared*

C. A. Reynolds (✉)
Department of Psychology, University of California-Riverside, Riverside, CA 92521, USA
e-mail: chandra.reynolds@ucr.edu

D. Finkel
School of Social Sciences, Indiana University,
4201 Grant Line Road, New Albany, IN 47150, USA
e-mail: dfinkel@ius.edu

C. Zavala
Department of Psychology, University of California-Riverside, Riverside, CA 92521, USA
e-mail: catalina.zavala@email.ucr.edu

D. Finkel, C. A. Reynolds (eds.), *Behavior Genetics of Cognition Across the Lifespan*,
Advances in Behavior Genetics 1, DOI 10.1007/978-1-4614-7447-0_6,

environmental variance (Johansson et al. 2004; McGue and Christensen 2002; Reynolds et al. 2005). Some exceptions to this pattern exist: working and episodic memory traits display increases in both genetic and environmental variance (Reynolds 2008a; Reynolds et al. 2005).

Increasing nonshared environmental variance has important implications for investigations of gene by environment interplay. Indeed, if interactions between genes and nonshared environment (denoted G × E) exist but are not formally accounted for in analyses, as is typically the case, G × E effects become part of the nonshared environmental variance estimates (Falconer 1989). Patterns of increasing nonshared environmental variance suggest, therefore, emergent GE interactions. In addition to G × E interactions, correlative associations may arise among genes and environments (GE correlations), which if not accounted for become part of the genetic variance term in biometrical models (Falconer 1989).

In this chapter, we reinterpret existing theoretical models of GE interplay using a lifespan perspective, focusing on the changing nature of the environments that impact cognitive function throughout adulthood. We then evaluate the models in relationship to existing evidence for GE interplay in cognitive aging, including investigations that tap recent advances in genotyping and gene expression that allow researchers to examine GE interplay at molecular levels.

6.1 Models of Gene–Environment Interplay in Cognitive Aging

Both behavioral genetic and lifespan perspectives provide theoretical models of the interplay of genes and environments applicable to cognitive function in adulthood, both in terms of GE correlation and G × E interaction.

6.1.1 Gene–Environment Correlation

Scarr and McCartney (1983) placed three forms of GE correlation into an early-life developmental context (infancy through adolescence): passive, evocative, and active. First, *passive* GE correlation occurs because in nuclear families children receive both their genes and their rearing environment from the same source: their parents. Because of the limited impact of rearing environment on measures of cognitive function in adulthood (Pedersen et al. 1992), it is logical to conclude that passive GE correlation will play a minimal role in adult development and aging (see Fig. 6.1). One possible exception is education, which can play a large role in cognitive functioning in late life, particularly with regard to cognitive reserve hypotheses for dementia (Glymour et al. 2012; Scarmeas and Stern 2003; see Chap. 7). Factors promoting educational achievement are complex but are likely to involve parental education as both an environmental and genetic source of variance. Second, environments and other people react to our (genetically influenced) traits and behaviors,

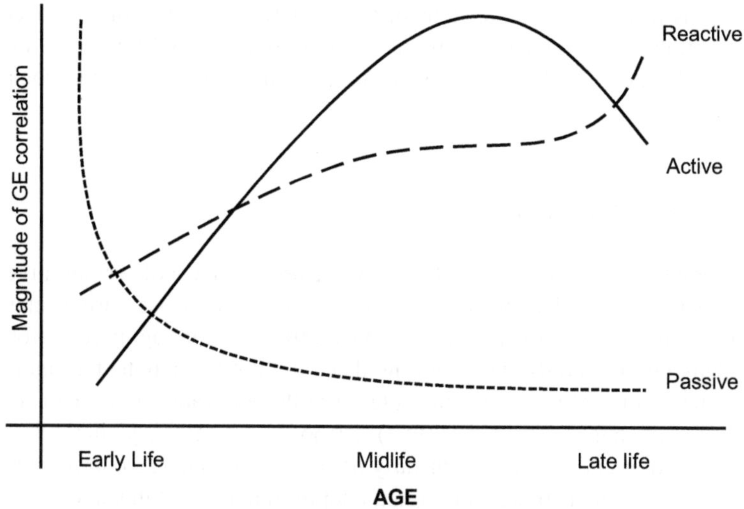

Fig. 6.1 Lifespan model of Gene–Environment (GE) correlation

creating *evocative* or reactive GE correlations. This process can only continue and perhaps even intensify as we move through adulthood experiencing changing societal expectations for cognitive function. Whereas society may expect, and thus promote, high levels of cognitive function in midlife, powerful stereotypes about cognitive decline in late adulthood may produce "social facilitation of the nonuse of competence" (Bieman-Copland et al. 1998). Because of a physical appearance of aging or frailty, older adults may evoke assumptions by others around them of cognitive frailty that inhibit attempts to maintain cognitive function. The competence–environmental press model emphasizes that functioning is maximized when the demands of the environment are sufficiently tailored to the individual's ability to promote stimulation and maintenance of competence, and even growth (Lawton and Nahemow 1973).

Third, evidence for increasing genetic variance in late adulthood and the acceleration of nonshared environmental variance suggests that the most powerful form of GE interplay in cognitive aging are likely to be *active* GE correlations. Our choices shape our environment and that environment in turn shapes us; moreover, our choices are—at least to some extent—influenced by our genetic make-up. The environments we choose for ourselves are, by definition, unique to each of us and thus act as sources of nonshared environmental influence on cognitive function. For example, we choose our occupations and our working environments in turn impact our cognitive functioning (Andel et al. 2005; Finkel et al. 2009; Schooler and Caplan 2008). In the old-old age period, however, we predict that evocative GE correlational effects surpass active GE correlation. Increasing frailty and reductions in function are necessarily associated with reduced control over one's environment (Rodin 1986, 1989) resulting in decreasing opportunities for active GE correlation.

Similarly, as frailty and visible signs of aging increase, the response evoked from the environment will intensify, resulting in not only decreased functional independence but also decreased expectations of functioning and narrowing of social contexts.

6.1.2 *G × E Interaction*

G × E interaction processes are another set of factors that may bolster cognitive maintenance or precipitate declines in later adulthood. As described earlier, increasing nonshared environmental variance, observed uniformly for cognitive performance across domains (Reynolds 2008a; Reynolds et al. 2005), is a potential indicator of G × E interactions. Given the extant literature three decades ago, Scarr and McCartney (Scarr and McCartney 1983) proposed a relatively limited role of G × E interaction in development in early life (as opposed to GE correlation), arguing that environmental "treatments" (e.g., adoption) that affect mean level (IQ) performance affect most individuals in like direction rather than altering individuals' rank ordering. However, we argue for an updated view of the saliency of G × E interaction based on: (1) the qualitatively different impacts of stress during particular developmental periods (Lupien et al. 2009), in particular the "brain maintenance" phase in late adulthood; (2) familiality of epigenetic alterations to gene expression (thus impacting individual differences and potentially rank-order; Boks et al. 2009; Coolen et al. 2011); (3) discordance of monozygotic (MZ) twins in cognitive decline and dementia (e.g., MZ differences in memory trajectories associated with differences in depressive symptoms and moderated by *APOE* genotype; Reynolds et al. 2007); and (4) growing epidemiological literatures on the *APOE* gene and measured risk factors (Gatz 2007; Reynolds 2008a, b; see Fig. 6.2).

6.1.3 *Epigenetic Landscape*

Lifespan models of GE interplay provide additional means for conceptualizing the environments that impact cognitive aging. For example, we can apply Waddington's epigenetic landscape (Waddington 1942) to cognitive functioning throughout the lifespan. Waddington emphasized that developmental pathways are shaped by evolution and thus are fairly robust to minor variations in environmental conditions. In contrast, Gottlieb (1991) stressed the influence of environmental variations on the genetic programming. Combining aspects of both genetic canalization (Waddington 1942) and experiential canalization (Gottlieb 1991), development occurs as genetically influenced pathways are impacted by environmental forces, giving rise to individual phenotypes. Although most research on cognition focuses on canalization processes that occur during childhood (e.g., see Chap. 2), there is no reason to assume that the process does not continue throughout the lifespan. Environmental forces (evoked, self-selected, or random) continue to impact cognitive functioning,

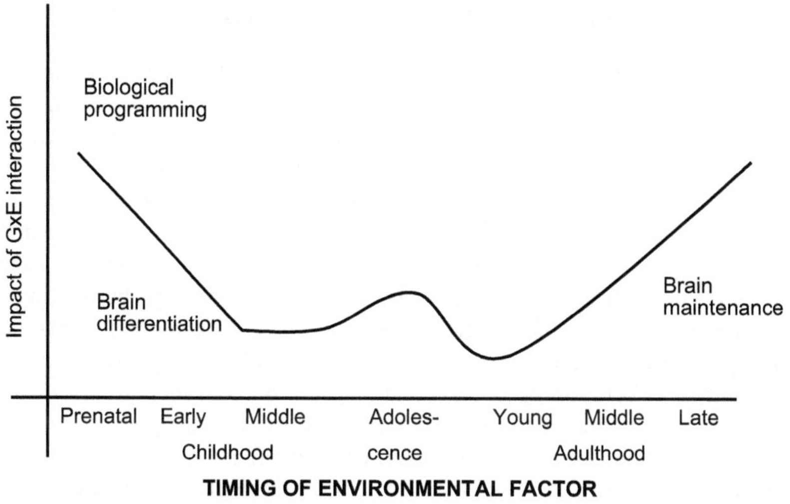

Fig. 6.2. Lifespan model of G × E interaction

pushing the individual into particular modes of functioning. Active and even reactive GE correlations can be seen in the context of the epigenetic landscape as *continued* canalization of cognitive function.

Genetic canalization may imply inflexibility and unmodifiability: once you are headed down a certain canal or path, there is no turning back. Genetic forces have impacted your phenotype and your behavioral options are thus limited. In contrast, with experiential canalization Gottlieb (1991) promoted the concepts of malleability and flexibility. Clearly, these ideas are appealing to researchers looking for interventions to slow or delay cognitive aging. Although Salthouse (2006) has questioned whether there is sufficient support for the disuse theory of cognitive aging, evidence for increases in environmental variance with aging provides hope that experiences may moderate genetic predispositions for cognitive functioning. The SOC model may provide an individual-centered adaptive framework for *selecting* appropriate goals for cognitive functioning, *optimizing* the allocation of internal and external resources, and *compensating* with additional (environmental) resources to counteract loss and decline (Baltes et al. 2006).

6.1.4 Characterizing the Environment

Clearly, a vital step in the application of any of these models is to determine the aspects of the environment that may have the most impact on the genetically influenced cognitive decline described in Chap. 5.

6.1.4.1 Compensating and Enhancing Environments

Balte's SOC model implies an individual-centric view of active compensation for aging-related change and declines. On the whole, whether by individual agency or otherwise, enriched environments may compensate for genetic vulnerabilities, such as the "social context as compensation" model (Shanahan and Hofer 2005). Empirical evidence of environmental compensation has been supported by mouse studies in which enriched environments have mitigated cognitive deficits (Markham and Greenough 2004) such as those due to gene knockouts (*NMDAR1*; Rampon et al. 2000) or to dietary deficiencies (Lee et al. 2012), perhaps by altering gene expression related to metabolic processes, e.g., *GLUT1* expression in cortex and CA1 region of the hippocampus (Harbeby et al. 2012). In humans, evidence for compensation effects can be observed in research on educational attainment and complex social environments on cognitive reserve. For example, social engagement in late adulthood may help to support cognitive functioning in late life despite the increasing presence of age-associated neural pathologies (Bennett et al. 2006), and higher educational attainment is associated with better cognitive performance and to a lesser extent smaller rates of decline (Glymour et al. 2012) and a lower risk of dementia (Ferrari et al. 2012). Moreover, it has been recently observed that high education and participation in leisure activities may lower the risk of dementia otherwise associated with carrying the *APOE* e4 allele (Ferrari et al. 2012), by serving to delay the onset of symptoms.

Enhancing environments refer to social contexts that interact with genes to promote higher levels of functioning (Shanahan and Hofer 2005). Evidence in support of this concept includes larger heritability estimates for educational attainment across recent generations (perhaps due to more open school access) and higher heritability of cognitive abilities with higher levels of socioeconomic status (SES) (cf., Shanahan and Hofer 2005; Chap. 2 in this volume). As we have noted earlier, heritability for general cognitive abilities increases with age to late adulthood, followed by significant downturns. Whether this pattern is a function of enhancing environments, i.e., reflecting G × Social Contexts (Shanahan and Hofer 2005), or active GE correlational processes (i.e., niche picking) remains to be established. We note, however, that with respect to lifespan development and aging, it is important to go beyond proportions of variance such as that conveyed by heritability statistics and consider "raw" genetic and environmental variances for a more accurate picture of changing genetic and environmental variance (Reynolds 2008a; Reynolds et al. 2005).

6.1.4.2 Benign Versus Adverse Environments

Both genetic factors and environmental influences may impact susceptibility to cognitive decline with aging. Investigations of GE interplay must be sensitive to the possibility of both benign and adverse effects (e.g., Boardman et al. 2012). For example, a genetic risk factor for cognitive decline (*APOE* e4) may interact with an environmental factor to exacerbate (head injury) or delay (education) changes in functioning (McArdle and Prescott 2010; Dardiotis et al. 2012). The "social trigger" model proposes that an environmental risk factor (lower education) may be required to elicit the impact of a genetic risk factor like *APOE* e4 (Reiss and Leve 2007; Shanahan and Hofer 2005). Alternatively, the "social push" model argues that disadvantageous environments may crowd out genetic effects on traits without necessarily playing a causal, or triggering, role (Raine 2002). Thus, according to the social push model, the impact of genetic influences on cognitive decline will only become apparent when disadvantageous social conditions (e.g., low education) are *minimized.* In contrast, the "social trigger" model posits that genetic influences on cognitive decline will be greater when unfavorable social conditions are maximized (Reiss and Leve 2007; Shanahan and Hofer 2005). It is important to note that the social push model was originally developed for antisocial behavior and related traits and the impact on discussions of cognitive aging has only begun to be explored (Boardman et al. 2012).

6.1.4.3 Perceived Versus Objective

Is perception reality? It is likely that individual perception moderates the impact of environmental factors. Psychologists attempt to collect objective measurements of environmental factors like SES and availability of social contact, but often the measures used are obtained via self-report; for example, recent evidence suggests that self-reported social participation may drive subsequent changes in perceptual speed (Lovden 2005). However, it is possible that individual perceptions of SES and satisfaction with social engagement color self-reports of ostensibly objective measures, but are equally important in their own rights. Indeed, perceived or subjective environments have been notably referred to as "effective" environments (Rutter 2012). Studies comparing the predictive value of subjective and objective SES for various health outcomes in older adults routinely report that subjective SES is the better predictor (e.g., Singh-Manoux et al. 2005); the same may be true for cognitive decline.

In particular, SES lends itself to the social comparisons inherent in subjective perceptions. As a social species, people are sensitive to the social interactions and disparities that occur within their various social settings. The work place is one example of a social setting in which individuals are organized in a ranked system from dominant to subordinate individuals. Yet, social interactions that highlight resource differences between individuals are prevalent not only in the workplace. At the societal level, social stratification of resources has been associated with a gradient in level of health, with average individual health improving at each level on the social ladder (Adler et al. 1994). This social stratification is not unique to our own species. Primate models suggest well-being may be impacted via hierarchical social systems within primate social groups even when environmental resources are held equal (Sapolsky 2005), an effect that is likely mirrored in our own species (Boyce 2007; Sapolsky 2004). Understanding how these social interactions may impact the development of subjective or "effective" environments should contribute to our understanding of the SES impact on cognitive aging. Social comparisons add a subjective facet to SES, contributing to an individual's perceptions of their own economic situation. These perceptions, although largely influenced by familial and nonfamilial environmental factors, likely reflect the impact of genetic factors as well (Lichtenstein et al. 1992). Individuals with fewer resources but equally impoverished peers may feel more financial satisfaction than individuals with relatively more resources but surrounded by wealthier peers (Liang and Fairchild 1979). Clearly, the distinction between objective and subjective measures can extend to many of the environmental factors commonly believed to impact cognitive decline: life events, health events, etc.

6.1.4.4 Proximal Versus Distal

A primary issue in uncovering GE interplay, whether environments are enhancing, compensating, adverse, perceived or otherwise, is one of timing: do proximal or distal environmental factors have a larger impact on cognitive aging? A strict application of Waddington's epigenetic model, for example, would hypothesize that environmental selection was more salient during early development than during aging; thus the theory would likely nominate distal (early) environmental factors for dominant roles in GE interplay in cognitive aging. Recent longitudinal analyses support this hypothesis, reporting that IQ at age 11 was the strongest predictor of IQ at age 79, over and above concurrent SES and education (Gow et al. 2011). Other research, however, indicates that late-life functioning has many unique aspects and may not relate to variables that predicted functioning even in midlife (e.g., Vaillant 2002). Thus, proximal (or concurrent) environmental factors may play large roles in cognitive aging and may underlie the reported increases in nonshared environmental variance. In mouse models, lifelong enriched environments benefited learning and memory processes across development, whereas enriched exposures appeared to improve performance when introduced in young and middle age, but not when introduced only in late life (Harati et al. 2012), suggesting the saliency of early life as well as later developmental periods.

6.2 Evidence for Gene–Environment Interplay on Cognitive Aging

Many methods exist for investigating these potential forms of GE interplay in cognitive aging, including analysis of variance components, experimental and epidemiological approaches, and methods that focus on SES and other social and potentially stressful environments. Our review of the literature indicates that some methods remain relatively untapped; thus there is significant potential for future developments.

6.2.1 Biometrical Approaches to G × E

Variance components approaches consider measured environmental factors as they moderate genetic variance for cognitive aging phenotypes. Typically, such models have been applied to twin data, evaluating information from both MZ and dizygotic (DZ) pairs in an extension of classic biometrical models (Purcell 2002). As described in Chap. 2, higher heritability estimates for childhood and adolescent cognitive ability have been observed as social status and prosperity levels increased suggesting the presence of G × E interaction (Harden et al. 2007). However, these findings have not been fully replicated, and the heterogeneity in findings may be due in part to a lack of consideration of the overlapping variance among SES and cognitive traits (cf., Johnson et al. 2009). Indeed, there is some suggestion that social status may moderate *environmental* not genetic variance (Hanscombe et al. 2012). Moreover, there have been inconsistent findings in adult twin samples, for example, as to whether parental education modifies heritability of cognitive performance and during what developmental period (i.e., early vs. middle adulthood Grant et al. 2010; Kremen et al. 2005). Applications of the variance components approach to cognitive aging suggest that education moderates both genetic and shared environmental influences in late adulthood (Johnson, et al. 2009). To date, no studies have applied this approach to the longitudinal case. Power of the G × E variance components approach has been evaluated with suggestions that minimally 5,000 twin pairs are required to detect moderation of genetic variance (Hanscombe et al. 2012), which may be a factor in the inconsistent findings. Recent methodological work suggests that false positives are a potential problem with typical G × E models and suggests expansions of the approach to include full appreciation of the genetic and environmental covariance structures of both the putative environmental moderator and phenotype of interest (van der Sluis et al. 2012). The G × E variance components model has been extended to consider gene candidates in an association context (van der Sluis et al. 2008) as well as for the case of ordinal and binary traits (Medland et al. 2009).

A second method for evaluating G × E interaction focuses on MZ twin similarities vs. differences (Fisher 1925; Jinks and Fulker 1970; Martin et al. 1983; van der Sluis, Dolan et al. 2006). Because MZ pairs share identical genotypes, any within-pair differences are attributed to environmental factors unique to the individual pair members, i.e., the nonshared environment. If associations exist between the shared genetic factors and the environmentally driven differences, then it is taken as support for G × E. In terms of cognitive aging, one may use such methods to compare differences in trajectories between MZ pairs and relate these differences to measured gene candidates and environmental exposures. The MZ-only approach inspired by Fisher (1925) considers first whether heterogeneity exists (Fisher 1925; Martin et al. 1983), i.e., whether there are mixtures of within-pair or intrapair difference distributions. If so, presence of gene candidates may then be measured in combination with environmental factors to consider whether they could contribute to the heterogeneity in MZ pair differences (Martin et al. 1983; Reynolds et al. 2007). To avoid false-positive tests of G × E, it is critical that the outcome traits are normally distributed (Jinks and Fulker 1970). A G × E analysis of MZ twins from the Swedish Adoption Twin Study of Aging (SATSA) suggested the presence of GE interactions for cognitive tasks that are particularly dependent upon semantic or episodic memory. Specifically, variations in genes regulating aspects of serotonin (*5HTT*, *HTR2A*), estrogen (*ESR1a*), and cholesterol (*APOE* e4) interacted with the exposure by those individuals to unique environmental factors that predicted differential semantic or episodic memory change (Reynolds et al. 2007). First, we evaluated and observed significant heterogeneity in within-pair differences, i.e., intrapair variability in memory trajectory features, including performance level at age 65, linear change at age 65, and nonlinear change across age. Moreover, the intrapair variability in memory trajectories differed by genotype whereby those MZ pairs who did not carry the risk alleles showed greater differences in semantic and episodic memory change than those who did carry risk alleles (e.g., *APOE* e4). Last, the intrapair variability of depression was shown to be associated with the intrapair variability of longitudinal memory change, however, only for *noncarriers* of either the *APOE* e4 allele or *ESR1a* rare allele (rs1801132). This result indicates that noncarriers of these risk alleles for dementia may have greater sensitivity to environmental sequelae that result from depressive symptoms and thereby show differential memory trajectories; however, carriers of risk alleles, who otherwise have an elevated risk of decline (especially vis-à-vis *APOE* e4), may be less impacted by environmental challenges posed by depression. The findings for *APOE* are consistent with work on cognitive health and dementia that suggest that non-e4 carriers may be relatively more sensitive to a variety of environmental factors than *APOE* e4 carriers, while *APOE* e4 carriers may be more sensitive to vascular risk factors (see Gatz 2007). Indeed, related cognitive aging findings also support this conclusion, as female *APOE* e4 homozygotes who were more aerobically fit showed better cognitive performance (Etnier et al. 2007). Subsequent studies of physical activity (see later), suggest a range of findings of enhanced effects, compensation effects or no appreciable moderation of physical activity and *APOE* e4 status on cognitive performance and brain phenotypes (Erickson et al. 2012) indicating further work is necessary to elucidate when and what type of GE interplay is at work.

6.2.2 Experimental and Epidemiological Approaches

Genetic–epidemiological approaches examining candidate gene variants and environmental exposures in unrelated individuals have also been used to identify G × E associations. For example, nondemented *APOE* e4 carrying adults ages 16 and 65 years who had sustained head injuries performed worse on verbal memory and attention and perceptual speed tasks 6 months postinjury (Ariza et al. 2006). Untreated hypertension in the presence of positive *APOE* e4 status was associated with poorer cognitive performance in nondemented women from the Nurses' Health Study (Kang et al. 2005).

Individuals in deprived neighborhoods are often exposed to more toxins, lower quality housing, and violence; these conditions subject individuals to a higher allostatic load (Evans 2004). Indeed, a study of neighborhood effects and *APOE* e4 status (Lee et al. 2011) indicated that living in a neighborhood rife with "psychosocial hazards" (e.g., higher crime rates, economic deprivation, familial disruptions, lower educational attainment, poorer infrastructure upkeep, etc.) coupled with carrying the *APOE* e4 allele predicted worse processing speed and executive performance in adults aged 50–70 years old enrolled in the Baltimore Memory Study. Memory abilities did not show a clear neighborhood environmental effect, although patterns of performance were suggestive of the expected *APOE* e4 effect. However, recent longitudinal evidence from the Chicago Health and Aging Project supports an interaction, suggesting that *APOE* e4 coupled with lower 'neighborhood social disorder' predicts change in general cognitive functioning (Boardman et al, 2012). Hence, more work is needed.

Physical activity may also interact with *APOE* e4 allele status to predict cognitive functioning (Erickson et al. 2012). As introduced earlier, a study of female *APOE* e4 homozygotes who performed better on an in-person aerobic fitness test demonstrated higher cognitive performance on a variety of cognitive measures including learning and attention tasks (Etnier et al. 2007). In a recent population-based study of 1,799 participants aged 60 years or older in the NHANES III study with available *APOE* genotyping (Obisesan et al. 2012), increased self-reported physical activity predicted better cognitive status performance in non-e4 carrying individuals but not e4 carrying individuals between the ages of 60 and 69, with adjustments for illnesses burden and mobility restrictions. However, in those older than 70 years, physical activity benefitted all individuals, including those who carried an e4 allele although the effect dropped when accounting for mobility restrictions (Obisesan et al. 2012). The physical exercise by *APOE* e4 genotype interaction was supported by recent brain imaging work suggesting that self-reported physical exercise engagement was associated with amyloid plaque deposition in a sample of adults between 45 and 88 years (Head et al. 2012). Sedentariness was most detrimental in terms of increased amyloid plaque deposition in those who were *APOE* e4 carriers (Head et al. 2012). Although the NHANES III study is cross-sectional, the findings of differential impact of physical activity on cognitive functioning by age suggest an age dependency of GE interplay effects with respect to *APOE*, perhaps due in part to selectivity resulting from morbidity or mortality. Whether physical activity is particularly beneficial to or merely mitigates risk for *APOE* e4 carriers is not yet clear (Erickson et al. 2012), but it is likely that a developmental framework taking into account age dependencies is important.

6.2.3 Socioeconomic Status and Cognitive Aging

In cognitive testing and measures of IQ, researchers have observed differences in cognitive abilities across levels of SES. Often, privileged individuals perform better on cognitive tasks compared to individuals of low SES. Initial explanations of these observed differences attempted to disentangle the impacts of genetic and environmental influences (to some controversy). One might be tempted to conclude that the apparent relationships between perceived SES and health and cognitive aging are generally due to environmental pathways. The etiological factors underlying cognitive performance, educational attainment, and cognitive dysfunction are indicative of the expected complexity. Although environmental pathways may largely influence the association between education and dementia risk (Gatz et al. 2007), one study indicated that the genetic factors that do underlie education and mental status performance among typically aging adults overlap completely with the genetic factors attributed to general cognitive ability (Pedersen et al. 1996). Moreover, Schooler and colleagues propose a person–environment pathway whereby individual difference traits (e.g., self-directedness) and occupational features contribute to later intellectual functioning (Schooler and Caplan 2008). Specifically, higher SES coupled with cognitive ability leads to more demanding and self-directing occupational contexts (Schooler and Caplan 2008), an example of active GE correlation. Such contexts thereby boost cognitive functioning, amplify early-life SES effects and mutually benefit self-directedness and intellectual flexibility. Based on a series of analyses of two-wave data collected 20 years apart, those with occupations that were high on substantive complexity, more self-directed, and less routine predicted positive reciprocal relationships with intellectual flexibility, i.e., mutually increased flexibility and self-directedness, accounting for baseline levels, respectively. Similarly, the complexity of household work may similarly impact intellectual flexibility (Caplan and Schooler 2006; Schooler and Caplan 2008). While these results are in no way definitive of GE processes, the findings provide candidate life course pathways to evaluate from a genetically informative perspective.

Studies of aging in rodent models underscore the positive effect of environmental complexity on dendritic growth, and these benefits seem to obtain throughout the lifespan of the aging rat (Greenough et al. 1986; Markham and Greenough 2004; Mohammed et al. 1993). Animal research continues to show evidence that the surrounding environment can alter the expression of genes in neurons (Harbeby et al. 2012; Mohammed, et al. 1993; Pinaud et al. 2002). With respect to cognitive phenotypes, lifelong enriched environments support maintenance of learning and memory processes, and introducing enriched exposures in young and middle age appear to be restorative though perhaps not in late life (Harati et al. 2012), suggesting that environmental interventions may have a more limited impact as plasticity wanes. While SES as a developmental context is much more complex than the experimental environments of lab animals, the overall implications of the epigenetic forces at play in cognitive development evident from this body of work should not be overlooked.

SES has become an important contextual marker in measuring environmental experiences as a proxy for exposure to toxins, nutrition, education, and leisure activities (Evans 2004). Much of the research in brain functioning and late-life cognition has focused on incidence of dementia (discussed in Chap. 6). Individuals from low SES are at higher risk for developing dementia. One of the theories posited for the relationship between SES, cognitive functioning, and dementia is the hypothesis of cognitive reserve (Staff et al. 2004), with evidence from Swedish studies supporting increased reserve largely via education and occupational complexity (Andel et al. 2005; Andel et al. 2007; Andel et al. 2006). Controlling for education, the complexity of one's occupation prior to retirement, particularly with respect to working with people (e.g., mentoring roles vs. subordinate roles), supported the relative maintenance of cognitive performance for verbal and spatial skills whereas in postretirement, spatial performance dropped (Finkel et al. 2009). In Sweden, age 65 is a mandatory retirement age and it is uncommon to work formally past this point, which constrains the extent to which GE correlational processes might otherwise play out with continued working. Taken together, empirical findings support the work context as a measureable environmental influence on cognitive aging. As a second example, we note findings that individuals with higher levels of educational attainment may still perform at preclinical levels on the Mini Mental Status Exam even when comparable amounts of brain atrophy are otherwise indicative of Alzheimer's disease in lower SES individuals (Fotenos et al. 2008). Most interesting are those individuals who remain undiagnosed as demented at time of death. The question stands: What aspects of education and higher SES have afforded these individuals protective cognitive resources that allowed them to function with otherwise biologically compromised brain structures? Additionally, what aspects of genetic endowment have contributed to healthy cognitive aging? To begin answering these questions, we look to current research in brain imaging for a preliminary conjecture until researchers further address these questions in older populations.

SES as a contextual marker of differing environmental conditions underscores the sensitivity of human cognitive aging to variations in environmental conditions. Moreover, *perception* of SES can augment the impact of objective SES, per se. For example, in a sample drawn from three longitudinal studies of aging in the Swedish Twin Registry, an individual's perceived SES was predicted of cognitive performance at age 75 for perceptual speed, spatial performance, verbal memory, and episodic memory (Zavala et al. 2013, in preparation). This was particularly true for the oldest cohort, perhaps suggesting the impact of early environments on perceptions of later environments. Overall, individual differences in cognitive performance and decline within and across SES environments highlight the fundamental biological nature of this sensitivity evident in individual brain structure and function. To gain a greater understanding of the mechanisms involved in cognitive aging across the SES spectrum, research in epigenetics may provide clues to possible GE interplay occurring within the human brain, especially with concern to individual differences in plasticity and susceptibility to environmental influences in neuronal gene expression as described later.

6.2.4 *Social/Stressful Environments*

We have given primacy to SES and related indices as observable, albeit global, markers of environmental contexts or exposures that may interact with genotype to lead to poorer or better cognitive aging. Physiological and psychological stress may be greater in lower SES contexts (Matthews and Gallo 2011). Recent work on gene expression and childhood SES (Miller et al. 2009) suggest that being raised in disadventageous childhood SES contexts may lead to differential gene programming that potentiates aging-associated dysfunction and disease. Specifically, findings suggested that adversity predicts elevated gene expression in the proinflammatory-immune pathways (CREB/ATF) and reduced expression of glucocorticoid receptor response elements (NF-κ), leading to greater production of stress markers such as cortisol and interleukin 6 (IL-6), a cytokine that is elevated at sites within the body given the presence of acute or chronic inflammation. There was also elevated expression of genes coding for inflammatory mediators (other cytokines or enzymes) such as *IL1A*, *CCL2*, *CXCL2*, *CCL20*, as well as such as *OLR1* and *GPR132*, which initiate inflammation processes such as macrophage scavenging of oxidized low-density lipoproteins that may lead to accumulations of atherosclerotic plaques and risk of myocardial infarction. The altered gene expression patterns due to exposure to early adversity are presumed to be initiated before adulthood, as controlling for current SES, lifestyle habits, and perceived stress did not alter the described findings (Miller et al. 2009). Inflammatory biomarkers, such as IL-6, as well as CRP, TNF1A, and ICAM-1, have been linked to cardiovascular disease risk and dementia (Dziedzic 2006), as well as normative cognitive aging performance and decline (Gimeno et al. 2008; Jordanova et al. 2007; Krabbe et al. 2009; Mooijaart et al. 2011; Rafnsson et al. 2007; Schram et al. 2007; van den Kommer et al. 2010). Some studies suggest that the presence of cardiovascular disease may be a moderator of the association between inflammatory biomarkers and cognitive decline (Hoth et al. 2008). Last, mouse models directly support the connection between early social adversity and age-associated impairments in spatial memory, which were associated with alterations in hippocampal *BDNF* expression and synaptophysin immunoreactivity, suggesting both structural and plasticity-related sequlae of the exposure to chronic social stress (Sterlemann et al. 2010).

A growing body of research in both human and animal literatures supports altered brain structures (particularly the hippocampus) and altered cognitive performance as a consequence of early adversity writ large, not only in terms of socioeconomic adversities but also in other forms of early adversity including childhood maltreatement, combat exposure, and other stress exposures (for review, see Pechtel and Pizzagalli 2011). Last, evidence suggests that perceived (but not objective) social isolation increases gene expression of an array of genes involved in inflammatory processes (Cole 2009). Thus, the perceptions of environmental adversity may be just as salient in some cases, or even more so, than objective adversities (which may indeed become "effective" environments; cf., Rutter 2012).

In sum, the emerging evidence on early adversity would suggest that adverse life experience, objective or perceived, leads to differential gene expression and downstream effects on brain structure and plasticity that may eventually show notable impacts on cognitive performance across the life course and differential impacts on cognitive decline in late life. However, the current findings on early adversity, differential gene expression, and adult outcomes are relatively slim as yet, let alone the findings for cognitive outcomes. Much work is needed from a prospective life course perspective to fully evaluate the direction of effects and extent of impact on adult cognitive performance and aging before strong conclusions can be reached.

6.3 Biomarkers of Gene–Environment Interplay

6.3.1 *Brain Morphology*

As described in Sect. 6.1.3, Gottlieb's theory of experiential canalization highlights the interaction of biological systems with the surrounding environment (Gottlieb 1991). Variation in brain structure and function among identical twins has been found, suggesting that structure and function are at least partly experience-dependent, and possibly reflective of GE interaction (Thompson et al. 2001). In an adult twin study, average age 48.2 (SD=3.4 years), 10 pairs of MZ twins (both male and female pairs) had higher similarities in quantity of frontal gray matter than the 10 pairs of DZ twins. Included among the regions examined were cortical language regions, i.e., Broca's and Wernick's area (Thompson et al. 2001). Predictably, individual differences in gray matter were related to individual differences in IQ. A subsequent study of MZ and DZ twins from the VETSA study, average age 55.8 (SD=2.6 years), that was 10-fold larger in sample size, suggested that heritability estimates varied within the frontal cortex, and findings were consistent with respect to a high heritability in Wernicke's area but not Brocas's area (Rimol et al. 2010). Changes in GE interplay influencing the frontal cortex and language areas of the brain across the lifespan would be consistent with heritability changes in cognition. The frontal cortex, in particular, may be subject to changes in heritability across the lifespan due to the protracted developmental timeframe (see Giedd et al. 2010). The extended developmental period typical of the frontal cortex may allow for individuals to influence their own development through active GE correlational processes as individuals seek out environments and experiences most consistent with their general cognitive abilities (such as noncompulsory higher education). The role of the environment as an enhancer of potential has been raised in interpreting findings of GE interaction on child and adolescence achievement and IQ as noted earlier (cf., Shanahan and Hofer 2005; Chap. 2 in this volume).

Evidence for continued plasticity within adult brain structure and function suggests an inherent framework by which dynamic genetic and environmental processes may play out to shape cognitive aging trajectories throughout the adult lifespan. In the brains of adult twins (42 years of age and older), magnetic resonance imaging reveals that heritability is not uniform across nor within brain structures (Chen et al. 2012; Giedd et al. 2010; Pfefferbaum et al. 2004; Sullivan et al. 2001; Thompson et al. 2001). In a twin sample of World War II veterans, brain mapping revealed a heritability of about 15–26%, while other brain structures exhibited evidence of greater genetic influences, including the bilateral temporal horn (38–47%) and the corpus callosum (46–48%; Sullivan et al. 2001). For MZ and DZ twin pairs, though genetic influences remained stable across 4 years follow-up, evidence suggested environmental influences on the lateral ventricles increased with age (Pfefferbaum et al. 2004). In particular, the plasticity of the hippocampus is one of the most well-documented phenomena in the study of brain morphology (for a review, see Neves et al. 2008). For example, changes in individual behaviors, such as exercise, can lead to changes in hippocampal brain volume in aging adults (an effect associated with the *BDNF* gene), leading to increased performance on memory tasks (Erickson et al. 2012). Furthermore, the role that the hippocampus plays in episodic memory (Chadwick et al. 2010) may help to partly explain changes in genetic and environmental variance in memory ability in late life (e.g., Reynolds et al. 2005). For an extended discussion of brain morphology and cognition, see Chap. 8.

6.3.2 Epigenetic Processes

An individual's genotype may provide a guide for the development of biological systems, but recent research supports the concept of probabilistic epigenesis (Gottlieb 2007): a cascade of feedback between genes and the environment that may result in changes in gene expression and cell senescence within the living organism that are not a result of DNA sequence variation or somatic mutations. Advancements particularly in mouse models, but including work on human cognitive disorders, have provided evidence that epigenetic modifications are important to cognition broadly, including learning and memory, and implicated in cognitive disorders such as dementia. Moreover, emerging evidence suggests that DNA biomarkers such as telomere length are associated with cognitive performance and risk of dementia in aging adults.

Epigenetic processes reflect ubiquitous forms of G × E interplay that occur at an environmental-by-molecular level. Epigenetic modifications include acetylation, phosphorylation, and methylation of histone proteins, components of the chromatin, as well as direct methylation of DNA (Day and Sweatt 2010, 2011, 2012). Particular combinations of histone modifications result either in activation or suppression of gene transcription (Day and Sweatt 2011, 2012). Moreover, the persistence of various histone modifications may be of brief duration, while methylation may have a relatively longer time-course (Day and Sweatt 2011). Of particular interest is 5-methylcytosine, i.e., methylation of cytosine-5, occurring at CpG (i.e., CG sequence) rich genomic regions denoted "islands" that occur in or near gene promoter regions. Such methylation has been demonstrated in a variety of human tissues, including brain, muscle, and leukocytes (Fernandez et al. 2012) and it is associated with aging and neurological disease (Boks et al. 2009; Christensen et al. 2009; Fernandez et al. 2012). Moreover, specific methylation patterns may be associated with some forms of neurological disease (e.g., dementia with Lewy bodies), although particular patterns for Alzheimer disease are thus far elusive in one of the largest studies of methylation across tissue types and across 1505 CpG sites (Fernandez et al. 2012).

Recent evidence suggests that histone modifications may be relevant to learning and memory processes as well to disease risk spanning "susceptibility to stress," depression, addictions, and cognitive disorders including Alzheimer's disease and Huntington's disease (Day and Sweatt 2012; Graff and Mansuy 2009), suggesting that such modifications may be important to cognitive aging. Epigenetic dysregulation of the amyloid precursor protein may explain beta-amyloid production or deposition (Maekawa and Watanabe 2007; Scarpa et al. 2006; Wang et al. 2008; Wu et al. 2008), processes implicated in Alzheimer's disease neuropathology. Moreover, epigenetic modifications including methylation have been linked to the formation of memories vis-à-vis alterations of hippocampal gene expression in mouse models (Day and Sweatt 2010), such as *BDNF* (Day and Sweatt 2010; Lubin et al. 2008). While methylation processes in the hippocampus appear to be relatively dynamic, relatively lasting methylation processes underlying remote memory storage in cortical regions have been implicated in the anterior cingulate cortex (Day and Sweatt 2012; Miller et al. 2010).

G × E interplay may be seen in the environmental factors that impact the extent of global DNA methylation. Again, while empirical evidence for cognitive aging outcomes is not yet available, the available findings suggest that such mechanisms could play a role. For example, Fraga et al. (2005) highlighted the increasing differences in DNA methylation profiles for identical twins in a cross-sectional study. The oldest twins with the most divergent self-reported health histories had more divergent acetylation of histones as well as indices of global methylation (Fraga et al. 2005). Moreover, the chromosomal locations of divergent methylation patterns in normal metaphase chromosomes included telomeric regions among twins who differed in global methylation (Fraga et al. 2005). Methylation of 88 gene loci assayed from saliva samples has been shown to be linearly related to chronological age and touted as a biomarker of biological age (Bocklandt et al. 2011). The primary analysis was conducted on 34 male twin pairs and replicated in unrelated male and female individuals. Results highlighted methylation in the promoter regions of the *EDARADD* and *TOM1L1* genes were strongly associated with age in both males and females. This emergent work suggests that environmental factors associated with loci-specific methylation may be important to consider for cognitive aging. Indeed, calorie restriction has been shown to relate to epigenetic processes in the hippocampus in mouse models (Chouliaras et al. 2010b). Additionally, physical exercise is proposed as a promising environmental factor given the numerous studies linking exercise and cognitive performance in older adults and mouse models showing altered expression of genes involved in learning and memory, including *BDNF* (Kaliman et al. 2011).

The extent to which epigenetic processes, particularly in basic learning and memory processes, are indicative of GE interplay for cognitive aging writ large remains to be addressed. While epigenetic processes are separate from DNA sequence variation by definition, genetic influences on methylation are evident from examinations of significant twin concordance for methylation (Coolen et al. 2011); indeed, heritability of DNA methylation patterns may be gene-specific (Boks et al. 2009). Hence, a full understanding of epigenetic mechanisms is not yet within grasp (Feil and Fraga 2012), including the extent to which epigenetic alterations promote or are a consequence of cognitive aging or dementing processes (Chouliaras et al. 2010a); this necessitates longitudinal measurement of DNA methylation or other epigenetic markers as well as cognitive performance.

6.3.3 Telomere Shortening

Telomeres are segments of DNA bases that cap the ends of chromosomes. Telomeres become shorter and shorter over the course of thousands of cell divisions and are associated with cellular senescence (Shawi and Autexier 2008). With the loss of telomere length, risk of somatic mutations and damage during cell division may increase (Aubert and Lansdorp 2008). With respect to cognitive aging, a study of female twins who averaged 50.6 years in age (range 19–78 years) from the Twins UK sample suggests that longer telomere lengths are associated with better working and episodic memory performance (Valdes et al. 2008). Moreover, in pairs discordant for telomere length, a shorter telomere length was associated with worse performance relative to the cotwin with longer telomere length (Valdes et al. 2008). However, not all studies find associations with telomere length and dementia risk or related neuropathologies (Lof-Ohlin et al. 2008; Lukens et al. 2009; Martin-Ruiz et al. 2006; Zekry et al. 2010; Zekry et al. 2008). In fact, a recent review of telomere lengths as a biomarker of aging suggests additional work is necessary, particularly from a longitudinal perspective, to ascertain its potential importance (Mather et al. 2010). Differences in findings may be attributable to what tissues are sampled, with a recent study suggesting that shorter telomere lengths measured from buccal or white cells were significantly associated with a Alzheimer's disease diagnosis, but *longer* telomere lengths among those with Alzheimer's disease were observed from hippocampal brain tissue samples (Thomas et al. 2008). Moreover, longer telomeres, as measured from leukocytes tissue, have been observed among nondemented adults (age range 41–81 years) who carried the *APOE* e4 allele than among noncarriers (Wikgren et al. 2010); this result was noted particularly among younger adults in the study. Last, longer telomere lengths among *APOE* e4 carriers predicted worse episodic memory performance (Wikgren et al. 2010). The study authors suggested that altered cell maintenance processes may be features of *APOE* e4 carriers. Taken together, tissue type and genotype may underlie the complexity of telomere length findings on cognitive aging traits. Moreover, variation in methods to assess telomere lengths may be a critical consideration as well (Vera and Blasco 2012).

Environmental factors that predict telomere shortening include SES, stress, and inflammation, all factors that are associated with more rapid cognitive aging (see Chap. 5). First, differences in twins' telomere lengths can be seen as evidence that phenotypic differences in biomarkers cannot be solely attributed to differences in genetic factors. Second, telomere shortening may occur due to exposures to both psychosocial and physical stressors. For example, shorter telomere lengths are associated with greater perceived stress (Epel 2009; Epel et al. 2004), mood disorder (Epel 2009; Epel et al. 2004), and low SES (e.g., Cherkas et al. 2006). Telomere length is related to physical stressors as well, such as cancer, CVD (Gilley et al. 2008), inflammation (Carrero et al. 2008), and oxidative stress (Houben et al. 2008). Of particular interest, Cherkas et al. (2006) illustrated that female identical twins divergent on SES had significantly different telomere lengths after controlling for BMI, physical activity, and smoking profiles. One implication of such research is that low SES may be a salient risk factor for biological aging as well as cognitive aging. Preliminary longitudinal evidence appears to bear out patterns from cross-sectional findings (Biegler et al. 2012), but it is clear that more work remains to be done, particularly with cognitive aging outcomes.

6.4 Conclusions and Future Directions

In the course of reviewing a diverse set of literatures on GE interplay on cognitive aging, it is apparent that while many threads suggest the potential importance of GE correlation or G × E interaction on cognitive aging there remains a dearth of studies dedicated to addressing these processes directly, particularly from an informative behavioral genetic perspective that can evaluate the etiologies of phenotypes and "environments." Theories of development and aging suggesting the pertinence of GE processes have been in place while the empirical data are more or less wanting, particularly with respect to normative cognitive aging. For example, as described in Sect. 6.1.3, Baltes' SOC model (see Baltes et al. 2006) can be framed as an individual-specific active GE model whereby individuals adapt and reinvest energies in order to maximize or maintain (cognitive) skills in the face of increasing functional loss with age. It is also the case that as individuals lose function, their own personal agency decreases and evocative environmental GE correlations may become increasingly important (see Fig. 6.1). Nevertheless, the extent to which genetic factors are actually correlated with environments that provide more or less support for cognitive skills is unknown (e.g., social interaction vs. isolation). Moreover, it is behavior that mediates the relationships between genes and environments (Rutter 2012), and thus genetically mediated behaviors that appropriate or evoke particular environmental contexts in late life, conducive or not to cognitive maintenance, are perhaps ripe for deeper examinations of GE correlational processes. Educational and occupational attainment as well as leisure activities and physical exercise may reflect GE correlational processes and indeed explain why heritability of cognitive ability increases up to late life.

The extent to which environmental interventions in late life support or improve cognitive function, particularly for those predisposed to cognitive decline (or dementia) due to risk genotypes, such as *APOE* e4, is not yet clear. Emerging evidence is perhaps encouraging: higher education and participation in leisure and physical activities may lower the risk of cognitive decline or dementia otherwise posed by carrying the *APOE* e4 allele (Ferrari et al. 2012). However, *when* one begins to engage in beneficial pursuits may matter: interventions introduced in young or middle age may be beneficial but late-life interventions may be met with more limited success (Harati et al. 2012). In contrast, we should not discount the fact that some environmental factors seem to be more pertinent in late old age than earlier (e.g., physical activity; Head et al. 2012; Obisesan et al. 2012). Moreover, GE interplay may differ for *APOE* across the life course and *APOE* genotype may relate to which environmental factors are salient (Gatz 2007; Reynolds et al. 2007). Last, the unique impacts of particular activities on cognitive aging outcomes remain to be elucidated, using appropriate and rigorous control groups. The social and cognitive features of particular physical activities may be relevant to unpack to determine the underlying bases of the associations with cognitive performance and change (Miller et al. 2012).

This chapter has focused only on selected gene candidates beyond *APOE* that may interact with environmental factors and impact cognitive or brain aging, e.g., *BDNF* as well as others in estrogen or serotonergic neurotransmitter pathways. This focus largely reflects the extant literature. Indeed, epigenetic processes—potential biomarkers of GE interplay—are implicated in basic memory formation and maintenance (e.g., *BDNF*; Day and Sweatt 2012; Graff and Mansuy 2009), and thus may be critical to day-to-day and even moment-to-moment adaptations to environments. Familiality of methylation levels and telomere lengths (Bakaysa et al. 2007; Bischoff et al. 2005, but see Huda et al. 2007; Bocklandt et al. 2011; Boks et al. 2009; Coolen et al. 2011) suggests that genetically driven differential sensitivities to environments (e.g., stress) may be important to individual differences in cognitive aging. However, evidence from MZ differences or discordancy approaches suggests interaction of G with nonshared E may also be salient (Reynolds et al. 2007; Valdes et al. 2008), although much work remains to be done. Indeed, the consideration of biomarkers of GE interplay processes is relatively recent with a lot of suggestive findings, but not yet a lot of data, particularly for cognitive aging outcomes.

Importantly, the most recent work in gene-finding efforts using genome-wide association (GWAS) approaches affirms the polygenic nature of general intelligence traits (Davies et al. 2011) and cognitive decline (Davies et al. submitted), with few "hits" beyond *APOE* and neighboring genes such as *TOMM40*. That is, genes of very small effect contribute to cognitive abilities, with up to 51 % of the variance in spatial/fluid cross-sectional performance accounted for by thousands of SNPs included in the GWAS (Davies et al. 2011). Moreover, 24 % of genetic influences on general cognitive ability may differ (i.e., new genes) between childhood and late adulthood (Deary et al. 2012), which would be consistent with previous SATSA longitudinal work reporting evidence of increasing genetic variation up to age 65 (Reynolds et al. 2005). That working and episodic memory traits display both increasing genetic and environmental variance (Reynolds et al. 2005; Reynolds 2008a) is consistent with the putative time-dependent impact of *APOE* on dementia risk as well as the variety of significant *APOE* × E effects. Moreover, it has been argued that *APOE* is one example of a "plasticity" gene (Chen et al. 2010; Holtzman and Fagan 1998; Myers and Nemeroff 2012; Nichol et al. 2009; Petit-Turcotte et al. 2005; Teter 2004; Weeber et al. 2002); *BDNF* (Fritsch et al. 2010; Li Voti et al. 2011) and dopaminergic or serotonergic candidates (cf., Belsky and Beaver 2011; Belsky et al. 2009; Rutter 2012) could be added to the argument. Certainly, pleiotropic effects appear to be evident. Nonetheless, taking a broader approach to consider multiple genetic variants within and across genes will be more informative, compared to single markers, given the current status of genetics research to date (Dick 2011). The paucity of strong genetic signals from candidates, apart from *APOE*, likely a consequence of SNP-by-SNP evaluations, may also suggest the importance of considering GE interplay, albeit with much caution in an essentially postGWAS era (Dick 2011).

It is also important to consider differential GE interplay for men and women, particularly in the timing of GE associations. All too often sex is treated as a covariate to be controlled rather than considered as a moderator. Recent studies suggest men may be at greater risk than women for mild cognitive impairments (Roberts et al. 2012), while it has been long established that the prevalence of dementia among women is greater (Alzheimer's Association 2012). Twin studies have not found appreciable evidence for differential sex effects on longitudinal trajectories for normative cognitive aging of most traits, with the exception of verbal ability (Finkel et al. 2006). However, risk factors such as serum lipids may differentially predict cognitive decline after age 50 (Reynolds et al. 2010), which may be in part attributable to differential life course trajectories in cholesterol and other lipids and lipoproteins. Thus, distinctive age-related risk profiles may be important to consider in evaluating G × E interplay for cognitive aging for men and women.

Large-scale efforts to study G × E influences on aging outcomes are afoot that will add to emerging literature on GE interplay using behavioral genetics methods. The IGEMS project (Interplay of Genes and Environment across Multiple Studies) is a new collaboration among nine existing longitudinal twin and family studies in Sweden, Denmark, and the United States (Pedersen et al. 2013). The central focus is to harmonize social–environmental data that can be related to physical functioning and health, psychological well-being (emotional stability/depression), and cognitive health outcomes in midlife and old age in order to address GE interplay, both GE correlation and G × E interactions. Through this and similar efforts, the several threads suggestive of GE interplay may become clearer in the near future as research begins to illuminate the dynamic pathways to variation in cognitive maintenance and aging in late life.

Acknowledgments The authors gratefully acknowledge that a portion of this work was supported by the National Institutes of Aging (R01 AG037985).

References

Adler, N. E., Boyce, T., Chesney, M. A., Cohen, S., Folkman, S., Kahn, R. L., & Syme, S. L. (1994). Socioeconomic status and health: The challenge of the gradient. *American Psychologist, 49*(1), 15–24. doi:10.1037/0003-066x.49.1.15.

Andel, R., Crowe, M., Pedersen, N. L., Mortimer, J., Crimmins, E., Johansson, B., & Gatz, M. (2005). Complexity of work and risk of Alzheimer's disease: a population-based study of Swedish twins. (Research Support, N.I.H., Extramural Research Support, Non-U.S. Gov't Research Support, U.S. Gov't, P.H.S. Twin Study). *The Journals of Gerontology. Series B, Psychological Sciences and Social Sciences, 60*(5). P251–258.

Andel, R., Vigen, C., Mack, W. J., Clark, L. J., & Gatz, M. (2006). The effect of education and occupational complexity on rate of cognitive decline in Alzheimer's patients. (Research Support, N.I.H., Extramural). *Journal of the International Neuropsychological Society, 12*(1), 147–152. doi:10.1017/S1355617706060206.

Andel, R., Kareholt, I., Parker, M. G., Thorslund, M., & Gatz, M. (2007). Complexity of primary lifetime occupation and cognition in advanced old age. (Research Support, Non-U.S. Gov't). *Journal of Aging and Health, 19*(3), 397–415. doi:10.1177/0898264307300171.

Ariza, M., Pueyo, R., Matarin Mdel, M., Junque, C., Mataro, M., Clemente, I., et al. (2006). Influence of APOE polymorphism on cognitive and behavioural outcome in moderate and severe traumatic brain injury. (Research Support, Non-U.S. Gov't). *Journal of Neurology, Neurosurgery, and Psychiatry, 77*(10), 1191–1193. doi:10.1136/jnnp.2005.085167.

Alzheimer's Association (2012). Alzheimer's disease facts and figures. *Alzheimer's and Dementia: The Journal of the Alzheimer's Association, 8,* 131–168.

Aubert, G., & Lansdorp, P. M. (2008). Telomeres and aging. *Physiological Reviews, 88*(2), 557–579. doi:10.1152/physrev.00026.2007.

Bakaysa, S. L., Mucci, L. A., Slagboom, P. E., Boomsma, D. I., McClearn, G. E., Johansson, B., & Pedersen, N. L. (2007). Telomere length predicts survival independent of genetic influences. (Research Support, N.I.H., Extramural Research Support, Non-U.S. Gov't Twin Study). *Aging Cell, 6*(6), 769–774. doi:10.1111/j.1474-9726.2007.00340.x.

Baltes, P. B., Lindenberger, U., & Staudinger, U. M. (2006). Life span theory in developmental psychology. In R. M. Lerner & W. Damon (Eds.), *Handbook of Child Psychology: Vol. 1, Theoretical Models of Human Development* (6th ed, pp. 569–664). Hoboken: Wiley.

Belsky, J., & Beaver, K. M. (2011). Cumulative-genetic plasticity, parenting and adolescent self-regulation. (Research Support, N.I.H., Extramural Research Support, Non-U.S. Gov't). *Journal of Child Psychology and Psychiatry, and Allied Disciplines, 52*(5), 619–626. doi:10.1111/j.1469-7610.2010.02327.x.

Belsky, J., Jonassaint, C., Pluess, M., Stanton, M., Brummett, B., & Williams, R. (2009). Vulnerability genes or plasticity genes? (Research Support, N.I.H., Extramural Research Support, Non-U.S. Gov'tReview). *Molecular Psychiatry, 14*(8), 746–754. doi:10.1038/mp.2009.44.

Bennett, D. A., Schneider, J. A., Tang, Y., Arnold, S. E., & Wilson, R. S. (2006). The effect of social networks on the relation between Alzheimer's disease pathology and level of cognitive function in old people: A longitudinal cohort study. (Research Support, N.I.H., Extramural). *Lancet Neurology, 5*(5), 406–412. doi:10.1016/S1474-4422(06)70417–3.

Biegler, K. A., Anderson, A. K., Wenzel, L. B., Osann, K., & Nelson, E. L. (2012). Longitudinal change in telomere length and the chronic stress response in a randomized pilot biobehavioral clinical study: Implications for cancer prevention. (Research Support, N.I.H., Extramural). *Cancer Prevention Research, 5*(10), 1173–1182. doi:10.1158/1940-6207.CAPR-12-0008.

Bieman-Copland, S., Ryan, E. B., & Cassano, J. (1998). Responding to challenges of late life. In D. M. Stack & D. R. White (Eds.), *Improving competence across the lifespan: Building interventions based on theory and research* (pp. 141–157). New York: Plenum.

Bischoff, C., Graakjaer, J., Petersen, H. C., Hjelmborg, J. V. B., Vaupel, J. W., Bohr, V., et al. (2005). The heritability of telomere length among the elderly and oldest-old. *Twin Research and Human Genetics, 8*(5), 433–439. doi:10.1375/183242705774310141.

Boardman, J. D., Barnes, L. L., Wilson, R. S., Evans, D. A., & de Leon, C. F. M. (2012). Social disorder, apoe-e4 genotype, and change in cognitive function among older adults living in Chicago. *Social Science & Medicine, 74*(10), 1584–1590. doi: http://dx.doi.org/10.1016/j.socscimed.2012.02.012.

Bocklandt, S., Lin, W., Sehl, M. E., Sanchez, F. J., Sinsheimer, J. S., Horvath, S., & Vilain, E. (2011). Epigenetic predictor of age. (Twin Study). *PLoS ONE, 6*(6), e14821. doi:10.1371/journal.pone.0014821.

Boks, M. P., Derks, E. M., Weisenberger, D. J., Strengman, E., Janson, E., Sommer, I. E., Ophoff, R. A., et al. (2009). The relationship of DNA methylation with age, gender and genotype in twins and healthy controls. (Research Support, N.I.H., Extramural Research Support, Non-U.S. Gov't Twin Study). *PLoS ONE, 4*(8), e6767. doi:10.1371/journal.pone.0006767.

Boyce, W. T. (2007). A biology of misfortune: Stress reactivity, social context, and the ontogeny of psychopathology in early life. In A. Masten (Ed.), *Multilevel dynamics in developmental psychopathology: Pathways to the future* (34th ed., pp. 45–82). Minneapolis: University of Minnesota.

Buetnner, D. (2008). *The Blue Zones: Lessons for living longer from the people who've lived the longest*. Washington, DC: National Geographic Press.

Caplan, L. J., & Schooler, C. (2006). Household work complexity, intellectual functioning, and self-esteem in men and women. *Journal of Marriage and Family, 68*, 883–900.

Carrero, J. J., Stenvinkel, P., Fellstrom, B., Qureshi, A. R., Lamb, K., Heimburger, O., et al. (2008). Telomere attrition is associated with inflammation, low fetuin-A levels and high mortality in prevalent haemodialysis patients. (Research Support, Non-U.S. Gov't). *Journal of Internal Medicine, 263*(3), 302–312. doi:10.1111/j.1365-2796.2007.01890.x.

Chadwick, M. J., Hassabis, D., Weiskopf, N., & Maguire, E. A. (2010). Decoding individual episodic memory traces in the human hippocampus. (Research Support, Non-U.S. Gov't). *Current Biology, 20*(6), 544–547. doi:10.1016/j.cub.2010.01.053.

Chen, Y., Durakoglugil, M. S., Xian, X., & Herz, J. (2010). ApoE4 reduces glutamate receptor function and synaptic plasticity by selectively impairing ApoE receptor recycling. (In Vitro Research Support, Non-U.S. Gov't). *Proceedings of the National Academy of Sciences of the United States of America, 107*(26), 12011–12016. doi:10.1073/pnas.0914984107.

Chen, C.-H., Gutierrez, E. D., Thompson, W., Panizzon, M. S., Jernigan, T. L., Eyler, L. T., et al. (2012). Hierarchical genetic organization of human cortical surface area. *Science, 335*(6076), 1634–1636. doi:10.1126/science.1215330.

Cherkas, L. F., Aviv, A., Valdes, A. M., Hunkin, J. L., Gardner, J. P., Surdulescu, G. L., et al. (2006). The effects of social status on biological aging as measured by white-blood-cell telomere length. (Comparative Study Research Support, N.I.H., Extramural Research Support, Non-U.S. Gov't). *Aging Cell, 5*(5), 361–365. doi:10.1111/j.1474-9726.2006.00222.x.

Chouliaras, L., Rutten, B. P., Kenis, G., Peerbooms, O., Visser, P. J., Verhey, F., et al. (2010a). Epigenetic regulation in the pathophysiology of Alzheimer's disease. *Progress in Neurobiology, 90*(4), 498–510. doi:S0301-0082(10)00013-4 [pii] 10.1016/j.pneurobio. 2010. 01.002.

Chouliaras, L., van den Hove, D. L., Kenis, G., Dela Cruz, J., Lemmens, M. A., van Os, J., et al. (2010b). Caloric restriction attenuates age-related changes of DNA methyltransferase 3a in mouse hippocampus. *Brain, Behavior, and Immunity, 25*(4), 616–623. doi:S0889-1591(10)00571-4 [pii] 10.1016/j.bbi.2010.11.016.

Christensen, H. (2001). What cognitive changes can be expected with normal ageing? (Research Support, Non-U.S. Gov't Review). *The Australian and New Zealand Journal of Psychiatry, 35*(6), 768–775.

Christensen, B. C., Houseman, E. A., Marsit, C. J., Zheng, S., Wrensch, M. R., Wiemels, J. L., et al. (2009). Aging and environmental exposures alter tissue-specific DNA methylation dependent upon CpG island context. (Research Support, N.I.H., Extramural Research Support, Non-U.S. Gov't). *PLoS Genetics, 5*(8), e1000602. doi:10.1371/journal.pgen.1000602.

Cole, S. W. (2009). Social regulation of human gene expression. *Current Directions in Psychological Science, 18*(3), 132–137. doi:10.1111/j.1467-8721.2009.01623.x.

Coolen, M. W., Statham, A. L., Qu, W., Campbell, M. J., Henders, A. K., Montgomery, G. W., et al. (2011). Impact of the genome on the epigenome is manifested in DNA methylation patterns of imprinted regions in monozygotic and dizygotic twins. (Research Support, Non-U.S. Gov't Twin Study). *PLoS ONE, 6*(10), e25590. doi:10.1371/journal.pone.0025590.

Dardiotis, E., Grigoriadis, S., & Hadjigeorgiou, G. M. (2012). Genetic factors influencing outcome from neurotrauma. (Review). *Current Opinion in Psychiatry, 25*(3), 231–238. doi:10.1097/YCO.0b013e3283523c0e.

Davies, G., Tenesa, A., Payton, A., Yang, J., Harris, S. E., Liewald, D., et al. (2011). Genome-wide association studies establish that human intelligence is highly heritable and polygenic. (Meta-Analysis Research Support, Non-U.S. Gov't). *Molecular Psychiatry, 16*(10), 996–1005. doi:10.1038/mp.2011.85.

Day, J. J., & Sweatt, J. D. (2010). DNA methylation and memory formation. *Nature Neuroscience, 13*(11), 1319–1323. doi:10.1038/nn.2666.

Day, J. J., & Sweatt, J. D. (2011). Epigenetic mechanisms in cognition. *Neuron, 70*(5), 813–829. doi:S0896-6273(11)00433-8 [pii] 10.1016/j.neuron.2011.05.019.

Day, J. J., & Sweatt, J. D. (2012). Epigenetic treatments for cognitive impairments. *Neuropsychopharmacology, 37*(1), 247–260. doi:10.1038/npp.2011.85 npp201185 [pii].

Deary, I. J., Yang, J., Davies, G., Harris, S. E., Tenesa, A., Liewald, D., Visscher, P. M., et al. (2012). Genetic contributions to stability and change in intelligence from childhood to old age. (Research Support, Non-U.S. Gov't). *Nature, 482*(7384), 212–215. doi:10.1038/nature10781.

Dick, D. M. (2011). Gene-environment interaction in psychological traits and disorders. (Research Support, N.I.H., Extramural Review). *Annual Review of Clinical Psychology, 7,* 383–409. doi:10.1146/annurev-clinpsy-032210-104518.

Dziedzic, T. (2006). Systemic inflammatory markers and risk of dementia. *American Journal of Alzheimer's Disease and Other Dementias, 21*(4), 258–262.

Epel, E. S. (2009). Psychological and metabolic stress: A recipe for accelerated cellular aging? (Research Support, N.I.H., Extramural Research Support, Non-U.S. Gov't Review). *Hormones, 8*(1), 7–22.

Epel, E. S., Blackburn, E. H., Lin, J., Dhabhar, F. S., Adler, N. E., Morrow, J. D., & Cawthon, R. M. (2004). Accelerated telomere shortening in response to life stress. (Research Support, Non-U.S. Gov't Research Support, U.S. Gov't, P.H.S.). *Proceedings of the National Academy of Sciences of the United States of America, 101*(49), 17312–17315. doi:10.1073/pnas.0407162101.

Erickson, K. I., Miller, D. L., Weinstein, A. M., Akl, S. L., & Banducci, S. (2012). *Physical activity and brain plasticity in late adulthood: A conceptual and comprehensive review* (Vol. 3).

Etnier, J. L., Caselli, R. J., Reiman, E. M., Alexander, G. E., Sibley, B. A., Tessier, D., & McLemore, E. C. (2007). Cognitive performance in older women relative to ApoE-epsilon4 genotype and aerobic fitness. (Research Support, N.I.H., Extramural Research Support, Non-U.S. Gov't). *Medicine and Science in Sports and Exercise, 39*(1), 199–207. doi:10.1249/01.mss.0000239399.85955.5e.

Evans, G. W. (2004). The environment of childhood poverty. (Research Support, Non-U.S. Gov't Research Support, U.S. Gov't, P.H.S. Review). *The American Psychologist, 59*(2), 77–92. doi:10.1037/0003-066X.59.2.77.

Falconer, D. S. (1989). *Introduction to Quantitative Genetics*. (3rd ed.). London: Longman, Scientific & Technical.

Feil, R., & Fraga, M. F. (2012). Epigenetics and the environment: Emerging patterns and implications. (10.1038/nrg3142). *Nature Reviews Genetics, 13*(2), 97–109.

Fernandez, A. F., Assenov, Y., Martin-Subero, J. I., Balint, B., Siebert, R., Taniguchi, H., & Esteller, M. (2012). A DNA methylation fingerprint of 1628 human samples. (Research Support, Non-U.S. Gov't). *Genome Research, 22*(2), 407–419. doi:10.1101/gr.119867.110.

Ferrari, C., Xu, W.-L., Wang, H.-X., Winblad, B., Sorbi, S., Qiu, C., & Fratiglioni, L. (2012). How can elderly apolipoprotein E Œμ4 carriers remain free from dementia? *Neurobiology of Aging, 34*(1), 13–21. doi:10.1016/j.neurobiolaging.2012.03.003.

Finkel, D., & Reynolds, C. A. (2009). Behavioral genetic investigations of cognitive aging. In Y.-K. Kim (Ed.), *Handbook of Behavior Genetics* (pp. 101–112). New York: Springer.

Finkel, D., Reynolds, C. A., Berg, S., & Pedersen, N. L. (2006). Surprising lack of sex differences in normal cognitive aging in twins. *International Journal of Aging and Human Development, 62*(4), 335–357.

Finkel, D., Andel, R., Gatz, M., & Pedersen, N. L. (2009). The role of occupational complexity in trajectories of cognitive aging before and after retirement. (Research Support, N.I.H., Extramural Research Support, Non-U.S. Gov't Twin Study). *Psychology and Aging, 24*(3), 563–573. doi:10.1037/a0015511.

Fisher, R. A. (1925). The resemblance between twins, a statistical examination of Lauterbach's Measurements. *Genetics, 10*(6), 569–579.

Fotenos, A. F., Mintun, M. A., Snyder, A. Z., Morris, J. C., & Buckner, R. L. (2008). Brain volume decline in aging: evidence for a relation between socioeconomic status, preclinical Alzheimer disease, and reserve. (Research Support, N.I.H., Extramural Research Support, Non-U.S. Gov't). *Archives of Neurology, 65*(1), 113–120. doi:10.1001/archneurol.2007.27.

Fozard, J. L., Reitsema, J., Bouma, H., & Graafmans, J. A. M. (2000). Gerontechnology: Creating enabling environments for the challenges and opportunities of aging. *Educational Gerontology, 26,* 331–344.

Fraga, M. F., Ballestar, E., Paz, M. F., Ropero, S., Setien, F., Ballestar, M. L., et al. (2005). Epigenetic differences arise during the lifetime of monozygotic twins. *Proceedings of the National Academy of Sciences, 102*(30), 10604–10609.

Fritsch, B., Reis, J., Martinowich, K., Schambra, H. M., Ji, Y., Cohen, L. G., & Lu, B. (2010). Direct current stimulation promotes BDNF-dependent synaptic plasticity: Potential implications for motor learning. (Research Support, N.I.H., Intramural Research Support, Non-U.S. Gov't). *Neuron, 66*(2), 198–204. doi:10.1016/j.neuron.2010.03.035.

Gatz, M. (2007). Genetics, dementia, and the elderly. *Current Directions in Psychological Science, 16,* 123–127.

Gatz, M., Mortimer, J. A., Fratiglioni, L., Johansson, B., Berg, S., Andel, R., et al. (2007). Accounting for the relationship between low education and dementia: A twin study. (Research Support, N.I.H., Extramural Research Support, Non-U.S. Gov't Twin Study). *Physiology & Behavior, 92*(1–2), 232–237. doi:10.1016/j.physbeh.2007.05.042.

Giedd, J., Stockman, M., Weddle, C., Liverpool, M., Alexander-Bloch, A., Wallace, G., et al. (2010). Anatomic magnetic resonance imaging of the developing child and adolescent brain and effects of genetic variation. *Neuropsychology Review, 20*(4), 349–361. doi:10.1007/s11065-010-9151-9.

Gilley, D., Herbert, B. S., Huda, N., Tanaka, H., & Reed, T. (2008). Factors impacting human telomere homeostasis and age-related disease. (Research Support, N.I.H., Extramural Research Support, Non-U.S. Gov't Review). *Mechanisms of Ageing and Development, 129*(1–2), 27–34. doi:10.1016/j.mad.2007.10.010.

Gimeno, D., Marmot, M. G., & Singh-Manoux, A. (2008). Inflammatory markers and cognitive function in middle-aged adults: The Whitehall II study. *Psychoneuroendocrinology, 33*(10), 1322–1334.

Glymour, M. M., Tzourio, C., & Dufouil, C. (2012). Is cognitive aging predicted by one's own or one's parents' educational level? Results from the three-city study. *American Journal of Epidemiology, 175*(8), 750–759. doi:10.1093/aje/kwr509.

Gottlieb, G. (1991). Experiential canalization of behavioral development: Theory. *Developmental Psychology, 27,* 4–13.

Gottlieb, G. (2007). Probabilistic epigenesis. Developmental Science, 10(1), 1-11. doi: DESC556 [pii] 10.1111/j.1467-7687.2007.00556.x

Gow, A. J., Johnson, W., Pattie, A., Brett, C. E., Roberts, B., Starr, J. M., & Deary, I. J. (2011). Stability and change in intelligence from age 11 to ages 70, 79, and 87: The Lothian Birth Cohorts of 1921 and 1936. (Research Support, Non-U.S. Gov't). *Psychology and Aging, 26*(1), 232–240. doi:10.1037/a0021072.

Graff, J., & Mansuy, I. M. (2009). Epigenetic dysregulation in cognitive disorders. *European Journal of Neuroscience, 30*(1), 1–8. doi:EJN6787 [pii] 10.1111/j.1460-9568.2009.06787.x.

Grant, M. D., Kremen, W. S., Jacobson, K. C., Franz, C., Xian, H., Eisen, S. A., et al. (2010). Does parental education have a moderating effect on the genetic and environmental influences of general cognitive ability in early adulthood? (Research Support, N.I.H., Extramural Research Support, U.S. Gov't, Non-P.H.S. Twin Study). *Behavior Genetics, 40*(4), 438–446. doi:10.1007/s10519-010-9351-3.

Greenough, W. T., McDonald, J. W., Parnisari, R. M., & Camel, J. E. (1986). Environmental conditions modulate degeneration and new dendrite growth in cerebellum of senescent rats. *Brain research, 380*(1), 136–143.

Hanscombe, K. B., Trzaskowski, M., Haworth, C. M., Davis, O. S., Dale, P. S., & Plomin, R. (2012). Socioeconomic status (SES) and children's intelligence (IQ): In a UK-representative sample SES moderates the environmental, not genetic, effect on IQ. (Research Support, N.I.H., Extramural Research Support, Non-U.S. Gov't Twin Study). *PLoS ONE, 7*(2), e30320. doi:10.1371/journal.pone.0030320.

Harati, H., Barbelivien, A., Herbeaux, K., Muller, M. A., Engeln, M., Kelche, C., et al. (2012). Lifelong environmental enrichment in rats: impact on emotional behavior, spatial memory vividness, and cholinergic neurons over the lifespan. *Age*. doi:10.1007/s11357-012-9424-8.

Harbeby, E., Jouin, M., Alessandri, J. M., Lallemand, M. S., Linard, A., Lavialle, M., et al. (2012). n-3 PUFA status affects expression of genes involved in neuroenergetics differently in the fronto-parietal cortex compared to the CA1 area of the hippocampus: Effect of rest and neuronal activation in the rat. *Prostaglandins, Leukotrienes, and Essential Fatty Acids, 86*(6), 211–220. doi:10.1016/j.plefa.2012.04.008.

Harden, K. P., Turkheimer, E., & Loehlin, J. C. (2007). Genotype by environment interaction in adolescents' cognitive aptitude. *Behavior Genetics, 37*(2), 273–283. doi:10.1007/s10519-006-9113-4.

Head, D., Bugg, J. M., Goate, A. M., Fagan, A. M., Mintun, M. A., Benzinger, T., et al. (2012). Exercise engagement as a moderator of the effects of APOE genotype on amyloid deposition. *Archives of neurology, 69*(5), 636–643. doi:10.1001/archneurol.2011.845.

Holtzman, D. M., & Fagan, A. M. (1998). Potential role of apoE in structural plasticity in the nervous system; implications for disorders of the central nervous system. *Trends in cardiovascular medicine, 8*(6), 250–255.

Hoth, K. F., Haley, A. P., Gunstad, J., Paul, R. H., Poppas, A., Jefferson, A. L., et al. (2008). Elevated C-reactive protein is related to cognitive decline in older adults with cardiovascular disease. *Journal of the American Geriatrics Society, 56*(10), 1898–1903.

Houben, J. M., Moonen, H. J., van Schooten, F. J., & Hageman, G. J. (2008). Telomere length assessment: Biomarker of chronic oxidative stress? (Review). *Free Radical Biology & Medicine, 44*(3), 235–246. doi:10.1016/j.freeradbiomed.2007.10.001.

Huda, N., Tanaka, H., Herbert, B. S., Reed, T., & Gilley, D. (2007). Shared environmental factors associated with telomere length maintenance in elderly male twins. (Research Support, N.I.H., Extramural Research Support, Non-U.S. Gov't Twin Study). *Aging Cell, 6*(5), 709–713. doi:10.1111/j.1474-9726.2007.00330.x.

Jinks, J. L., & Fulker, D. W. (1970). Comparison of the biometrical genetical, MAVA, and classical approaches to the analysis of human behavior. *Psychological bulletin, 73*(5), 311–349.

Johansson, B., Hofer, S. M., Allaire, J. C., Maldonado-Molina, M. M., Piccinin, A. M., Berg, S., et al. (2004). Change in cognitive capabilities in the oldest old: The effects of proximity to death in genetically related individuals over a 6-year period. (Research Support, U.S. Gov't, P.H.S. Twin Study). *Psychology and Aging, 19*(1), 145–156. doi:10.1037/0882-7974.19.1.145.

Johnson, W., Deary, I. J., McGue, M., & Christensen, K. (2009). Genetic and environmental transactions linking cognitive ability, physical fitness, and education in late life. *Psychology and Aging, 24*, 48–62.

Jordanova, V., Stewart, R., Davies, E., Sherwood, R., & Prince, M. (2007). Markers of inflammation and cognitive decline in an African-Caribbean population. *International Journal of Geriatric Psychiatry, 22*(10), 966–973.

Kaliman, P., Parrizas, M., Lalanza, J. F., Camins, A., Escorihuela, R. M., & Pallas, M. (2011). Neurophysiological and epigenetic effects of physical exercise on the aging process. *Ageing Research Reviews, 10*(4), 475–486. doi:S1568-1637(11)00046-8 [pii] 10.1016/j.arr.2011.05.002.

Kang, J. H., Logroscino, G., De Vivo, I., Hunter, D., & Grodstein, F. (2005). Apolipoprotein E, cardiovascular disease and cognitive function in aging women. *Neurobiology of Aging, 26*(4), 475–484. doi:10.1016/j.neurobiolaging.2004.05.003.

Krabbe, K. S., Mortensen, E. L., Avlund, K., Pilegaard, H., Christiansen, L., Pedersen, A. N., et al. (2009). Genetic priming of a proinflammatory profile predicts low IQ in octogenarians. *Neurobiology of Aging, 30*(5), 769–781.

Kremen, W. S., Jacobson, K. C., Xian, H., Eisen, S. A., Waterman, B., Toomey, R., et al. (2005). Heritability of word recognition in middle-aged men varies as a function of parental education. (Research Support, N.I.H., Extramural Research Support, U.S. Gov't, Non-P.H.S. Research Support, U.S. Gov't, P.H.S. Twin Study). *Behavior Genetics, 35*(4), 417–433. doi:10.1007/s10519-004-3876-2.

Lawton, M. P., & Nahemow, L. (1973). Ecology of the aging process. In C. Eisdorfer & M. P. Lawton (Eds.), *The psychology of adult development and aging* (pp. 619–674). Washington, DC: American Psychological Association.

Lee, B. K., Glass, T. A., James, B. D., Bandeen-Roche, K., & Schwartz, B. S. (2011). Neighborhood psychosocial environment, Apolipoprotein E genotype, and cognitive function in older adults. *Archives of General Psychiatry, 68*(3), 7.

Lee, S., Doulames, V., Donnelly, M., Levasseaur, J., & Shea, T. B. (2012). Environmental enrichment can prevent cognitive decline induced by dietary oxidative challenge. (Research Support, Non-U.S. Gov't). *Journal of Alzheimer's Disease, 28*(3), 497–501. doi:10.3233/JAD-2011-111562.

Li Voti, P., Conte, A., Suppa, A., Iezzi, E., Bologna, M., Aniello, M. S., et al. (2011). Correlation between cortical plasticity, motor learning and BDNF genotype in healthy subjects. (Comparative Study). *Experimental Brain Research. Experimentelle Hirnforschung. Experimentation Cerebrale, 212*(1), 91–99. doi:10.1007/s00221-011-2700-5.

Liang, J., & Fairchild, T. J. (1979). Relative deprivation and perception of financial adequacy among the aged. *Journal of Gerontology, 34*(5), 768–759.

Lichtenstein, P., Harris, J. R., Pedersen, N. L., & McClearn, G. E. (1992). Socioeconomic status and physical health, how are they related? An empirical study based on twins reared apart and twins reared together. *Social Science & Medicine, 36*(4), 441–450.

Lof-Ohlin, Z. M., Hagnelius, N. O., & Nilsson, T. K. (2008). Relative telomere length in patients with late-onset Alzheimer's dementia or vascular dementia. *Neuroreport, 19*(12), 1199–1202. doi:10.1097/WNR.0b013e3283089220 00001756-200808060-00006 [pii].

Lovden, M., Ghisletta, P., & Lindenberger, U. (2005). Social participation attenuates decline in perceptual speed in old and very old age. *Psychology and Aging, 20*(3), 423–434. doi:10.1037/0882-7974.20.3.423.

Lubin, F. D., Roth, T. L., & Sweatt, J. D. (2008). Epigenetic regulation of BDNF gene transcription in the consolidation of fear memory. *The Journal of Neuroscience, 28*(42), 10576–10586. doi:28/42/10576 [pii] 10.1523/JNEUROSCI.1786-08.2008.

Lukens, J. N., Van Deerlin, V., Clark, C. M., Xie, S. X., & Johnson, F. B. (2009). Comparisons of telomere lengths in peripheral blood and cerebellum in Alzheimer's disease. *Alzheimers Dement, 5*(6), 463–469. doi:S1552-5260(09)02014-7 [pii] 10.1016/j.jalz.2009.05.666.

Lupien, S. J., McEwen, B. S., Gunnar, M. R., & Heim, C. (2009). Effects of stress throughout the lifespan on the brain, behaviour and cognition. (Review). Nature reviews. *Neuroscience, 10*(6), 434–445. doi:10.1038/nrn2639.

Maekawa, M., & Watanabe, Y. (2007). Epigenetics: Relations to disease and laboratory findings. *Current Medicinal Chemistry, 14*(25), 2642–2653.

Markham, J. A., & Greenough, W. T. (2004). Experience-driven brain plasticity: Beyond the synapse. *Neuron Glia Biology, 1*(4), 351–363. doi:10.1017/s1740925×05000219.

Martin, N. G., Rowell, D. M., & Whitfield, J. B. (1983). Do the MN and Jk systems influence environmental variability in serum lipid levels? (Comparative Study). *Clinical Genetics, 24*(1), 1–14.

Martin-Ruiz, C., Dickinson, H. O., Keys, B., Rowan, E., Kenny, R. A., & Von Zglinicki, T. (2006). Telomere length predicts poststroke mortality, dementia, and cognitive decline. *Annals of Neurology, 60*(2), 174–180. doi:10.1002/ana.20869.

Mather, K. A., Jorm, A. F., Parslow, R. A., & Christensen, H. (2010). Is telomere length a biomarker of aging? A review. *The Journals of Gerontology Series A: Biological Sciences and Medical Sciences, 66*(2), 202–213. doi:glq180 [pii] 10.1093/gerona/glq180.

Matthews, K. A., & Gallo, L. C. (2011). Psychological perspectives on pathways linking socioeconomic status and physical health. (Research Support, N.I.H., Extramural Review). *Annual Review of Psychology, 62*, 501–530. doi:10.1146/annurev.psych.031809.130711.

McArdle, J. J., & Prescott, C. A. (2010). Contemporary modeling of Gene-by-Environment effects in randomized multivariate longitudinal studies. *Perspectives on Psychological Science: A Journal of the Association for Psychological Science, 5*(5), 606–621. doi:10.1177/1745691610383510.

McGue, M., & Christensen, K. (2002). The heritability of level and rate-of-change in cognitive functioning in Danish twins aged 70 years and older. *Experimental Aging Research, 28*(4), 435–451. doi:10.1080/03610730290080416.

Medland, S. E., Neale, M. C., Eaves, L. J., & Neale, B. M. (2009). A note on the parameterization of Purcell's G x E model for ordinal and binary data. (Research Support, N.I.H., Extramural Research Support, Non-U.S. Gov't). *Behavior Genetics, 39*(2), 220–229. doi:10.1007/s10519-008-9247-7.

Miller, C. A., Gavin, C. F., White, J. A., Parrish, R. R., Honasoge, A., Yancey, C. R., et al. (2010). Cortical DNA methylation maintains remote memory. *Nature Neuroscience, 13*(6), 664–666. doi:nn.2560 [pii] 10.1038/nn.2560.

Miller, G. E., Chen, E., Fok, A. K., Walker, H., Lim, A., Nicholls, E. F., et al. (2009). Low early-life social class leaves a biological residue manifested by decreased glucocorticoid and increased proinflammatory signaling. *PNAS Proceedings of the National Academy of Sciences of the United States of America, 106*(34), 14716–14721.

Miller, D. I., Taler, V., Davidson, P. S., & Messier, C. (2012). Measuring the impact of exercise on cognitive aging: Methodological issues. (Research Support, Non-U.S. Gov't Review). *Neurobiology of Aging, 33*(3), 622, e629–643. doi:10.1016/j.neurobiolaging.2011.02.020.

Mohammed, A. H., Henriksson, B. G., Söderström, S., Ebendal, T., Olsson, T., & Seckl, J. R. (1993). Environmental influences on the central nervous system and their implications for the aging rat. *Behavioural Brain Research, 57*(2), 183–191.

Mooijaart, S. P., Sattar, N., Trompet, S., Polisecki, E., de Craen, A. J. M., Schaefer, E. J., et al. (2011). C-reactive protein and genetic variants and cognitive decline in old age: The PROSPER Study. *PLoS One, 6*(9), e23890.

Morse, C. K. (1993). Does variability increase with age? An archival study of cognitive measures. *Psychology and Aging, 8*(2), 156–164.

Myers, A. J., & Nemeroff, C. B. (2012). APOE: A risk factor for multiple disorders. *The American Journal of Geriatric Psychiatry: Official Journal of the American Association for Geriatric Psychiatry, 20*(7), 545–548. doi:10.1097/JGP.0b013e318259b9a5.

Neves, G., Cooke, S. F., & Bliss, T. V. P. (2008). Synaptic plasticity, memory and the hippocampus: a neural network approach to causality. *Nature Reviews Neuroscience, 9*(1), 65–75.

Nichol, K., Deeny, S. P., Seif, J., Camaclang, K., & Cotman, C. W. (2009). Exercise improves cognition and hippocampal plasticity in APOE epsilon4 mice. (Research Support, N.I.H., Extramural). *Alzheimer's & Dementia: The Journal of the Alzheimer's Association, 5*(4), 287–294. doi:10.1016/j.jalz.2009.02.006.

Obisesan, T. O., Umar, N., Paluvoi, N., & Gillum, R. F. (2012). Association of leisure-time physical activity with cognition by apolipoprotein-E genotype in persons aged 60 years and over: The National Health and Nutrition Examination Survey (NHANES-III). (Comparative Study Research Support, N.I.H., Extramural). *Clinical Interventions in Aging, 7*, 35–43. doi:10.2147/CIA.S26794.

Pechtel, P., & Pizzagalli, D. A. (2011). Effects of early life stress on cognitive and affective function: an integrated review of human literature. (Research Support, N.I.H., Extramural Research Support, Non-U.S. Gov't Review). *Psychopharmacology, 214*(1), 55–70. doi:10.1007/s00213-010-2009-2.

Pedersen, N. L., Plomin, R., Nesselroade, J. R., & McClearn, G. E. (1992). A quantitative genetic analysis of cognitive abilities during the second half of the life span. *Psychological Science, 3*(6), 346–353.

Pedersen, N. L., Reynolds, C. A., & Gatz, M. (1996). Sources of covariation among Mini-Mental State Examination scores, education, and cognitive abilities. *The Journals of Gerontology Series B: Psychological Sciences and Social Sciences, 51*(2), P55–63.

Pedersen, N. L., Christensen, K., Dahl, A., Finkel, D., Franz, C. E., Gatz, M., Horwitz, B. N., Johansson, B., Johnson, W., Kremen, W. S., Lyons, M. J., Malmberg, B., McGue, M., Neiderhiser, J. M., Petersen, I., & Reynolds, C. A. (2013). IGEMS: The consortium on interplay of genes and environment across multiple studies. *Twin Research and Human Genetics, 16*, 481–489.

Petit-Turcotte, C., Aumont, N., Beffert, U., Dea, D., Herz, J., & Poirier, J. (2005). The APOE receptor apoER2 is involved in the maintenance of efficient synaptic plasticity. (Comparative Study Research Support, Non-U.S. Gov't Research Support, U.S. Gov't, P.H.S.). *Neurobiology of Aging, 26*(2), 195–206. doi:10.1016/j.neurobiolaging.2004.04.007.

Pfefferbaum, A., Sullivan, E. V., & Carmelli, D. (2004). Morphological changes in aging brain structures are differentially affected by time-linked environmental influences despite strong genetic stability. *Neurobiology of Aging, 25*(2), 175–183.

Pinaud, R., Tremere, L. A., Penner, M. R., Hess, F. F., Robertson, H. A., & Currie, R. W. (2002). Complexity of sensory environment drives the expression of candidate-plasticity gene, nerve growth factor induced-A. *Neuroscience, 112*(3), 573–582.

Plomin, R., DeFries, J. C., McClearn, G. E., & McGuffin, P. (2008). *Behavioral Genetics* (5th ed.). New York: Worth.

Purcell, S. (2002). Variance components models for gene-environment interaction in twin analysis. (Research Support, Non-U.S. Gov't Research Support, U.S. Gov't, P.H.S. Twin Study). *Twin research: The Official Journal of the International Society for Twin Studies, 5*(6), 554–571. doi:10.1375/136905202762342026.

Rafnsson, S. B., Deary, I. J., Smith, F. B., Whiteman, M. C., Rumley, A., Lowe, G. D. O., & Fowkes, F. G. R. (2007). Cognitive decline and markers of inflammation and hemostasis: The Edinburgh Artery Study. *Journal of the American Geriatrics Society, 55*(5), 700–707.

Raine, A. (2002). Biosocial studies of antisocial and violent behavior in children and adults: A review. *Journal of Abnormal Child Psychology, 30*, 311–326.

Rampon, C., Tang, Y. P., Goodhouse, J., Shimizu, E., Kyin, M., & Tsien, J. Z. (2000). Enrichment induces structural changes and recovery from nonspatial memory deficits in CA1 NMDAR1-knockout mice. (Research Support, Non-U.S. Gov't Research Support, U.S. Gov't, P.H.S.). *Nature Neuroscience, 3*(3), 238–244. doi:10.1038/72945.

Reiss, D., & Leve, L. D. (2007). Genetic expression outside the skin: Clues to mechanisms of Genotype x Environment interaction. (Research Support, N.I.H., extramural review). *Development and Psychopathology, 19*(4), 1005–1027. doi:10.1017/S0954579407000508.

Reynolds, C. A. (2008a). Genetic and environmental influences on cognitive change. In S. M. Hofer & D. F. Alwin (Eds.), *The Handbook on Cognitive Aging: Interdisciplinary Perspectives*. Thousand Oaks: Sage.

Reynolds, C. A. (2008b). Genetics of brain aging—twin aging. In L. Squire (Ed.), *Encyclopedia of Neuroscience* (Network Version ed., pp. 10500): Elsevier.

Reynolds, C. A., Finkel, D., McArdle, J. J., Gatz, M., Berg, S., & Pedersen, N. L. (2005). Quantitative genetic analysis of latent growth curve models of cognitive abilities in adulthood. *Developmental Psychology, 41*(1), 3–16.

Reynolds, C. A., Gatz, M., Berg, S., & Pedersen, N. L. (2007). Genotype-environment interactions: Cognitive aging and social factors. *Twin Research and Human Genetics, 10*(2), 241–254. doi:10.1375/twin.10.2.241.

Reynolds, C. A., Gatz, M., Prince, J. A., Berg, S., & Pedersen, N. L. (2010). Serum lipid levels and cognitive change in late life. *Journal of the American Geriatrics Society, 58*(3), 501–509. doi:JGS2739 [pii] 10.1111/j.1532-5415.2010.02739.x.

Rimol, L. M., Panizzon, M. S., Fennema-Notestine, C., Eyler, L. T., Fischl, B., Franz, C. E., et al. (2010). Cortical thickness is influenced by regionally specific genetic factors. (Research Support, N.I.H., Extramural Research Support, U.S. Gov't, Non-P.H.S. Twin Study). *Biological Psychiatry, 67*(5), 493–499. doi:10.1016/j.biopsych.2009.09.032.

Roberts, R. O., Geda, Y. E., Knopman, D. S., Cha, R. H., Pankratz, V. S., Boeve, B. F., Petersen, R. C., et al. (2012). The incidence of MCI differs by subtype and is higher in men: The Mayo Clinic Study of Aging. (Research Support, N.I.H., Extramural Research Support, Non-U.S. Gov't). *Neurology, 78*(5), 342–351. doi:10.1212/WNL.0b013e3182452862.

Rodin, J. (1986). Aging and health: Effects of sense of control. *Science, 233*, 1271–1276.

Rodin, J. (1989). Sense of control: Potentials for intervention. *Annals of the American Academy of Political and Social Science, 503*, 29–42.

Rowe, J. W., & Kahn, R. L. (1997). Successful aging. (Research Support, Non-U.S. Gov't Review). *The Gerontologist, 37*(4), 433–440.

Rutter, M. (2012). Gene–environment interdependence. *European Journal of Developmental Psychology, 9*(4), 391–412. doi:10.1080/17405629.2012.661174.

Salthouse, T. A. (2006). Mental exercise and mental aging: Evaluating the validity of the "use it or lose it" hypothesis. *Perspectives on Psychological Science, 1,* 68–87.

Sapolsky, R. M. (2004). Social status and health in humans and other animals. *Annual Review of Anthropology, 33,* 393–418.

Sapolsky, R. M. (2005). The influence of social hierarchy on primate health. *Science, 308,* 648–652.

Scarmeas, N., & Stern, Y. (2003). Cognitive reserve and lifestyle. *Journal of Clinical and Experimental Neuropsychology, 25,* 625–633.

Scarpa, S., Cavallaro, R. A., D'Anselmi, F., & Fuso, A. (2006). Gene silencing through methylation: An epigenetic intervention on Alzheimer disease. *Journal of Alzheimer's Disease, 9,* 407–414.

Scarr, S., & McCartney, K. (1983). How people make their own environments: A theory of genotype greater than environment effects. (Research Support, Non-U.S. Gov't Research Support, U.S. Gov't, P.H.S.). *Child Development, 54*(2), 424–435.

Schooler, C., & Caplan, L. J. (2008). Those who have, get: Social structure, environmental complexity, intellectual functioning, and self-directed orientations in the elderly. In K. W. Schaie & R. P. Abeles (Eds.), *Social Structures and Aging Individuals: Continuing Challenges* (Vol. 20, pp. 131–153). New York: Springer.

Schram, M. T., Euser, S. M., de Craen, A. J., Witteman, J. C., Frölich, M., Hofman, A., Westendorp, R. G. J. (2007). Systemic markers of inflammation and cognitive decline in old age. *Journal of the American Geriatrics Society, 55*(5), 708–716.

Shanahan, M. J., & Hofer, S. M. (2005). Social context in gene-environment interactions: Retrospect and prospect. (Review). *The Journals of Gerontology. Series B, Psychological Sciences and Social Sciences, 60*(Spec No 1), 65–76.

Shawi, M., & Autexier, C. (2008). Telomerase, senescence and ageing. (Research Support, Non-U.S. Gov't Review). *Mechanisms of Ageing and Development, 129*(1–2), 3–10. doi:10.1016/j.mad.2007.11.007.

Singh-Manoux, A., Marmot, M. G., & Adler, N. E. (2005). Does subjective social status predict health and change in health status better than objective status? *Psychosomatic Medicine, 67*(6), 855–861. doi:10.1097/01.psy.0000188434.52941.a0.

Staff, R. T., Murray, A. D., Deary, I. J., & Whalley, L. J. (2004). What provides cerebral reserve? *Brain, 127*(5), 1191–1199. doi:10.1093/brain/awh144.

Sterlemann, V., Rammes, G., Wolf, M., Liebl, C., Ganea, K., Muller, M. B., & Schmidt, M. V. (2010). Chronic social stress during adolescence induces cognitive impairment in aged mice. *Hippocampus, 20*(4), 540–549. doi:10.1002/hipo.20655.

Sullivan, E. V., Pfefferbaum, A., Swan, G. E., & Carmelli, D. (2001). Heritability of hippocampal size in elderly twin men: Equivalent influence from genes and environment. *Hippocampus, 11*(6), 754–762.

Teter, B. (2004). ApoE-dependent plasticity in Alzheimer's disease. (Review). *Journal of Molecular Neuroscience, 23*(3), 167–179. doi:10.1385/JMN:23:3:167.

Thomas, P., O'Callaghan, N. J., & Fenech, M. (2008). Telomere length in white blood cells, buccal cells and brain tissue and its variation with ageing and Alzheimer's disease. *Mechanisms Ageing and Development, 129*(4), 183–190. doi:S0047-6374(07)00194-7 [pii] 10.1016/j.mad.2007.12.004.

Thompson, P. M., Cannon, T. D., Narr, K. L., van Erp, T., Poutanen, V. P., Huttunen, M., Toga, A. W. (2001). Genetic influences on brain structure. (Research Support, U.S. Gov't, Non-P.H.S. Research Support, U.S. Gov't, P.H.S.). *Nature Neuroscience, 4*(12), 1253–1258. doi:10.1038/nn758.

Vaillant, G. E. (2002). *Aging well: Surprising guideposts to a happier life*. Boston: Little, Brown.

Valdes, A. M., Deary, I. J., Gardner, J., Kimura, M., Lu, X., Spector, T. D., Cherkas, L. F. (2008). Leukocyte telomere length is associated with cognitive performance in healthy women. *Neurobiology of Aging, 31*(6), 986–992. doi:S0197-4580(08)00256-X [pii] 10.1016/j.neurobiolaging.2008.07.012

van den Kommer, T. N., Dik, M. G., Comijs, H. C., Jonker, C., & Deeg, D. J. H. (2010). Homocysteine and inflammation: Predictors of cognitive decline in older persons? *Neurobiology of Aging, 31*(10), 1700–1709.

van der Sluis, S., Dolan, C. V., Neale, M. C., Boomsma, D. I., & Posthuma, D. (2006). Detecting genotype-environment Interaction in monozygotic twin data: Comparing the Jinks and Fulker test and a new test based on Marginal Maximum Likelihood estimation. *Twin Research and Human Genetics, 9*(3), 377–392. doi:10.1375/twin.9.3.377.

van der Sluis, S., Dolan, C. V., Neale, M. C., & Posthuma, D. (2008). A general test for gene-environment interaction in sib pair-based association analysis of quantitative traits. *Behavior Genetics, 38*(4), 372–389.

van der Sluis, S., Posthuma, D., & Dolan, C. (2012). A note on false positives and power in G × E modelling of twin data. *Behavior Genetics, 42*(1), 170–186. doi:10.1007/s10519-011-9480-3.

Vera, E., & Blasco, M. A. (2012). Beyond average: Potential for measurement of short telomeres. *Aging, 4*(6), 379–392.

Waddington, C. H. (1942). Canalization of development and the inheritance of acquired characters. *Nature, 150,* 563–565.

Wang, S. C., Oelze, B., & Schumacher, A. (2008). Age-specific epigenetic drift in late-onset Alzheimer's disease. *PLoS ONE, 3*(7), e2698. doi:10.1371/journal.pone.0002698.

Weeber, E. J., Beffert, U., Jones, C., Christian, J. M., Forster, E., Sweatt, J. D., & Herz, J. (2002). Reelin and ApoE receptors cooperate to enhance hippocampal synaptic plasticity and learning. (Research Support, Non-U.S. Gov't Research Support, U.S. Gov't, P.H.S.). *The Journal of Biological Chemistry, 277*(42), 39944–39952. doi:10.1074/jbc.M205147200.

Wikgren, M., Karlsson, T., Nilbrink, T., Nordfjall, K., Hultdin, J., Sleegers, K., et al. (2010). APOE epsilon4 is associated with longer telomeres, and longer telomeres among epsilon4 carriers predicts worse episodic memory. *Neurobiology of Aging, 33*(2), 335–344. doi:S0197-4580(10)00115-6 [pii] 10.1016/j.neurobiolaging.2010.03.004.

Wu, J., Basha, M., & Zawia, N. (2008). The environment, epigenetics and amyloidogenesis. *Journal of Molecular Neuroscience, 34*(1), 1–7.

Zavala, C., Gatz, M., Johansson, B., Malmberg, B., Pedersen, N. L., & Reynolds, C. A. (2013, in preparation). Perceived socioeconomic status predicts cognitive trajectories: A cohort comparison.

Zekry, D., Herrmann, F. R., Irminger-Finger, I., Ortolan, L., Genet, C., Vitale, A. M., et al. (2008). Telomere length is not predictive of dementia or MCI conversion in the oldest old. *Neurobiology of Aging, 31*(4), 719–720. doi:S0197-4580(08)00169-3 [pii] 10.1016/j.neurobiolaging.2008.05.016.

Zekry, D., Herrmann, F. R., Irminger-Finger, I., Graf, C., Genet, C., Vitale, A. M., et al. (2010). Telomere length and ApoE polymorphism in mild cognitive impairment, degenerative and vascular dementia. *Journal of the Neurological Sciences, 299*(1–2), 108–111. doi:S0022-510X(10)00342-4 [pii] 10.1016/j.jns.2010.07.019.

Chapter 7
Dementia: Genes, Environments, Interactions

Margaret Gatz, Jung Yun Jang, Ida K. Karlsson and Nancy L. Pedersen

Older adulthood is characterized by normative changes in cognition as described in Chaps. 5 and 6. At the same time, age is also the most important risk factor for nonnormative cognitive changes, or dementia. In this chapter, we briefly review the epidemiology of dementia then turn to behavior genetic research, molecular and genomic studies, environmental risk factors, and interactions of genetic and environmental risk factors. The field is rapidly growing, with new work on biomarkers, ever larger genome-wide association studies (GWAS) consortia, and yet more "omics" approaches; thus, we conclude by pointing to areas where new developments are likely to emerge.

7.1 Introduction to Dementia

7.1.1 Defining Dementia

Dementia refers to a group of disorders marked by progressive cognitive deterioration, primarily in old age. Persons with dementia show significant difficulties in performing everyday activities, which eventually lead to complete reliance on

M. Gatz (✉) · J. Y. Jang · N. L. Pedersen
Department of Psychology, University of Southern California, Los Angeles, CA 90089, USA
e-mail: gatz@usc.edu

J. Y. Jang
e-mail: jungyjan@usc.edu

N. L. Pedersen
e-mail: nancy.pedersen@ki.se

I. K. Karlsson · M. Gatz · N. L. Pedersen
Department of Medical Epidemiology and Biostatistics, Karolinska Institutet,
Stockholm 17177, Sweden
e-mail: ida.karlsson@ki.se

D. Finkel, C. A. Reynolds (eds.), *Behavior Genetics of Cognition Across the Lifespan*,
Advances in Behavior Genetics 1, DOI 10.1007/978-1-4614-7447-0_7,

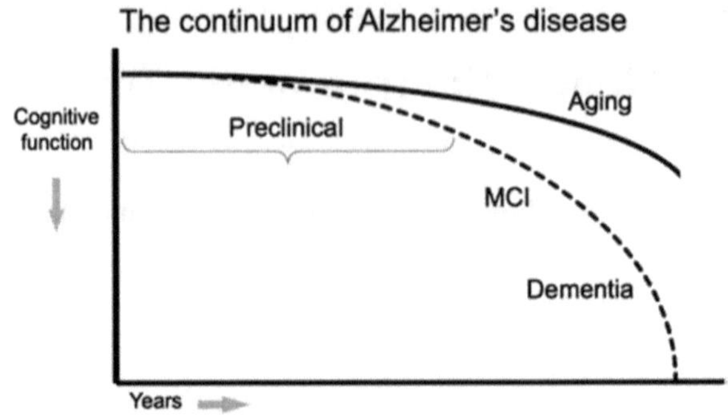

Fig. 7.1 Model of the clinical trajectory of Alzheimer disease (AD). The stage of preclinical AD precedes mild cognitive impairment (MCI) and encompasses the spectrum of presymptomatic autosomal dominant mutation carriers, asymptomatic biomarker-positive older individuals at risk for progression to MCI due to AD and AD dementia, as well as biomarker-positive individuals who have demonstrated subtle decline from their own baseline that exceeds that expected in typical aging, but would not yet meet criteria for MCI. Note that this diagram represents a hypothetical model for the pathological–clinical continuum of AD but does not imply that all individuals with biomarker evidence of AD-pathophysiological process will progress to the clinical phases of the illness. (Reprinted from Sperling et al. 2011, p. 283, Copyright 2011, with permission from Elsevier)

others in basic self-care. Current understanding places dementia on a spectrum, where a disease process may start to occur years before symptoms manifest and cause a mild cognitive decline before symptoms become severe enough to meet diagnostic criteria for dementia (Sperling et al. 2011; see Fig. 7.1). In an effort for early detection and intervention, in the past two decades, a large number of studies have been conducted to characterize mild cognitive impairment (MCI) in relation to normative cognitive aging (see Gauthier et al. 2006, for review). Acknowledging the recent advancement in understanding the continuum of dementia and its clinical utility, the new soon-to-be-released fifth edition of the Diagnostic and Statistical Manual of Mental Disorders (DSM-5) revised its definition of dementia and proposed the new terms “minor neurocognitive disorder” and “major neurocognitive disorder.” According to the DSM-5, a neurocognitive disorder may be broadly defined as a decline from a previously attained level of cognitive functioning in one or more domains (Jeste et al. 2010). Cognitive domains that may be affected include complex attention (sustained and divided attention, processing speed, and selective attention), executive ability (planning, decision-making, working memory, and mental flexibility), learning and memory (immediate and recent episodic memory), language (expressive and receptive language), visuoconstructional–perceptual ability (construction and visual perception), and social cognition (emotion recognition and behavioral regulation). Major neurocognitive disorder indicates sufficient severity of impairment in these domains and loss of independence in daily functioning to be consistent with dementia (Reiman et al. 2011). Recognizing the pattern of specific cognitive domains affected may be helpful to further diagnose subtypes of dementia, with the subtypes representing different etiologies.

7.1.2 *Dementia Prevalence*

Reports on dementia prevalence use different age classifications and assessment approaches, making comparisons difficult. Further, prevalence reflects a combination of incidence and survival, and survival rates among the nondemented vary widely in different parts of the world. That said, the following is a summary of the most recent, most comprehensive numbers.

As of 2010, the number of people with dementia above 60 years of age worldwide was estimated to be 35.6 million, with a projection of almost twofold increase by 2030 (Ferri et al. 2005; Wimo and Prince 2010). Estimated crude prevalence of dementia among those aged 60 years and older was higher among developed countries, with approximately 7% in North America and Western Europe, than among developing countries; whereas the rate of increase in prevalence was far higher among developing countries, including Latin American nations, China, and India (Ferri et al. 2005). Across all population-based studies in different regions of the world, prevalence of dementia consistently increases with age, with the highest percentage of affected people in the population aged 85 years or older (Berr et al. 2005; Ferri et al. 2005; Plassman et al. 2007). Thus, the projected increase in number of people with dementia directly reflects increased life expectancy.

7.1.3 *Dementia Subtypes*

By far, the most prevalent subtype is dementia due to AD, a degenerative process that accelerates neuronal death in the brain. Population-based studies of the prevalence of dementia show that AD accounted for 70% of all cases of dementia in the USA (Plassman et al. 2007); 64% in Canada (Canadian Study of Health and Aging 1994); 54% across eight European countries (Lobo et al. 2000); and 60% in developing countries (Kalaria et al. 2008).

Early clinical presentation of typical AD has progressive short-term memory deficits at its core (Cummings and Cole 2002). Histopathological markers of AD observed postmortem include: extracellular amyloid plaques and intracellular neurofibrillary tangles (NFTs). The National Institute on Aging and the Alzheimer's Association (NIA-AA) recently published the new diagnostic criteria for AD (McKhann et al. 2011). To meet diagnostic criteria for "probable dementia due to AD," an individual should have (1) a clinical diagnosis of dementia, including impairment in multiple domains that interfere with functional independence, (2) gradual cognitive decline with insidious onset, (3) either amnestic (learning and recall) or nonamnestic (language, visuospatial, and executive dysfunction) cognitive deficits, and (4) no prominent features of other dementia subtypes. Additionally, corroborative evidence that improves diagnostic confidence includes documented decline based on informant report and neuropsychological assessments, presence of known AD genetic mutations, and abnormal pathophysiological biomarkers on imaging or cerebrospinal fluid assays. Of note, the new NIA-AA criteria expanded the definition of AD to include nonmemory types of AD, recognizing that some AD cases may not present memory deficits early on and instead show deficits in other cognitive domains (Lopez et al. 2011). With respect to the continuum of dementia, the new NIA-AA criteria specify diagnostic features of "MCI due to AD" and "preclinical AD." Criteria for MCI due to AD include self- or informant-reported changes in cognition, education- and age-inappropriate cognitive impairment in one or more cognitive domains, and slight decline in performing functional tasks with intact functional independence (Albert et al. 2011). Notably, preclinical AD is proposed as a category for research only, not for clinical use. Criteria for preclinical AD include asymptomatic individuals with positive biomarkers and presumed at risk for developing either MCI due to AD or AD dementia, or individuals with positive biomarkers and subtle age-incongruent cognitive decline (Sperling et al. 2011).

Hypotheses regarding the cause of AD are relevant to possible genetic pathways, environmental exposures, and treatments. The two most characteristic neuropathological features of the disease are extracellular plaques composed of amyloid beta (Aβ) peptide and neurofibrillary tangles (NFTs) composed of abnormal tau protein (Hyman et al. 2012). The amyloid cascade hypothesis (Hardy 2006) has provided major contributions to understanding the pathophysiology of AD. The amyloid hypothesis as it has evolved posits overaccumulation of Aβ in the form of soluble oligomers and insoluble fibrils that aggregate as plaques. Currently, it is thought that the oligomers instigate the sequence of events, including chronic inflammation that results in neuronal injury and death (White et al. 2005). The tau hypothesis postulates that disruptions in tau–microtubule binding by increased phosphorylation promote abnormal aggregation of "free" tau proteins, eventually leading to the formation of NFTs, which facilitate neuronal injury and death (Mudher and Lovestone 2002). The two hypotheses are not independent; it has been observed that an increased Aβ concentration triggers abnormal changes in tau protein and resultant formation of NFTs (Oddo et al. 2006), with measures of tau thus representing a more "downstream" biomarker of neuronal injury (Albert et al. 2011).

Vascular dementia (VaD) is the second most common subtype of dementia in the elderly. Prevalence estimates range from 16 to 24 % of all dementia cases (Canadian Study of Health and Aging 1994; Kalaria et al. 2008; Lobo et al. 2000; Plassman et al. 2007).

A diagnosis of VaD requires (1) clinical symptomatology of dementia, (2) evidence of ischemic or hemorrhagic cerebrovascular disease (CVD) or hypoperfusive ischemic cerebral infarcts resulting from cardiovascular and circulatory disorders, and (3) close temporal association between dementia and vascular etiology (Chui et al. 1992; Román et al. 1993). Several versions of diagnostic criteria for VaD exist to date, with lack of consensus among them (Chui 2006; Wiederkehr et al. 2008). In particular, it involves considerable challenge to characterize the profile of cognitive impairment in VaD because of the heterogeneous nature of underlying cerebral lesions in terms of number, size, and location (Moorhouse and Rockwood 2008). Increasingly, researchers prefer to use the term vascular cognitive impairment (VCI) (Hachinski and Bowler 1993; O'Brien et al. 2003), which incorporates a range of cognitive disorders with presumed vascular implications in order to encompass significant cognitive decline that does not meet criteria for dementia (Moorhouse and Rockwood 2008). Moreover, VCI construct includes recognition of the interplay between vascular disease and neurodegenerative pathology. Post-mortem studies often find that mixed neuropathology, including plaques and tangles characteristic of AD and vascular infarcts characteristic of VaD, is common (Kalaria and Ballard 1999; Schneider et al. 2007), and most experts view these pathologies as additive (Schneider and Bennett 2010).

After AD, dementia with Lewy bodies (DLB) is the second most prevalent neurodegenerative dementia. A systematic review concluded that it accounts for 0–22 % of all dementia cases, with the large range reflecting a need for more studies and greater use of consensus diagnostic criteria (Zaccai et al. 2005). Early cognitive features of DLB include decline in attentional, visuospatial, and executive abilities with relative memory preservation, compared with AD (Mrak and Griffin 2007). In addition to a clinical diagnosis of dementia, DLB is characterized by (1) fluctuating cognition with pronounced variation in attention and alertness, (2) recurrent visual hallucinations, and (3) spontaneous features of parkinsonism (McKeith et al. 2005). In terms of pathophysiology, DLB is marked by the presence of abnormally aggregated protein known as Lewy bodies throughout the whole brain, including neocortical areas and paralimbic structures. Progressive cognitive impairment may also occur in patients with Parkinson disease, called Parkinson disease with dementia (PDD). Both DLB and PDD are Lewy body dementias (LBD), with Lewy bodies comprising clumps of alpha-synuclein protein in the brain (Ballard 2004). The two LBDs are generally distinguished by which symptoms come first, motor (PDD) or cognitive (DLB). Overlapping pathology is not uncommon, with Lewy bodies often observed in patients with AD (Bonifati 2008).

Another subtype is frontotemporal lobar degeneration (FTLD), of which Pick disease is one rare clinical syndrome. FTLD is characterized by early manifestations of deficits in behavior, personality, executive functioning, and language (Rabinovici and Miller 2010). With respect to neuropathology and pathophysiology, many FTLD cases have tau protein deposits, a portion of which comprises Pick bodies, while a number of cases who were not tau-positive can show ubiquitin inclusions (Forman et al. 2006).

Finally, individuals may develop dementia symptomatology secondary to many diseases that affect the immune or metabolic system (World Health Organization, 2010).

7.2 Familial Influences and Estimating Heritability

Heritability is defined as the relative percentage of variance in a phenotype explained by genetic influences compared with the percentage of variance explained by environmental influences, within the population under study. The heritability of dementia carries both clinical and scientific importance and must be considered with regard to the context of each study. Clinically, the information helps relatives of dementia patients to understand their own risk for dementia. Most research attention has been given to investigating heritability in studies of AD, with reports typically specifying AD or combining across all dementias.

7.2.1 Family Studies

Research findings have consistently shown elevated risk for developing AD in first-degree relatives of AD patients. Among AD probands, various studies have reported that 34–42 % had a positive family history of AD (Green et al. 2002; Lautenschlager et al. 1996; Silverman et al. 1994). Taking a different approach to characterizing the importance of family history, cumulative risk for AD among those with a positive family history ranges from 30 to 39 % (Lautenschlager et al. 1996; Silverman et al. 1994). These figures can be compared with a risk for AD of 12 % among first-degree relatives of normal controls (Silverman et al. 1994) or to an overall estimated lifetime risk for AD of 19 % (Plassman and Breitner 1997).

There is some suggestion that African-American first-degree relatives and normal controls may be at higher risk than respective samples of Whites (Green et al. 2002), and that female first-degree relatives are at greater risk of developing dementia than their male counterparts, even after accounting for the difference in longevity (Lautenschlager et al. 1996; Van Duijn et al. 1993).

Some family studies stratified first-degree relatives of AD patients according to the proband's age of onset (Li et al. 1995; Lautenschlager et al. 1996; Silverman et al. 2003). From these studies, Li et al. (1995) concluded that earlier age of onset in the case may increase risk of earlier onset AD in the relative but not their total lifetime risk. For example, Lautenschlager et al. (1996) reported that relatives of cases with onset before age 72 years had increased risk of developing AD, but only until they themselves reached age 82 years.

In a population-based study with 74 FTLD probands and 561 age- and gender-matched controls, Stevens et al. (1998) reported that the risk for developing dementia before age 80 years among 411 relatives of FTLD probands was 22 %, compared with 11 % among 2,934 relatives of controls.

7.2.2 Twin Studies

As the most basic descriptive step, comparing monozygotic (MZ) and dizygotic (DZ) twins in their concordance rates of AD provides evidence with respect to genetic influence on liability to AD. Probandwise concordance rates are based on the ratio of the number of affected twin partners of independently ascertained probands to the total number of probands. Four different population-based twin studies were launched in the mid-1990s. Probandwise concordance rates for all dementia and for AD alone are shown in Table 7.1; all four studies report higher concordance for AD among MZ pairs than DZ pairs, although concordance rates and estimates of heritability vary across studies.

Bergem et al. (1997) identified dementia probands in long-term care facilities in Norway and located their cotwins using the Norwegian Twin Registry. Heritability for AD in this study was estimated to be 60% (Bergem and Lannfelt 1997).

Breitner et al. (1995) ascertained all twins with dementia from the National Academy of Sciences-National Research Council (NAS-NRC) Registry of Aging Twin Veterans, largely from World War II. Twins in this study were relatively young, with their ages ranging from 62 and 73 years at the time of screening, prior to the age at which many would likely develop dementia, hence reducing concordance and heritability. Heritability was estimated to be 28% (Plassman and Breitner 1997). When liability to disease was modeled using age of onset rather than disease risk, heritability was estimated to be 37% (Meyer and Breitner 1998). As the cohort has aged, concordance has increased (Plassman et al. 2004; Steffens et al. 2000). Additionally, Steffens et al. found more cases of AD among first-degree relatives of the concordant twin pairs compared with first-degree relatives of the discordant twin pairs.

In the Finnish Twin Registry consisting of all same-sex twin pairs in Finland, Räihä et al. (1996) identified twins with dementia through matching the twin registry to the national hospital discharge database. On the basis of the data in the article, Plassman and Breitner (1997) computed heritability to be 45% in this sample. Use of the discharge registry to identify cases likely resulted in underascertainment, which would depress heritability (Gatz and Pedersen 1996).

Gatz et al. (1997) identified all cases of dementia in the Swedish Adoption/Twin Study of Aging (SATSA) sample, which comprises a subset of Swedish Twin Registry (STR). Using standard biometrical models, heritability of AD was estimated to be 74%, and heritability of all dementias, 43%. However, the difference between age of onset for twins in concordant MZ pairs was as large as 16 years. Pedersen et al. (2001) estimated heritability of AD in this same sample using multiple thresholds reflecting age of onset rather than disease risk. Using this approach, heritability estimated using a threshold fixed to age-based population prevalence was 78%, whereas using a model that allows for varying thresholds derived from observed data produced an estimate of 57%, accounting for mortality and likelihood of follow-up to a certain age in the data set.

Subsequently, Gatz et al. (2006) screened all twin pairs in the STR aged 65 years and older for cognitive impairment, and referred those who screened positive and their cotwins for a complete diagnostic work-up. A total of 11,884 twin pairs were included in the study. Heritability estimates for AD were 58% in an age-adjusted full model, including genetic, shared, and nonshared environmental factors, and 79% in the age-adjusted best fitting model, excluding shared environmental effects. Probandwise concordance rates for all dementia were 44% in MZ and 25% in DZ pairs for men and 58% in MZ and 45% in DZ pairs for women. As a point of comparison, Gatz (2007) created unrelated pairs matched by sex and year of birth; the

Table 7.1 Summary of concordance rates from twin studies of dementia and AD

Authors	Registry	Dementia			AD			
		Pairs where one or both were affected	Probandwise concordance		Pairs where one or both were affected	Probandwise concordance		Heritability of AD (%)
			MZ (%)	DZ (%)		MZ (%)	DZ (%)	
Bergem et al. (1997)	Norwegian Twin Registry	72	57	33	38	83	46	60
Breitner et al. (1995)	NAS-NRC	–	–	–	37	21	11	28
Meyer and Breitner (1998)	NAS-NRC							38
Steffens et al. (2000)	NAS-NRC plus volunteers	–	–	–	94	36	18	
Plassman et al. (2004)	NAS-NRC					51	21	
Räihä et al. (1996)	Finnish Twin Registry	–	–	–	84	31	9	45
Gatz et al. (1997)	Swedish Twin Registry: SATSA	65	50	30	40	67	22	74
Pedersen et al. (2001)	SATSA, multiple thresholds							57–78
Gatz et al. (2006)	Swedish Twin Registry, age-adjusted thresholds	562	54	34	392	59	32	58–79

SATSA Swedish Adoption/Twin Study of Aging, *NAS-NRC* National Academy of Sciences-National Research Council, *MZ* monozygotic, *DZ* dizygotic, *AD* Alzheimer disease estimated concordance for dementia over a series of random unrelated pairs was 12% for men and 21% for women.

Few findings are available for dementias other than AD. In the Norwegian study, concordance for VaD was 29% among both MZ and DZ twins (Bergem et al. 1997); in the Finnish study, the figures were 31% for MZ and 12% for DZ twins. Wang et al. (2009) examined autopsy-confirmed DLB in the NAS-NRC twins. In 17 pairs, only one MZ pair was concordant for DLB. Four additional pairs were concordant for dementia, but only one twin in each pair was diagnosed with DLB. One newer twin study, the Vietnam Era Twin Study of Aging (VETSA), a longitudinal study of cognitive and brain aging beginning in midlife (Kremen et al. 2006), will soon be able to report on concordance and heritability of MCI.

An important contribution of twin studies to knowledge about AD lies in their indicating the extent to which genes likely play a role in liability to the disease, providing a context for the search for specific risk genes. Although findings show variability, the variability is less in samples of similar age, and it is clear that heritability of liability for AD is substantial. On the basis of MZ twin similarity alone, Roberts et al. (2012) estimated the predictive capacity of knowing an individual's personal genome. For most diseases, predictive capacity was low. However, for AD, those who hypothetically received a positive genetic test would have a markedly elevated risk of eventually developing the disease, whereas a negative genetic test would indicate a substantially lowered risk.

7.3 Molecular Studies

Once there is evidence for familial aggregation of a disorder, such as the increased risk in first-degree relatives, differential concordances in MZ and DZ twins, or heritability estimates, the next logical step is to try to identify which genes are contributing to the disorder.

7.3.1 Family Linkage Studies and Rare Mutations

The earliest attempts to identify genes that could be responsible for dementia in general and AD in particular were classical linkage studies of relatively large pedigrees, in which multiple family members were identified with the disease (often affected sib pairs). Chromosome 21 was long thought to be a likely chromosome with loci that could be important for dementia, as those with Down syndrome often develop AD-typical plaques. The first gene with mutations linked to early-onset AD was amyloid precursor protein (*APP*) on chromosome 21 (Goate et al. 1991). Identifying mutations in this gene (now up to 29) and subsequent work with understanding the mechanisms by which mutations change the protein product of this gene did much for developing the amyloid cascade hypothesis of AD (Hardy 2006).

Two other genes also have mutations that are linked with familial, early-onset AD: presenilin 1 (*PSEN1*) on chromosome 14 (Sherrington et al. 1995) and presenilin 2 (*PSEN2*) on chromosome 1 (Levy-Lahad et al. 1995; Rogaev et al. 1995). Mutations in all three of these genes are completely penetrant, with an autosomal dominant mode of transmission. All are involved in production or processing of APP, hence leading to Aβ deposition and increases in the Aβ42:Aβ40 ratio (Tanzi 2012). Mutations in *PSEN1* are the most common (185 to date); nevertheless, mutations in these three genes are relatively rare, and they account for less than 5 % of all AD cases (Cummings and Cole 2002). Notably, the vast majority of AD cases do not carry mutations in any of these genes.

A rare form of VaD called cerebral autosomal dominant arteriopathy with subcortical infarcts and leukoencephalopathy (CADASIL) has similarly been attributed to a mutation in a single gene (Chabriat et al. 2009). Mutations in five genes have been associated with autosomal dominant FTLD, accounting for about 10 % of all FTLD (Rabinovici and Miller 2010). At least two genes have been identified in autosomal dominant LBD, implicated in both PD and DLB, but explaining only a small minority of cases (Bonifati 2008; Forman et al. 2005).

7.3.2 Association Studies of Candidate Genes

Prior to the technological advancements that enabled large-scale GWAS, most other efforts to identify genes for AD or any dementia were candidate gene association studies. Some leads in the late 1990s revolved around linkage findings on chromosomes 10 and 12. However, several dozen loci on other chromosomes were also considered. The majority of studies focused on candidate genes that were hypothesized to infer increased susceptibility due to what was known of their function, such as being part of certain pathogenic pathways. Many were considered because they were involved in APP processing and Aβ production, clearance, and degradation. Others were considered because of potential roles in the formation of NFTs, whereas others because of their role in inflammation, oxidative stress, or cerebrovascular events (Bertram and Tanzi 2008).

Genetic association studies for AD had their most important breakthrough in 1993 when the epsilon 4 (ε4) allele of apolipoprotein E (*APOE*) was associated to both late-onset familial and sporadic AD (Corder et al. 1993; Saunders et al. 1993; Strittmatter et al. 1993). Meta-analyses of 38 case-control studies indicated that carriers of the ε4 allele had an odds ratio (OR) of 3.68 compared with carriers of the "wild-type" ε3 allele, whereas ε2 carriers are protected (OR=0.62) (http://www.alzgene.org). Another meta-analysis indicates that ε4 homozygotes have an OR of 14.9 in Caucasian populations and 33.1 among Japanese (Farrer et al. 1997). *APOE* genotypes are actually haplotypes of two single nucleotide polymorphisms (SNPs), rs7412 and rs429358. The latter, which is essential for defining the ε4 allele, is solely responsible for the association of *APOE* and dementia, and this association is mediated predominantly through its effect on Aβ42 levels in the central nervous system (Bennet et al. 2010). *APOE* as a susceptibility gene for AD is the most robust genetic association for any complex disorder known today. Not only is the association consistently found across studies and ethnicities (although the effect size in African-Americans needs some clarification), the effect size is several orders of magnitude greater than those typically found for most candidate genes and even findings from GWAS studies described later.

Findings from genetic association studies of candidate genes have been systematically catalogued and reviewed by Bertram and others (Bertram 2011; Bertram and Tanzi 2001, 2008) and are publically available with meta-analyses in the AlzGene database (http://www.alzgene.org) (Bertram et al. 2007; Bertram et al. 2011). From a handful of studies in the early 1990s, there has been an explosion in the number of reports. Through 2001, approximately 450 association studies were published (Bertram 2011). In the most recent update of AlzGene 10 years later (as of April 18, 2011), there were nearly 1,400 studies reporting on nearly 3,000 polymorphisms in 700 genes. Despite the large number of reports and genes evaluated, only 40 genes show significant risk effects in meta-analyses. A few genes that reached significance but were not signals in GWAS include *SORL1* (sortilin-related receptor), *ACE* (angiotensin-converting enzyme), *IL8* (interleukin 8), and *LDLR* (low-density lipoprotein receptor) (Tanzi 2012). Many of the studies suffer from the classic perils of gene discovery studies, such as small sample sizes, publication bias, and insufficient attention to appropriate covariates and confounders.

7.3.3 Genome-Wide Association Studies of AD

Attempts to find genes involved in the pathogenesis of AD have now shifted over to high-density GWAS with the first reports by Coon et al. (2007), Grupe et al. (2007), and Reiman et al. (2007) (see Table 7.2). As of mid-2012, there were 28 published GWAS reported in recent summaries from http://www.alzgene.org and the catalogue available through the National Human Genome Research Institute website (http://www.genome.gov/gwastudies/index.cfm?pageid=26525384#searchForm), with more than 60 loci implicated as potential modifiers of susceptibility to AD or age at onset for AD. In most cases, *APOE* comes out as the most significant finding and with the largest effect size. Many of the findings from these studies have yet to be replicated in other samples. Nevertheless, nine other genes have sufficient replication or significant meta-analytic results to be considered real associations: *BIN1* (bridging integrator 1), *CLU* (clusterin), *ABCA7* (ATP-binding cassette subfamily A member 7), *CR1* (complement receptor 1), *PICALM* (phosphatidylinositol-binding clathrin assembly protein), *MS4A6A/MS4A4E* (membrane-spanning 4-domains, subfamily A, member 6A/4E), *CD33* (myeloid cell surface antigen CD33 isoform 2 precursor), and *CD2AP* (CD2 associated protein). In contrast to *APOE*, for which the meta-analytic OR for ε4 versus ε3 is 3.68, the ORs for these loci are much smaller, ranging from 1.11 to 1.23. Another way of putting the importance of *APOE* into perspective is to consider the population attributable fraction, which is the proportion of disease burden attributable to a factor, in this case, an allele (Levine 2007). Using OR and minor allele frequencies, Bertram (2011) estimated that the population attributable fraction for *APOE* was 27 %, whereas the combined

Table 7.2 Summary of GWAS findings

GWAS	Design	Population	SNPs	AD GWAS (Follow-up)	Control GWAS (Follow-up)	"Featured" genes
Grupe et al. (2007)	Case-control	USA and UK	17,343	380 (1,428)	396 (1,666)	APOE, ACAN, BCR, CTSS, EBF3, FAM63A, GALP, GWA_14q32.13, GWA_7p15.2, LMNA, LOC651924, MYH13, PCK1, PGBD1, TNK1, TRAK2, UBD
Coon et al. (2007)	Case-control	USA and Netherlands	502,627	664 (–)	422 (–)	APOE
Reiman et al. (2007)	Case-control	USA and Netherlands	312,316	446 (415)	290 (260)	GAB2
Li et al. (2008)	Case-control	Canada and UK	469,438	753 (418)	736 (249)	APOE, GOLM1, GWA_15q21.2, GWA_9p24.3
Abraham et al. (2008)	Case-control	UK	561,494	1,082 (–)	1,239 (1,400)	APOE, LRAT
Bertram et al. (2008)	Family-based	USA	484,522	941 (1,767)	404 (838)	APOE, ATXN1, CD33, GWA_14q31
Poduslo et al. (2009)	Family-based and Case-control	USA	489,218	9 (199)	10 (225)	TRPC4AP
Potkin et al. (2009)	Case-control	USA (ADNI)	516,645	172 (–)	209 (–)	APOE, ARSB, CAND1, EFNA5, MAGI2, PRUNE2, TOMM40
Beecham et al. (2009)	Case-control	USA	532	492 (238)	496 (220)	APOE, FAM113B
Carrasquillo et al. (2009)	Case-control	USA	313,504	844 (1,547)	1,255 (1,209)	APOE, PCDH11X
Lambert et al. (2009)	Case-control	Europe	~540,000	2,035 (3,978)	5,328 (3,297)	APOE, CLU(APOJ), CR1
Harold et al. (2009)	Case-control	USA and Europe	~610,000	3,941 (2,023)	7,848 (2,340)	APOE, CLU(APOJ), PICALM
Heinzen et al. (2010) (CNV[a])	Case-control	USA	not given	331 (–)	368 (–)	APOE, CHRNA7
Seshadri et al. (2010)	Case-control	Europe and USA	~2,540,000	3,006 (6,505)	22,604 (13,532)	APOE, BIN1, CLU (APOJ), EXOC3L2, PICALM
Naj et al. (2010)	Case-control	USA and Europe	483,399	931 (1,338)	1,104 (2,003)	APOE, MTHFD1L

Table 7.2 (continued)

GWAS	Design	Population	SNPs	AD GWAS (Follow-up)	Control GWAS (Follow-up)	"Featured" genes
Sherva (2011)	Case-control	Israel	~2,540,000	124 142		AGPAT1, ATP6V0A4, GLOD4, RGS6, TMEM132C
Wijsman et al. (2011)	Family-based and case-control	USA	565,336	1,848 (617)	1,991 (573)	APOE, CELF2
Hu et al. (2011)	Case-control	USA	509,376	1,831 (751)	1,764 (751)	APOE, B1N1
Lee et al. (2011)	Case-control	USA	627,380	549 (2,449)	544 (1,390)	DGKB, GWA_10q23.1. GWA_18q23, GWA_3q25.2, HPCAL1
Hollingworth et al. (2011)	Case-control	USA and Europe	496,763	6,688 (13,182)	13,685 (26,261)	ABCA7, B1N1, CD2AP, CD33, CR1, EPHA1, MS4A4E, MS4A6A
Naj et al. (2011)	Case-control	USA	2,324,889	8,309 (3,531)	7,366 (3,565)	APOE, B1N1, CD2AP, CD33, CLU, CR1, EPHA1, MS4A4A, PICALM
Antunez et al. (2011)	Case-control	Europe	696,707	319 (2,690)	769 (2,237)	MS4A
Logue et al. (2011)	Case-control	African-American	2,505,093	513 (–)	496 (–)	APOE, CLU, PICALM, BIN1, EPHA1, MS4A, ABCA7, CD33, PROX1, CNTNAP2, CNTN5, STK24, POLN
Lambert et al. (2013)	Case-control	USA and Europe	~7,700,000	2,025 (7,913)	5,328 (10,417)	FRMD4A

Modified after content on the AlzGene website (http://www.alzgene.org) and the National Human Genome Research Institute website (http://www.genome.gov/gwastudies/index.cfm?pageid=26525384#searchForm)

AD Alzheimer disease, *GWAS* genome-wide association studies, *SNP* single nucleotide polymorphism

[a] Copy number variants

attributable fraction for *BIN1*, *CLU*, *CR1*, and *PICALM* was only 19.3% with no single locus being greater than 6.5%. Similarly, many of the large consortia with GWAS data have applied genetic risk prediction models to their findings; all report that the addition of these genes minimally improved prediction of incident AD beyond age, sex, and *APOE* ε4 (see Seshadri et al. 2010, for example). GWAS findings reinforce the futility of using individual genetic risk profiling for AD beyond having information on age, sex, family history, and *APOE* status (Pedersen 2010).

At the same time, family history and twin heritability studies indicate that there is genetic risk not yet accounted for. Findings of relatively few replicable genes, each with very modest effect sizes (beyond *APOE*), suggest that there is considerable genetic heterogeneity for a complex disorder such as AD. It is not surprising that GWAS efforts are finding genes with relatively small effect sizes, as this would be compatible with a polygenic model of inheritance. Larger and larger consortia are combining their findings in the hunt for discovering new associations with AD. Power will increase to find genome-wide significant associations, all with effect sizes as small as those previously reported. Some consortia are implementing genome-wide gene-based approaches to find associations (Lambert et al. 2013). Others are focusing on whole-genome sequencing to identify rare variants that may contribute to late-onset AD. Recently, Jonsson et al. (2012) discovered that a rare variant in the *APP* gene (frequency of 0.2–0.5%) is protective of AD and cognitive decline in the oldest old. This finding is important as it gives further insight into the role of β-cleavage in *APP* and may lead to advances in finding therapeutic interventions.

7.3.4 Gene–Gene interactions

Moving beyond gene identification requires, focusing on multiple genes, including additive and interactive effects, and incorporating information on environmental risk and protective factors is required rather than further pursuit of gene identification or replication. Many cohorts that have contributed information to the GWAS analyses have at least some information on selected risk factors other than age, sex, and *APOE* genotype. Perhaps the greatest challenge for AD geneticists will be to evaluate both gene–gene interactions as well as gene–environment interactions.

Early evidence for potential interactions between genes at different loci (known as epistasis or gene–gene interaction) for AD came from candidate gene studies that found evidence for association of a candidate gene only when *APOE* was taken into account. Using synergy factor analysis, Combarros et al. (2009) evaluated 100 "claims or suggestions of epistasis" in AD. They found 27 gene–gene interactions in networks involving cholesterol metabolism, β-amyloid metabolism, inflammation, oxidative stress, and other networks. The vast majority of the interactions were with *APOE*. Most of the interactions were synergistic, such that the effect of another gene was considerably stronger in the presence of *APOE* ε4. Nevertheless, some interactions were antagonistic, with ε4 presence masking the effect of another gene. The Epistasis Project, a consortium of seven AD research groups with 1,757 AD cases and 6,294 controls, is systematically replicating interactions that have been reported in AD (Bullock et al. 2013; Heun et al. 2012; Kölsch et al. 2012) and has focused on genes involved in inflammation and glucose metabolism.

Given the strong association of *APOE* with AD, genome-wide studies that have not adjusted for *APOE* appropriately may find both false-positive and false-negative results (Wijsman et al. 2011). Indeed, the early GWAS finding for *GAB2* required post hoc stratification by APOE to reach significance (Reiman et al. 2007). Nevertheless, gene–gene interactions may explain some of the heritability of AD (Heun et al. 2012), although no attempts to quantify the contribution have been made.

7.4 Environmental Influences and Gene–Environment Interactions

7.4.1 Environmental Factors

Findings from twin studies provide evidence for a significant role of environmental influences on liability to dementia. Researchers have made vigorous efforts to identify potential environmental factors that can contribute to higher or lower risk of AD or dementia more generally. The focus has largely been directed to lifestyle choices and medical conditions. Recently, a group of experts was commissioned by National Institutes of Health (NIH) to provide an evidence report with regard to risk-modifying factors of AD (Williams et al. 2010). Key findings from that report, including 25 systematic reviews and 250 primary research studies mainly from developed countries, are featured here, while noting additional research evidence for possible underlying mechanisms.

7.4.1.1 Education, Occupational Complexity, and Cognitive Engagement

One of the most studied variables is level of education and the related factors of cognitively challenging occupational and leisure activities. The preponderance of evidence from prospective cohort studies indicates that fewer years of education is associated with increased risk of AD (Williams et al. 2010). Low education remains a significant risk in discordant MZ twin pairs, and bivariate twin modeling indicates that the association between low education and dementia is not genetically mediated (Gatz et al. 2007).

The Williams et al. (2010) report did not find sufficient evidence for a significant protective effect of occupation beyond the effect of education, but did conclude that more frequent participation during one's leisure in activities that are cognitively engaging is associated with reduced risk of AD. In twin studies, both complexity of work with people and midlife participation in cognitively engaging activities have been found to be protective (Andel et al. 2005; Carlson et al. 2008). Not yet resolved is the extent to which the mechanism accounting for the association of complex cognitive activities with lower rates of dementia is neuroprotection or improved compensation.

7.4.1.2 Physical Activity

Williams et al. (2010) reported a significant association between a high level of physical activity and decreased incident AD. On the basis of animal (Cotman and Berchtold 2002) and human (Erickson et al. 2011) studies, researchers posit that exercise induces increased levels of brain-derived neurotrophic factor (BDNF), important in facilitating neuronal growth.

7.4.1.3 Diet

Williams et al. (2010) conclude that high adherence to Mediterranean-type diet, typically involving higher consumption of fish, fruits, vegetables, and unsaturated fatty acids (e.g., olive oil), may be beneficial in lowering the risk of AD. Findings also seem to suggest a reliable association between low baseline serum folate levels and increased risk for AD and dementia. No other findings were judged conclusive with respect to demonstrating a role for any other dietary or nutritional factor.

7.4.1.4 Smoking

Evidence consistently indicates an elevated risk of AD for current smokers, compared with those who never smoked (Cataldo et al. 2010; Williams et al. 2010). The association is somewhat difficult to specify for former smokers because of variability in the length of tobacco use and abstinence (Williams et al. 2010).

7.4.1.5 Vascular Factors

The preponderance of research evidence establishes an increased risk of AD among persons with diabetes mellitus, with some evidence for increased risk of AD associated with elevated cholesterol in midlife (Williams et al. 2010). Statin use is associated with a moderately reduced risk for AD (Williams et al. 2010). Inconsistencies were found with respect to other vascular factors, including hypertension, antihypertensive use, and obesity. In twins, we find that obesity and overweight in midlife, but not in old age, are risk factors for dementia (Xu et al. 2011), possibly explaining some inconsistencies. One hypothesis posits that the insulin resistance syndrome may selectively affect the hippocampus and medial temporal cortex in the brain, areas affected in the AD patients (Craft 2009). Cerebrovascular changes associated with vascular risk factors may act additively with AD pathology in impairing brain function, giving additional importance to the role of vascular risk profiles for stroke (Gorelick et al. 2011).

7.4.1.6 Depression

Reviewed studies have found a reliable association between a history of clinical depression and incident AD (Williams et al. 2010). More studies than not, including studies of twins, have found that the association between depression and dementia only holds for depression that occurs for the first time, close in time to the age of onset of dementia (Brommelhoff et al. 2009). These findings are consistent with the hypothesis proposed by Alexopoulos (2005) that, at least for some individuals, there is disruption of frontal-striatal and frontal-limbic brain pathways that potentiates both late-onset depression and dementia.

7.4.1.7 Traumatic Brain Injury (TBI)

Some evidence suggests a heightened risk of AD for individuals with a history of TBI (Williams et al. 2010). It appears that the risk may be greater for males than for females, although this apparent difference may reflect the greater chance that males will be exposed, or inclusion of more males in studies reviewed.

7.4.1.8 Estrogen

Prospective cohort studies generally indicate a protective role for estrogen exposure and for estrogen replacement therapy (Williams et al. 2010). In contrast, in clinical trials with estrogen with or without progesterone, there is either no effect on risk of AD, all dementia, or MCI, or a slightly increased risk of AD (Williams et al. 2010).

7.4.2 Interaction between genes and environments

Environmental risk factors for dementia may have differential effects on individuals as a function of their genetic status, and vice versa. Understanding the interaction between genetic and environmental influences may be important for understanding disease mechanisms, treatment, and prevention. Studies to date focused almost entirely on the interactions with *APOE* status.

One of the earliest reports described a synergistic effect in which head injury significantly increased risk of AD only in the presence of *APOE* ε4 (Mayeux et al. 1995). However, a more recent review of subsequent studies determined that evidence for the interactive role of *APOE* and head injury in the development of AD was inconclusive (Van Den Heuvel et al. 2007).

Much attention has been devoted to whether cerebrovascular risk factors are potentiated in *APOE* ε4 carriers. Eriksson et al. (2010) found that nonstroke cardiovascular disease increased risk of AD in *APOE* ε4 carriers, but not in noncarriers. Similarly, previous stroke or transient ischemic attack predicted increased risk of developing AD only in *APOE* ε4 carriers, and not in noncarriers (Johnston et al. 2000). Peila et al. (2001) showed a synergistic effect of *APOE* ε4 and midlife hypertension on cognitive impairment in old age, where elevated systolic blood pressure had a greater adverse effect on cognition in *APOE* ε4 carriers than in noncarriers. Both Peila et al. (2002) and Irie et al. (2008a) reported that *APOE* ε4 increases the risk for AD in individuals with diabetes mellitus beyond an additive effect of the separate risks.

A minority of research findings suggest the opposite pattern of interaction between *APOE* ε4 and cerebrovascular risk factors. In population-based studies of African-Americans and Nigerians, researchers found that increased level of cholesterol was associated with increased risk of AD in noncarriers of *APOE* ε4, but not in carriers (Evans et al. 2000; Hall et al. 2006).

For other risk factors, predominantly nonvascular, the risk factor is more prominent among non-*APOE* ε4 carriers, or *APOE* status made no difference. For smoking, the increased risk of AD for current smokers is limited to those with no *APOE* ε4 alleles, compared with those with one or two ε4 alleles (Aggarwal et al. 2006; Reitz et al. 2007). These findings were similar for African-American and non-African-American respondents. Two studies have looked at fish intake and *APOE* ε4 status, finding that more than weekly consumption was associated with reduced risk of AD, but only in *APOE* ε4 noncarriers (Barberger-Gateau et al. 2007; Huang et al. 2005).

Researchers have also looked at the interaction between *APOE* ε4 and depression in the development of AD. Steffens et al. (1997) found no evidence for an interaction. Synergistic interactions were reported by Irie et al. (2008b) for AD and by Geda et al. (2006) for MCI, with elevated risk among individuals with both depression and *APOE* ε4.

Results are mixed for the interaction between physical activity and *APOE* genotype. In one study, low rate of leisure-time physical activity appears more deleterious among *APOE* ε4 carriers than noncarriers (Rovio et al. 2005). In another study, the relationship between higher physical activity and reduced risk for dementia was limited to noncarriers, and no such relationship was found among carriers (Podewils et al. 2005).

There are either null and inconsistent findings, or an absence of evidence, with respect to interactions between *APOE* and education or cognitive activity. Carlson et al. (2008), for example, reported that the protective effect of midlife participation in cognitively engaging activities was significant for *APOE* ε4 carriers but not for noncarriers, whereas Wilson et al. (2002) reported no difference in the protective effect by *APOE* status. Finally, using a community sample of older women, one study investigated the interaction between *APOE* ε4 and estrogen in cognitive impairment and found an association between the current estrogen use and attenuated risk of cognitive impairment only in noncarriers (Yaffe et al. 2000).

7.5 Current Directions

Numerous efforts continue to attempt to identify associations between gene variants and AD, primarily through large GWAS consortia and sequencing efforts to find rare variants (Jonsson et al. 2012). These gene discovery studies are being complemented by replication studies of previous gene candidates, often using additional information about detailed phenotyping, such as that gained through neuroimaging (Meda et al. 2012) or metabolomics. Gene–environment interactions are being pursued through classic epidemiological designs, where specific genes and environmental risk factors are evaluated in the same models, as described above in Sect. 7.4.2. A recent complement to this line of investigation is to evaluate the extent to which epigenetic changes may be induced by environmental risk factors for AD and hence account for gene–environment interactions (Chouliaras et al. 2010).

7.5.1 Metabolomics and Dementia

Metabolomics refer to the study of small molecules and metabolites in cells, tissues, and body fluids. It is now possible to identify and quantify hundreds to thousands of metabolites simultaneously. Hopes are that these metabolites will represent new biomarkers for disease detection (beyond Aβ and tau in cerebral spinal fluid), disease progression, and identification of networks implicated in disease pathogenesis, as envisioned by the new NIA-AA diagnostic criteria. Complex mathematical models are applied to detect differences in metabolic signatures between diseased and healthy individuals. Like other "omics" approaches, metabolomics is a hypothesis-free method of studying the state of the organism at the global level rather than studying one or a few potential biomarkers (Quinones and Kaddurah-Daouk 2009).

Studies investigating metabolomic changes in dementia are still rather sparse, but the field is rapidly growing. One of the first studies was conducted already in 1995, when Shonk et al. (1995) were able to demonstrate that AD patients had lower levels of N-acetylaspartate (NAA). These results have since been confirmed by several other studies (Adalsteinsson et al. 2000; Block et al. 2002; Rami et al. 2007). More recently, Kaddurah-Daouk et al. (2011) performed a pilot-study to assess the feasibility of identifying AD patients through metabolites in cerebrospinal fluid samples. They found that a model including levels of tryptophan, norepinephrine, and indoleacetic acid was able to completely separate the AD patient group from the control group. Moreover, they were also able to identify important differences between AD patients and controls in the levels of several metabolites related to the norepinephrine and serotonergic pathways. The largest difference was found in the level of norepinephrine, which was significantly decreased in patients with AD.

Focusing on lipidomics, Han et al. (2011) studied the levels of over 800 molecular lipid species in plasma from 26 AD patients and 26 cognitively normal controls. They found significantly reduced levels in eight molecular species of sphingomyelin and significantly increased levels of two ceramide species in AD patients compared with controls. Furthermore, they showed that the ratios of ceramide to sphingomyelin species better discriminated between AD patients and controls compared with either metabolite alone.

Although the field is still in its beginning years, metabolomics is providing dementia research several interesting new directions for further investigations. If new biomarkers for early disease detection and diagnosis can be identified, metabolomics could be of great importance for dementia, since the disease has such a long preclinical phase and is very difficult to diagnose. By using metabolic signatures rather than single biomarkers, it is also possible to capture a more comprehensive picture about the pathology of complex diseases.

7.5.2 Epigenetics and Dementia

Epigenetics refer to the regulation of gene expression through reversible mechanisms, mainly changes in DNA methylation and chromatin structure (epigenetics is described in more detail in Chap. 6). Several lines of evidence suggest that epigenetic mechanisms are involved in dementia, including the higher frequency of sporadic cases over familial cases, the non-Mendelian inheritance pattern, and the late age of onset (Bihaqi et al. 2012). The following sections provide examples of specific epigenetic mechanisms related to AD risk.

7.5.2.1 Dysregulation in Epigenetic Mechanisms

Deficient dietary intake of vitamins B6 and B12 and folic acid, which has been implicated in AD (Chouliaras et al. 2010), has been shown to influence the methylation regulatory pathway, specifically through a gene-encoding methylenetetrahydrofolate reductase (MTHFR). In turn, a polymorphism of MTHFR is associated with AD (Wang et al. 2008). Thus, the process by which B6, B12, and folic acid deficiency increases risk for AD may be through dysregulation of this epigenetic mechanism (see Kwok 2010 review).

The methylation status of repetitive elements, such as Line1, Alu, and SAT-α, is also thought to be important for global DNA methylation. Bollati et al. (2011) studied methylation in repetitive elements of AD patients and healthy controls and found a significant increase in methylation status for the transposable element LINE1.

7.5.2.2 Differences in Methylation of Specific Genes

Tissue-specific methylation patterns (both hypo- and hypermethylation) are associated with cancers, autoimmune diseases, and some neurological disorders, such as DLB (Fernandez et al. 2012), although no significant differences in patterns at

1,505 CpG sites could be detected in a small sample (n = 11) of AD brain tissues. Nevertheless, other studies have found that several genes already implicated in AD show dysregulation in methylation status. *APP*, the gene most commonly mutated in familial AD, has been shown to be hypomethylated in AD patients compared with healthy controls (West et al. 1995). Cell culture studies have shown that *PSEN1*, the second gene often mutated in familial AD, is overexpressed in response to alteration in methylation, leading to increase in Aβ production (Wang et al. 2008). Finally, the *APOE* gene has a hypomethylated CpG-poor promoter and a fully methylated 3′-CpG-island, that contains the sequences for the ε4-haplotype. Aberrant epigenetic control in this CpG-island may contribute to late-onset AD. Wang et al. (2008) showed hypermethylation of the *APOE* promoter in cells both from postmortem prefrontal cortex and lymphocytes of AD patients compared with controls. Without appropriate longitudinal samples, it is impossible to know whether these differences are a cause or a consequence of the AD pathology (Chouliaras et al. 2010).

7.5.2.3 Epigenetics as a Mechanism for Environmental Risks and for Gene–Environment Interaction

Environmental risk factors for dementia may act by inducing epigenetic changes, for example, deficiency of vitamin B12, B6, and folate, as discussed above. Head injury is another risk factor for dementia that has been shown to induce epigenetic changes (Chouliaras et al. 2010). Further work is necessary to determine whether epigenetic changes may also underlie gene–environment interactions (Iraola-Guzmán, et al. 2011). For example, it has been suggested that the methylated 3′-CpG-island in *APOE* may be dysregulated by exposure to environmental triggers, thus lending ε4 carriers more susceptible to developing AD pathology (Wang et al. 2008).

In 1989, Barker et al. proposed the "Fetal Basis of Adult Disease" hypothesis, postulating that many adult diseases actually have fetal origin, where insult at a critical period of development may result in changes in gene expression leading to functional deficits later in life (Barker et al. 1989). Along the same line, Lahiri et al. (2008) proposed the "Latent Early-Life Associated Regulation" (LEARn) model for AD, stating that environmental factors early in life can lead to latent expression of specific genes later in life. According to the model, environmental agents (such as heavy metals, cytokines, or dietary factors) can induce epigenetic changes in a gene, leading to changes in gene expression either immediately or after a period of latency in response to a secondary trigger. Animal studies support this hypothesis. Basha et al. (2005) showed that lead exposure in rodents led to a delayed overexpression of *APP* 20 months later. In contrast, no change in *APP* expression could be seen in response to lead exposure during old age. Further work is called for examining longitudinal differences in total methylation and gene-specific epigenetic dysregulation in concert with information about early and midlife exposures.

7.6 Summary

Dementia is one of the most common disorders in older adults, affecting an estimated 35.6 million people worldwide (or about 5 % of those aged 60 and older). Prevalence increases markedly with age; the number affected will increase as the proportion of the population aged 60 years and older, and especially aged 80 years and older, climbs sharply in both developed and developing countries. The pathophysiology of AD suggests hypotheses about genetic bases for the disorder, that is, that pathways concerning deposition of Aβ may be of importance. Twin and family studies demonstrate that AD is one of the most heritable disorders, with genetic factors accounting for as much as 79 % of the variation in AD.

Mutations in three genes, *APP*, *PSEN1*, and *PSEN2*, all related to Aβ, are highly penetrant, follow Mendelian transmission, but account for a tiny fraction of all AD cases, and mostly those with a relatively early age of onset. *APOE* continues to be the most important susceptibility gene for AD. Yet, the population attributable fraction for *APOE* is estimated at approximately 25 %, indicating that a great deal of the heritability for AD must be found in other genes of smaller effect size.

Even with GWAS, we have not succeeded at accounting for all genetic influences. Genes identified through GWAS have very small effect sizes, and little if anything will be gained from further gene discovery efforts. Thus, we are far from the point of personalized genetic risk profiling beyond using information on age, family history, and *APOE* status.

This situation leads us to pose two possibilities: might it be that AD is not just polygenic but also the result of risk alleles in a cluster of genes (most often including *APOE*), where some constellations of risk alleles are important in some individuals while other combinations are important in other individuals? Or might different combinations of risk alleles and environmental triggers (manifested as gene–environment interactions) characterize different individuals and thus thwart the ability to predict genetic risk? There are very few strong "environmental" risk (or protective) factors, and there is evidence that many of these work together with genes. The most consistent findings point to the importance of vascular risks in combination with *APOE*, which is related to cholesterol transport and Aβ42 levels. We urge further work to understand the extent and nature of gene–gene and gene–environment interactions and their role in the pathogenesis of AD. For example, one promising line of research may be in exploring the role of epigenetic mechanisms in explaining how environmental factors may impinge on genetic predisposition and trigger development of the disease.

Acknowledgments Grants R01 AG037985 and P50 AG05142.

References

Abraham, R., Moskvina, V., Sims, R., Hollingworth, P., Morgan, A., Georgieva, L., et al. (2008). A genome-wide association study for late-onset Alzheimer's disease using DNA pooling. *BMC Medical Genomics, 1,* 44.

Adalsteinsson, E., Sullivan, E. V., Kleinhans, N., Spielman, D. M., & Pfefferbaum, A. (2000). Longitudinal decline of the neuronal marker N-acetyl aspartate in Alzheimer's disease. *Lancet, 355*(9216), 1696–1697.

Aggarwal, N. T., Bienias, J. L., Bennett, D. A., Wilson, R. S., Morris, M. C., Schneider, J. A., et al. (2006). The relation of cigarette smoking to incident Alzheimer's disease in a biracial urban community population. *Neuroepidemiology, 26*(3), 140–146.

Albert, M. S., DeKosky, S. T., Dickson, D., Dubois, B., Feldman, H. H., Fox, N. C., et al. (2011). The diagnosis of mild cognitive impairment due to Alzheimer's disease: Recommendations from the National Institute on Aging-Alzheimer's Association workgroups on diagnostic guidelines for Alzheimer's disease. *Alzheimer's and Dementia, 7*(3), 270–279.

Alexopoulos, G. S. (2005). Depression in the elderly. *Lancet, 365,* 1961–1970.

Andel, R., Crowe, M., Pedersen, N. L., Mortimer, J., Crimmins, E., Johansson, B., & Gatz, M. (2005). Complexity of work and risk of Alzheimer's disease: A population-based study of Swedish twins. *Journals of Gerontology: Series B: Psychological Sciences, 60,* P251–P258.

Antunez, C., Boada, M., Gonzalez-Perez, A., Gayan, J., Ramirez-Lorca, R., Marin, J., et al. (2011). The membrane-spanning 4-domains, subfamily A (MS4A) gene cluster contains a common variant associated with Alzheimer's disease. *Genome Medicine, 3*(5), 33.

Ballard, C. G. (2004). Definition and diagnosis of dementia with Lewy bodies. *Dementia and Geriatric Cognitive Disorders, 17,* 15–24.

Barberger-Gateau, P., Raffaitin, C., Letenneur, L., Berr, C., Tzourio, C., Dartigues, J. F., & Alpérovitch, A. (2007). Dietary patterns and risk of dementia: the Three-City cohort study. *Neurology, 69*(20), 1921–1930.

Barker, D. J., Winter, P. D., Osmond, C., Margetts, B., & Simmonds, S. J. (1989). Weight in infancy and death from ischaemic heart disease. *Lancet, 2*(8663), 577–580.

Basha, M. R., Wei, W., Bakheet, S. A., Benitez, N., Siddiqi, H. K., Ge, Y.-W., et al. (2005). The fetal basis of amyloidogenesis: Exposure to lead and latent overexpression of amyloid precursor protein and beta-amyloid in the aging brain. *Journal of Neuroscience, 25*(4), 823–829.

Beecham, G. W., Martin, E. R., Li, Y. J., Slifer, M. A., Gilbert, J. R., Haines, J. L., & Pericak-Vance, M. A. (2009). Genome-wide association study implicates a chromosome 12 risk locus for late-onset Alzheimer disease. *American Journal of Human Genetics, 84*(1), 35–43.

Bennet, A. M., Reynolds, C. A., Gatz, M., Blennow, K., Pedersen, N. L., & Prince, J. A. (2010). Pleiotropy in the presence of allelic heterogeneity: Alternative genetic models for the influence of APOE on serum LDL, CSF amyloid-β42, and dementia. *Journal of Alzheimer's Disease, 22*(1), 129–134.

Bergem, A. L., & Lannfelt, L. (1997). Apolipoprotein E type 4 allele, heritability, and age at onset in twins with Alzheimer's disease and vascular dementia. *Clinical Genetics, 52*(5), 408–413.

Bergem, A. L. M., Engedal, K., & Kringlen, E. (1997). The role of heredity in late-onset Alzheimer disease and vascular dementia: A twin study. *Archives of General Psychiatry, 54*(3), 264–270.

Berr, C., Wancata, J., & Ritchie, K. (2005). Prevalence of dementia in the elderly in Europe. *European Neuropsychopharmacology, 15*(4), 463–471.

Bertram, L. (2011). Alzheimer's genetics in the GWAS era: A continuing story of 'replications and refutations'. *Current Neurology and Neuroscience Reports, 11*(3), 246–253.

Bertram, L., & Tanzi, R. E. (2001). Of replications and refutations: The status of Alzheimer's disease genetic research. *Current Neurology and Neuroscience Reports, 1*(5), 442–450.

Bertram, L., & Tanzi, R. E. (2008). Thirty years of Alzheimer's disease genetics: The implications of systematic meta-analyses. *Nature Reviews Neuroscience, 9*(10), 768–778.

Bertram, L., McQueen, M. B., Mullin, K., Blacker, D., & Tanzi, R. E. (2007). Systematic meta-analyses of Alzheimer disease genetic association studies: The AlzGene database. *Nature Genetics, 39*(1), 17–23.

Bertram, L., Lange, C., Mullin, K., Parkinson, M., Hsiao, M., Hogan, M. F., et al. (2008). Genome-wide association analysis reveals putative Alzheimer's disease susceptibility loci in addition to APOE. *American Journal of Human Genetics, 83*(5), 623–632.

Bertram, L., McQueen, M., Mullin, K., Blacker, D., & Tanzi, R. (2011, April 18). The AlzGene Database. Retrieved from http://www.alzgene.org. Accessed 23 May 2012.

Bihaqi, S. W., Schumacher, A., Maloney, B., Zhang, Y., Lahiri, D. K., & Zawia, N. H. (2012). Do epigenetic pathways initiate late onset Alzheimer disease (LOAD): Towards a new paradigm. *Current Alzheimer Research, 9*(5), 574–588.

Block, W., Jessen, F., Traber, F., Flacke, S., Manka, C., Lamerichs, R., et al. (2002). Regional N-acetylaspartate reduction in the hippocampus detected with fast proton magnetic resonance spectroscopic imaging in patients with Alzheimer disease. *Archives of Neurology, 59*(5), 828–834.

Bollati, V., Galimberti, D., Pergoli, L., Dalla Valle, E., Barretta, F., Cortini, F., et al. (2011). DNA methylation in repetitive elements and Alzheimer disease. *Brain, Behavior, and Immunity, 25*(6), 1078–1083.

Bonifati, V. (2008). Recent advances in the genetics of dementia with Lewy bodies. *Current Neurology and Neuroscience Reports, 8,* 187–189.

Breitner, J. C. S., Welsh, K. A., Gau, B. A., McDonald, W. M., Steffens, D. C., Saunders, A. M., et al. (1995). Alzheimer's disease in the National Academy of Sciences—National Research Council Registry of aging twin veterans: III. Detection of cases, longitudinal results, and observations on twin concordance. *Archives of Neurology, 52,* 763–771.

Brommelhoff, J. A., Gatz, M., Johansson, B., McArdle, J. J., Fratiglioni, L., & Pedersen, N. L. (2009). Depression as a risk factor or prodomal feature for dementia? Findings in a population-based sample of Swedish twins. *Psychology and Aging, 24,* 373–384.

Bullock, J. M., Medway, C., Cortina-Borja, M., Turton, J. C., Prince, J. A., Ibrahim-Verbaas, C. A. et al. (2013). Discovery by the Epistasis Project of an epistatic interaction between the GSTM3 gene and the HHEX/IDE/KIF11 locus in the risk of Alzheimer's disease. *Neurobiology of Aging, 34(4)*, 1309e1–7.

Canadian Study of Health and Aging Workshop Group (1994). Canadian study of health and aging: Study methods and prevalence of dementia. *Canadian Medical Association Journal, 150,* 899–913.

Carlson, M. C., Helms, M. J., Steffens, D. C., Burke, J. R., Potter, G. G., & Plassman, B. L. (2008). Midlife activity predicts risk of dementia in older male twin pairs. *Alzheimer's and Dementia, 4*(5), 324–331.

Carrasquillo, M. M., Zou, F., Pankratz, V. S., Wilcox, S. L., Ma, L., Walker, L. P., et al. (2009). Genetic variation in PCDH11X is associated with susceptibility to late-onset Alzheimer's disease. *Nature Genetics, 41*(2), 192–198.

Cataldo, J. K., Prochaska, J. J., & Glantz, S. A. (2010). Cigarette smoking is a risk factor for Alzheimer's disease: An analysis controlling for tobacco industry affiliation. *Journal of Alzheimers Disease, 19*(2), 465–480.

Chabriat, H., Joutel, A., Dichgans, M., Tournier-Lasserve, E., & Bousser, M.-G. (2009). Cadasil. *The Lancet Neurology, 8*(7), 643–653.

Chouliaras, L., Rutten, B. P., Kenis, G., Peerbooms, O., Visser, P. J., Verhey, F., et al. (2010). Epigenetic regulation in the pathophysiology of Alzheimer's disease. *Progress in Neurobiology, 90*(4), 498–510.

Chui, H. C. (2006). Vascular cognitive impairment: Today and tomorrow. *Alzheimer's and Dementia, 2*(3), 185–194.

Chui, H. C., Victoroff, J. I., Margolin, D., Jagust, W., Shankle, R., & Katzman, R. (1992). Criteria for the diagnosis of ischemic vascular dementia proposed by the State of California Alzheimer's disease diagnostic and treatment centers. *Neurology, 42*(3), 473–480.

Combarros, O., Cortina-Borja, M., Smith, A. D., & Lehmann, D. J. (2009). Epistasis in sporadic Alzheimer's disease. *Neurobiology of Aging, 30*(9), 1333–1349.

Coon, K. D., Myers, A. J., Craig, D. W., Webster, J. A., Pearson, J. V., Lince, D. H., et al. (2007). A high-density whole-genome association study reveals that APOE is the major susceptibility gene for sporadic late-onset Alzheimer's disease. *The Journal of Clinical Psychiatry, 68*(4), 613–618.

Corder, E. H., Saunders, A. M., Strittmatter, W. J., Schmechel, D. E., Gaskell, P. C., Small, G. W., et al. (1993). Gene dose of apolipoprotein E type 4 allele and the risk of Alzheimer's disease in late onset families. *Science, 261*(5123), 921–923.

Cotman, C. W., & Berchtold, N. C. (2002). Exercise: A behavioral intervention to enhance brain health and plasticity. *Trends in Neurosciences, 25*(6), 295–301.

Craft, S. (2009). The role of metabolic disorders in Alzheimer disease and vascular dementia: Two roads converged. *Archives of Neurology, 66*(3), 300–305.

Cummings, J. L., & Cole, G. (2002). Alzheimer disease. *JAMA: The Journal of the American Medical Association, 287,* 2335–2338.

Erickson, K. I., Voss, M. W., Prakash, R. S., Basak, C., Szabo, A., Chaddock, L., et al. (2011). Exercise training increases size of hippocampus and improves memory. *Proceedings of National Academy of Sciences of the United States of America, 108*(7), 3017–3022.

Eriksson, U. K., Bennet, A. M., Gatz, M., Dickman, P. W., & Pedersen, N. L. (2010). Non-stroke cardiovascular disease and risk of Alzheimer's disease and dementia. *Alzheimer Disease and Associated Disorders, 24,* 213–219.

Evans, R. M., Emsley, C. L., Gao, S., Sahota, A., Hall, K. S., Farlow, M. R., & Hendrie, H. (2000). Serum cholesterol, APOE genotype, and the risk of Alzheimer's disease: A population-based study of African Americans. *Neurology, 54*(1), 240–242.

Farrer, L. A., Cupples, L. A., Haines, J. L., Hyman, B., Kulkull, W. A., Mayeux, R., et al. (1997). Effects of age, sex and ethnicity on the association between apolipoprotein E genotype and Alzheimer disease. A meta analysis. *Journal of the American Medical Association, 278*(16), 1349–1356.

Fernandez, A. F., Assenov, Y., Martin-Subero, J. I., Balint, B., Siebert, R., Taniguchi, H., et al. (2012). A DNA methylation fingerprint of 1628 human samples. *Genome Research, 22*(2), 407–419.

Ferri, C. P., Prince, M., Brayne, C., Brodaty, H., Fratiglioni, L., Ganguli, M., et al. (2005). Global prevalence of dementia: A Delphi consensus study. *Lancet, 366*(9503), 2112–2117.

Forman, M. S., Lee, V. M.-Y., & Trojanowski, J. Q. (2005). Nosology of Parkinson's disease: Looking for the way out of a quackmire. *Neuron, 47*(4), 479–482.

Forman, M. S., Farmer, J., Johnson, J. K., Clark, C. M., Arnold, S. E., Coslett, H. B., et al. (2006). Frontotemporal dementia: Clinicopathological correlations. *Annals of Neurology, 59*(6), 952–962.

Gatz, M. (2007). Genetics, dementia and the elderly. *Current Directions in Psychological Science, 16,* 123–127.

Gatz, M., & Pedersen, N. L. (1996). Use of twin samples to estimate the heritability of Alzheimer's disease: A methodological note. *Alzheimer's Research, 2,* 229–231.

Gatz, M., Pedersen, N. L.., Berg, S., Johansson, B., Johansson, K., Mortimer, J. A., et al. (1997). Heritability for Alzheimer's disease: The study of dementia in Swedish twins. *The Journals of Gerontology: Series A: Biological Sciences and Medical Sciences, 52A*(2), M117–M125.

Gatz, M., Reynolds, C. A., Fratiglioni, L., Johansson, B., Mortimer, J. A., Berg, S., et al. (2006). Role of genes and environments for explaining Alzheimer disease. *Archives of General Psychiatry, 63*(2), 168–174.

Gatz, M., Mortimer, J. A. Fratiglioni, L., Johansson, B., Berg, S., Andel, R., et al. (2007). Accounting for the relationship between low education and dementia: A twin study. *Physiology & Behavior, 92,* 232–237.

Gauthier, S., Reisberg, B., Zaudig, M., Petersen, R. C., Ritchie, K., Broich, K., et al. (2006). Mild cognitive impairment. *Lancet, 367*(9518), 1262–1270.

Geda, Y. E., Knopman, D. S., Mrazek, D. A., Jicha, G. A., Smith, G. E., Negash, S., et al. (2006). Depression, apolipoprotein E genotype, and the incident of mild cognitive impairment: A prospective cohort study. *Archives of Neurology, 63,* 435–440.

Goate, A., Chartier-Harlin, M.-C., Mullan, M., Brown, J., Crawford, F., Fidani, L., et al. (1991). Segregation of a missense mutation in the amyloid precursor protein gene with familial Alzheimer's disease. *Nature, 349,* 704–706.

Gorelick, P. B., Scuteri, A., Black, S. E., DeCarli, C., Greenberg, S. M., Iadccola, C., et al. (2011). Vascular contributions to cognitive impairment and dementia: A statement for healthcare professionals from the American Heart Association/American Stroke Association. *Stroke, 42*(9), 2672–2713.

Green, R. C., Cupples, L. A., Go, R., Benke, K. S., Edeki, T., Griffith, P. A. (2002). Risk of dementia among White and African American relatives of patients with Alzheimer disease. *JAMA: The Journal of the American Medical Association, 287*(3), 329–336.

Grupe, A., Abraham, R., Li, Y., Rowland, C., Hollingworth, P., Morgan, A., et al. (2007). Evidence for novel susceptibility genes for late-onset Alzheimer's disease from a genome-wide association study of putative functional variants. *Human Molecular Genetics, 16*(8), 865–873.

Hachinski, V. C., & Bowler, J. V. (1993). Vascular dementia. *Neurology, 43*(10), 2159–2160.

Hall, K., Murrell, J., Ogunniyi, A., Deeg, M., Baiyewu, O., Gao, S., et al. (2006). Cholesterol, APOE genotype, and Alzheimer disease: An epidemiologic study of Nigerian Yoruba. *Neurology, 66,* 223–227.

Han, X., Rozen, S., Boyle, S. H., Hellegers, C., Cheng, H., Burke, J. R., et al. (2011). Metabolomics in early Alzheimer's disease: Identification of altered plasma sphingolipidome using shotgun lipidomics. *PLoS One, 6*(7), e21643.

Hardy, J. (2006). Alzheimer's disease: The amyloid cascade hypothesis: An update and reappraisal. *Journal of Alzheimer's Disease, 9,* 151–153.

Harold, D., Abraham, R., Hollingworth, P., Sims, R., Gerrish, A., Hamshere, M. L., et al. (2009). Genome-wide association study identifies variants at CLU and PICALM associated with Alzheimer's disease. *Nature Genetics, 41*(10), 1088–1093.

Heinzen, E. L., Need, A. C., Hayden, K. M., Chiba-Falek, O., Roses, A. D., Strittmatter, W. J., et al. (2010). Genome-wide scan of copy number variation in late-onset Alzheimer's disease. *Journal of Alzheimer's Disease, 19*(1), 69–77.

Heun, R., Kölsch, H., Ibrahim-Verbaas, C. A., Combarros, O., Aulchenko, Y. S., Breteler, M., et al. (2012). Interactions between PPAR-α and inflammation-related cytokine genes on the development of Alzheimer's disease, observed by the Epistasis Project. *International Journal of Molecular Epidemiology and Genetics, 3*(1), 39–47.

Hollingworth, P., Harold, D., Sims, R., Gerrish, A., Lambert, J. C., Carrasquillo, M. M., et al. (2011). Common variants at ABCA7, MS4A6A/MS4A4E, EPHA1, CD33 and CD2AP are associated with Alzheimer's disease. *Nature Genetics, 43*(5), 429–435.

Hu, X., Pickering, E., Liu, Y. C., Hall, S., Fournier, H., Katz, E., et al. (2011). Meta-analysis for genome-wide association study identifies multiple variants at the BIN1 locus associated with late-onset Alzheimer's disease. *PloS One, 6*(2), e16616.

Huang, T. L., Zandi, P. P., Tucker, K. L., Fitzpatrick, A. L., Kuller, L. H., Fried, L. P., et al. (2005). Benefits of fatty fish on dementia risk are stronger for those without APOE epsilon4. *Neurology, 65*(9), 1409–1414.

Hyman, B. T., Phelps, C. H., Beach, T. G., Bigio, E. H., Cairns, N. J., Carrillo, M. C., et al. (2012). National Institute on Aging–Alzheimer's Association guidelines for the neuropathologic assessment of Alzheimer's disease. *Alzheimer's and Dementia, 8*(1), 1–13.

Iraola-Guzmán, S., Extivill, X., & Rabionet, R. (2011). DNA methylation in neurodegenerative disorders: A missing link between genome and environment? *Clinical Genetics., 80,* 1–14.

Irie, F., Fitzpatrick, A. L., Lopez, O. L., Kuller, L. H., Peila, R., Newman, A. B., & Launer, L. J. (2008a). Enhanced risk for Alzheimer disease in persons with type 2 diabetes and APOE ε4: The Cardiovascular Health Study Cognition Study. *Archives of Neurology, 65*(1), 89–93.

Irie, F., Masaki, K. H., Petrovitch, H., Abbott, R. D., Ross, G. W., Taaffe, D. R., et al. (2008b). Apolipoprotein E ε4 allele genotype and the effect of depressive symptoms on the risk of dementia in men: The Honolulu-Asia Aging Study. *Archives of General Psychiatry, 65*(8), 906–912.

Jeste, D., Blacker, D., Blazer, D., Ganguli, M., Grant, I., Paulsen, J., et al. (2010, January 7). Neurocognitive disorders: A proposal from the DSM-5 neurocognitive disorders working group. DSM-5 Neurocognitive Criteria. http://www.dsm5.org/Proposed%20Revision%20Attachments/APA%20Neurocognitive%20Disorders%20Proposal%20for%20DSM-5.pdf. Accessed 21 Oct. 2010.

Johnston, J. M., Nazar-Stewart, V., Kelsey, S. F., Kamboh, M. I., & Ganguli, M. (2000). Relationships between cerebrovascular events, APOE polymorphism and Alzheimer's disease in a community sample. *Neuroepidemiology, 19,* 320–326.

Jonsson, T., Atwal, J. K., Steinberg, S., Snaedal, J., Jonsson, P. V., Bjornsson, S., et al. (2012). A mutation in APP protects against Alzheimer's disease and age-related cognitive decline. *Nature, 488*, 96–99.

Kaddurah-Daouk, R., Rozen, S., Matson, W., Han, X., Hulette, C. M., Burke, J. R., et al. (2011). Metabolomic changes in autopsy-confirmed Alzheimer's disease. *Alzheimer's and Dementia, 7*(3), 309–317.

Kalaria, R. N., & Ballard, C. (1999). Overlap between pathology of Alzheimer disease and vascular dementia. *Alzheimer Disease and Associated Disorders, 13*(3), S115–S123.

Kalaria, R. N., Maestre, G. E., Arizaga, R., Friedland, R. P., Galasko, D., Hall, K., et al. (2008). Alzheimer's disease and vascular dementia in developing countries: Prevalence, management, and risk factors. *The Lancet Neurology, 7*(9), 812–826.

Kölsch, H., Lehmann, D. J., Ibrahim-Verbaas, C. A., Combarros, O., van Duijn, C. M., Hammond, N., et al. (2012). Interaction of insulin and PPAR-α genes in Alzheimer's disease: The Epistasis Project. *Journal of Neural Transmission, 119*(4), 473–479.

Kremen, W. S., Thompson-Brenner, H., Leung, Y. J., Grant, M. D., Franz, C. E., Eisen, S. A., et al. (2006). Genes, environment, and time: The Vietnam Era Twin Study of Aging (VETSA). *Twin Research and Human Genetics, 9*(6), 1009–1022.

Kwok, J. B. J. (2010). Role of epigenetics in Alzheimer's and Parkinson's disease. *Epigenomics, 2*(5), 671–682.

Lahiri, D. K., Zawia, N. H., Greig, N. H., Sambamurti, K., & Maloney, B. (2008). Early-life events may trigger biochemical pathways for Alzheimer's disease: The "LEARn" model. *Biogerontology, 9*(6), 375–379.

Lambert, J. C., Heath, S., Even, G., Campion, D., Sleegers, K., Hiltunen, M., et al. (2009). Genome-wide association study identifies variants at CLU and CR1 associated with Alzheimer's disease. *Nature Genetics, 41*(10), 1094–1099.

Lambert, J. C., Grenier-Boley, B., Harold, D., Zelenika, D., Chouraki, V., Kamatani, Y., et al. (2013). Genome-wide haplotype association study identifies the FRMD4A gene as a risk locus for Alzheimer's disease. *Molecular Psychiatry, 18*(4), 461–470.

Lautenschlager, N. T., Cupples, L. A., Rao, V. S., Auerbach, S. A., Becker, R., Burke, J., et al. (1996). Risk of dementia among relatives of Alzheimer's disease patients in the MIRAGE study: What is in store for old? *Neurology, 46*(3), 641–650.

Lee, J. H., Cheng, R., Barral, S., Reitz, C., Medrano, M., Lantigua, R., et al. (2011). Identification of novel loci for Alzheimer disease and replication of CLU, PICALM, and BIN1 in Caribbean Hispanic individuals. *Archives of Neurology, 68*(3), 320–328.

Levine, B. (2007). What does the population attributable fraction mean? *Preventing Chronic Disease, 4*(1), 1–5.

Levy-Lahad, E., Wijsman, E. M., Nemens, E., Anderson, L., Goddard, K. A., Weber, J. L., et al. (1995). A familial Alzheimer's disease locus on chromosome 1. *Science, 269*(5226), 970–973.

Li, G., Silverman, J. M., Smith, C. J., Zaccario, M. L., Schmeidler, J., Mohs, R. C., & Davis, K. L. (1995). Age at onset and familial risk in Alzheimer's disease. *The American Journal of Psychiatry, 152*(3), 424–430.

Li, H., Wetten, S., Li, L., St Jean, P. L., Upmanyu, R., Surh, L., et al. (2008). Candidate single-nucleotide polymorphisms from a genomewide association study of Alzheimer disease. *Archives of Neurology, 65*(1), 45–53.

Lobo, A., Launer, L. J., Fratiglioni, L., Andersen, K., Di Carlo, A., Breteler, M. M. B., et al. (2000). Prevalence of dementia and major subtypes in Europe: A collaborative study of population-based cohorts. *Neurology, 54*(11), S4–S9.

Logue, M. W., Schu, M., Vardarajan, B. N., Buros, J., Green, R. C., Go, R. C., et al. (2011). A comprehensive genetic association study of Alzheimer disease in African Americans. *Archives of Neurology, 68*(12), 1569–1579.

Lopez, O. L., McDade, E., Riverol, M., & Becker, J. T. (2011). Evolution of the diagnostic criteria for degenerative and cognitive disorders. *Current Opinion in Neurology, 24*(6), 532–541.

Mayeux, R., Ottman, R., Maestre, G., Ngai, C., Tang, M. X., Ginsberg, H., et al. (1995). Synergistic effects of traumatic head injury and apolipoprotein-epsilon 4 in patients with Alzheimer's disease. *Neurology, 45,* 555–557.

McKeith, I. G., Dickson, D. W., Lowe, J., Emre, M., O'Brien, J. T., Feldman, H., et al. (2005). Diagnosis and management of dementia with Lewy bodies: Third report of the DLB consortium. *Neurology, 65*(12), 1863–1872.

McKhann, G. M., Knopman, D. S., Chertkow, H., Hyman, B. T., Jack, C. R., Kawas, C. H., et al. (2011). The diagnosis of dementia due to Alzheimer's disease: Recommendations from the National Institute on Aging-Alzheimer's Association workgroups on diagnostic guidelines for Alzheimer's disease. *Alzheimer's and Dementia, 7*(3), 263–269.

Meda, S. A., Narayanan, B., Liu, J., Perrone-Bizzozero, N. I., Stevens, M. C., Calhoun, V. D., et al. (2012). A large scale multivariate parallel ICA method reveals novel imaging-genetic relationships for Alzheimer's disease in the ADNI cohort. *Neuroimage, 60*(3), 1608–1621.

Meyer, J. M., & Breitner, J. C. S. (1998). Multiple threshold model for the onset of Alzheimer's disease in the NAS-NRC Twin Panel. *American Journal of Medical Genetics (Neuropsychiatric Genetics), 81,* 92–97.

Moorhouse, P., & Rockwood, K. (2008). Vascular cognitive impairment: Current concepts and clinical developments. *The Lancet Neurology, 7*(3), 246–255.

Mrak, R., & Griffin, W. S. T. (2007). Dementia with Lewy bodies: Definition, diagnosis, and pathogenic relationship to Alzheimer's disease. *Neuropsychiatric Diseases and Treatment, 3*(5), 619–625.

Mudher, A., & Lovestone, S. (2002). Alzheimer's disease—Do tauists and baptists finally shake hands? *Trends in Neurosciences, 25*(1), 22–26.

Naj, A. C., Beecham, G. W., Martin, E. R., Gallins, P. J., Powell, E. H., Konidari, I., et al. (2010). Dementia revealed: Novel chromosome 6 locus for late-onset Alzheimer disease provides genetic evidence for folate-pathway abnormalities. *PLoS Genetics, 6*(9).

Naj, A. C., Jun, G., Beecham, G. W., Wang, L. S., Vardarajan, B. N., Buros, J., et al. (2011). Common variants at MS4A4/MS4A6E, CD2AP, CD33 and EPHA1 are associated with late-onset Alzheimer's disease. *Nature Genetics, 43*(5), 436–441.

O'Brien, J. T., Erkinjuntti, T., Reisberg, B., Román, G., Sawada, T., Pantoni, L., et al. (2003). Vascular cognitive impairment. *The Lancet Neurology, 2,* 89–98.

Oddo, S., Caccamo, A., Tran, L., Lambert, M. P., Glabe, C. G., Klein, W. L., & LaFerla, F. M. (2006). Temporal profile of amyloid-beta (Abeta) oligomerization in an in vivo model of Alzheimer disease: A link between Abeta and tau pathology. *Journal of Biological Chemistry, 281*(3), 1599–1604.

Pedersen, N. L. (2010). Reaching the limits of genome-wide significance in Alzheimer disease: Back to the environment. *JAMA: The Journal of the American Medical Association, 303*(18), 1864–1865.

Pedersen, N. L., Posner, S. F., & Gatz, M. (2001). Multiple-threshold models for genetic influences on age of onset for Alzheimer disease: Findings in Swedish twins. *American Journal of Medical Genetics Part B: Neuropsychiatric Genetics, 105,* 724–728.

Peila, R., White, L. R., Petrovich, H., Masaki, K., Ross, G. W., Havlik, R. J., & Launer, L. J. (2001). Joint effect of the ApoE gene and midlife systolic blood pressure on late-life cognitive impairment: The Honolulu-Asia Aging Study. *Stroke, 32,* 2882–2889.

Peila, R., Rodriguez, B. L., & Launer, L. J. (2002). Type 2 diabetes, APOE gene, and the risk for dementia and related pathologies: The Honolulu-Asia Aging Study. *Diabetes, 51,* 1256–1262.

Plassman, B. L., & Breitner, J. C. S. (1997). The genetics of dementia in late life. *Psychiatric Clinics of North America, 20*(1), 60–76.

Plassman, B. L., Steffens, D. C., Burke, J. R., Welsh-Bohmer, K. A., Helms, M. J., & Breitner, J. C. S. (2004). Concordance for Alzheimer's disease in the NAS-NRC Twin Registry of WWII Veterans. *Neurobiology of Aging, 25,* S514.

Plassman, B. L., Langa, K. M., Fisher, G. G., Heeringa, S. G., Weir, D. R., Ofstedal, M. B., et al. (2007). Prevalence of dementia in the United States: The aging, demographics, and memory study. *Neuroepidemiology, 29*(1–2), 125–132.

Podewils, L. J., Guallar, E., Kuller, L. H., Fried, L. P., Lopez, O. L., Carlson, M., & Lyketsos, C. G. (2005). Physical activity, APOE genotype, and dementia risk: Findings from the Cardiovascular Health Cognition Study. *American Journal of Epidemiology, 161*(7), 639–651.

Poduslo, S. E., Huang, R., Huang, J., & Smith, S. (2009). Genome screen of late-onset Alzheimer's extended pedigrees identifies TRPC4AP by haplotype analysis. *American Journal of Medical Genetics Part B: Neuropsychiatric Genetics, 150B*(1), 50–55.

Potkin, S. G., Guffanti, G., Lakatos, A., Turner, J. A., Kruggel, F., Fallon, J. H., et al. (2009). Hippocampal atrophy as a quantitative trait in a genome-wide association study identifying novel susceptibility genes for Alzheimer's disease. *PloS One, 4*(8), e6501.

Quinones, M. P., & Kaddurah-Daouk, R. (2009). Metabolomics tools for identifying biomarkers for neuropsychiatric diseases. *Neurobiology of Disease, 35*(2), 165–176.

Rabinovici, G. D., & Miller, B. L. (2010). Frontotemporal lobar degeneration. *CNS Drugs, 24*(5), 375–398.

Räihä, I., Kaprio, J., Koskenvuo, M., Rajala, T., & Sourander, L. (1996). Alzheimer's disease in Finnish twins. *Lancet, 347,* 573–578.

Rami, L., Gomez-Anson, B., Bosch, B., Sanchez-Valle, R., Monte, G. C., Villar, A., & Molinuevo, J. L. (2007). Cortical brain metabolism as measured by proton spectroscopy is related to memory performance in patients with amnestic mild cognitive impairment and Alzheimer's disease. *Dementia and Geriatric Cognitive Disorders, 24*(4), 274–279.

Reiman, E. M., Webster, J. A., Myers, A. J., Hardy, J., Dunckley, T., Zismann, V. L., et al. (2007). GAB2 alleles modify Alzheimer's risk in APOE epsilon4 carriers. *Neuron, 54*(5), 713–720.

Reiman, E. M., McKhann, G. M., Albert, M. S., Sperling, R. A., Petersen, R. C., & Blacker, D. (2011). Alzheimer's disease: Implications of the updated diagnostic and research criteria. *The Journal of Clinical Psychiatry, 72*(9), 1190–1196.

Reitz, C., den Heijer, T., van Duijn, C., Hofman, A., & Breteler, M. M. (2007). Relation between smoking and risk of dementia and Alzheimer disease: The Rotterdam Study. *Neurology, 69*(10), 998–1005.

Roberts, N. J., Vogelstein, J. T., Parmigiani, G., Kinzler, K. W., Vogelstein, B., & Velculescu, V. E. (2012). The predictive capacity of personal genome sequencing. *Science Translational Medicine, 4*(133), 1–9.

Rogaev, E. I., Sherrington, R., Rogaeva, E. A., Levesque, G., Ikeda, M., Liang, Y., et al. (1995). Familial Alzheimer's disease in kindreds with missense mutations in a gene on chromosome 1 related to the Alzheimer's type 3 gene. *Nature, 376*(6543), 775–778.

Román, G. C., Tatemichi, T. K., Erkinjuntti, T., Cummings, J. L., Masdeu, J. C., Garcia, J. H., et al. (1993). Vascular dementia: Diagnostic criteria for research studies. Report of the NINDS-AIREN International Workshop. *Neurology, 43*(2), 250–260.

Rovio, S., Kåreholt, I., Helkala, E.-L., Viitanen, M., Vinblad, B., Tuomilehto, J., et al. (2005). Leisure-time physical activity at midlife and the risk of dementia and Alzheimer's disease. *The Lancet Neurology, 4,* 705–711.

Saunders, A. M., Strittmatter, W. J., Schmechel, D., St. George-Hyslop, P. H., Pericak-Vance, M. A., Joo, S. H., et al. (1993). Association of apolipoprotein E allele ε4 with late-onset familial and sporadic Alzheimer's disease. *Neurology, 43*(1), 1467–1472.

Schneider, J. A., & Bennett, D. A. (2010). Where vascular meets neurodegenerative disease. *Stroke, 41*(10), S144–S146.

Schneider, J. A., Arvanitakis, Z., Bang, W., & Bennett, D. A. (2007). Mixed brain pathologies account for most dementia cases in community-dwelling older persons. *Neurology, 69,* 2197–2204.

Seshadri, S., Fitzpatrick, A. L., Ikram, M. A., DeStefano, A. L., Gudnason, V., Boada, M., et al. (2010). Genome-wide analysis of genetic loci associated with Alzheimer disease. *JAMA: The Journal of the American Medical Association, 303*(18), 1832–1840.

Sherrington, R., Rogaev, E. I., Liang, Y., Rogaeva, E. A., Levesque, G., Ikeda, M., et al. (1995). Cloning of a gene bearing missense mutations in early-onset familial Alzheimer's disease. *Nature, 375*(6534), 754–760.

Sherva, R., Baldwin, C. T., Inzelberg, R., Vardarajan, B., Cupples, L. A., Lunetta, K., et al. (2011). Identification of novel candidate genes for Alzheimer's disease by autozygosity mapping using genome wide SNP data. *Journal of Alzheimer's Disease: JAD, 23*(2), 349–359.

Shonk, T. K., Moats, R. A., Gifford, P., Michaelis, T., Mandigo, J. C., Izumi, J., & Ross, B. D. (1995). Probable Alzheimer disease: Diagnosis with proton MR spectroscopy. *Radiology, 195*(1), 65–72.

Silverman, J. M., Li, G., Zaccario, M. L., Smith, C. J., Schmeidler, J., Mohs, R. C., & Davis, K. L. (1994). Patterns of risk in first-degree relatives of patients with Alzheimer's disease. *Archives of General Psychiatry, 51,* 577–586.

Silverman, J. M., Smith, C. J., Marin, D. B., Mohs, R. C., & Propper, C. B. (2003). Familial patterns of risk in very late-onset Alzheimer disease. *Archives of General Psychiatry, 60*(2), 190–197.

Sperling, R. A., Aisen, P. S., Beckett, L. A., Bennett, D. A., Craft, S., Fagan, A. M., et al. (2011). Toward defining the preclinical stages of Alzheimer's disease: Recommendations from the National Institute on Aging-Alzheimer's Association workgroups on diagnostic guidelines for Alzheimer's disease. *Alzheimer's and Dementia, 7*(3), 280–292.

Steffens, D. C., Plassman, B. L., Helms, M. J., Welsh-Bohmer, K. A., Saunders, A. M., & Breitner, J. C. (1997). A twin study of late-onset depression and apolipoprotein E epsilon 4 as risk factors for Alzheimer's disease. *Biological Psychiatry, 41,* 851–856.

Steffens, D. C., Plassman, B. L., Helms, M. J., Welsh-Bohmer, K. A., Newman, T. T., & Breitner, J. C. (2000). APOE and AD concordance in twin pairs as predictors of AD in first-degree relatives. *Neurology, 54*(3), 593–598.

Stevens, M., van Duijn, C. M., Kamphorst, W., de Knijff, P., Heutink, P., van Gool, W. A., et al. (1998). Familial aggregation in frontotemporal dementia. *Neurology, 50*(6), 1541–1545.

Strittmatter, W. J., Saunders, A. M., Schmechel, D., Pericak-Vance, M., Enghild, J., Salvesen, G. S., & Roses, A. D. (1993). Apolipoprotein E: High-avidity binding to beta-amyloid and increased frequency of type 4 allele in late-onset familial Alzheimer disease. *Proceedings of the National Academy of Sciences of the United States of America, 90*(5), 1977–1981.

Tanzi, R. E. (2012). The genetics of Alzheimer disease. *Cold Spring Harbor Perspectives in Medicine, 2*(10). doi:10.1101/cshperspect.a006296.

Van Den Heuvel, C., Thornton, E., & Vink, R. (2007). Traumatic brain injury and Alzheimer's disease: A review. *Progress in Brain Research, 161,* 303–316.

Van Duijn, C. M., Farrer, L. A., Cupples, A., & Hofman, A. (1993). Genetic transmission of Alzheimer's disease among families in a Dutch population based study. *Journal of Medical Genetics, 30*(8), 640–646.

Wang, S. C., Oelze, B., & Schumacher, A. (2008). Age-specific epigenetic drift in late-onset Alzheimer's disease. *PLoS One, 3*(7), e2698

Wang, C. S. M., Burke, J. R., Steffens, D. C., Hulette, C. M., Breitner, J. C. S., & Plassman, B. L. (2009). Twin pairs discordant for neuropathologically confirmed Lewy body dementia. *Journal of Neurology, Neurosurgery, and Psychiatry, 80*(5), 562–565.

West, R. L., Lee, J. M., & Maroun, L. E. (1995). Hypomethylation of the amyloid precursor protein gene in the brain of an Alzheimer's disease patient. *Journal of Molecular Neuroscience, 6*(2), 141–146.

White, J. A., Manelli, A. M., Holmberg, K. H., Van Eldik, L. J., & LaDu, M. J. (2005). Differential effects of oligomeric and fibrillar amyloid-β1-42 on astrocyte-mediated inflammation. *Neurobiology of Disease, 18*(3), 459–465.

Wiederkehr, S., Simard, M., Fortin, C., & van Reekum, R. (2008). Comparability of the clinical diagnostic criteria for vascular dementia: A critical review. Part I. *Journal of Neuropsychiatry and Clinical Neuroscience, 20*(2), 150–161.

Wijsman, E. M., Pankratz, N. D., Choi, Y., Rothstein, J. H., Faber, K. M., Cheng, R., et al. (2011). Genome-wide association of familial late-onset Alzheimer's disease replicates BIN1 and CLU and nominates CUGBP2 in interaction with APOE. *PLoS Genetics, 7*(2), e1001308.

Williams, J. W., Plassman, B. L., Burke, J., Holsinger, T., & Benjamin, S. (2010, April). Preventing Alzheimer's disease and cognitive decline: Evidence report/technology assessment No. 193. (AHRQ Publication No. 10-E005). Rockville, MD: Agency for Healthcare Research and Quality. http://www.ahrq.gov/downloads/pub/evidence/pdf/alzheimers/alzcog.pdf. Accessed 18 July 2011.

Wilson, R. S., Mendes De Leon, C. F., Barnes, L. L.., Schneider, J. A., Bienias, J. L., Evans, D. A., & Bennett, D. A. (2002). Participation in cognitively stimulating activities and risk of incident Alzheimer disease. *JAMA: The Journal of the American Medical Association, 287*(6), 742–748.

Wimo, A., & Prince, M. (2010, September 21). World Alzheimer report 2010: The global economic impact of dementia. Alzheimer Disease International website: http://www.alz.co.uk/research/files/WorldAlzheimerReport2010.pdf. Accessed 19 Mar. 2012.

World Health Organization. (2010). International statistical classification of diseases and related health problems 10th revision. http://apps.who.int/classifications/icd10/browse/2010/en. Accessed 6 Apr. 2012.

Xu, W., Atti, A. R., Gatz, M., Pedersen, N. L., Johansson, B., & Fratiglioni, L. (2011). Midlife overweight and obesity increase late-life dementia risk: A population-based twin study. *Neurology, 76,* 1568–1574.

Yaffe, K., Haan, M., Byers, A., Tangen, C., & Kuller, L. (2000). Estrogen use, APOE, and cognitive decline: Evidence of gene-environment interaction. *Neurology, 54*(10), 1949–1954.

Zaccai, J., McCracken, C., & Brayne, C. (2005). A systematic review of prevalence and incidence of dementia with Lewy bodies. *Age and Ageing, 34*(6), 561–566.

Part III
Biology and Neurobiology

Chapter 8
Brain Imaging and Cognition

Iván Chavarría-Siles, Guillén Fernández and Danielle Posthuma

During the last 3 decades, the study of the biological underpinnings of cognition has shown an exponential growth. The introduction of several brain imaging methods that allow studying the morphology and function of the brain have certainly contributed to this exponential growth of knowledge. Initially, functional imaging methods were expensive and invasive (such as positron and single–photon emission tomography); these methods were rapidly replaced with more cost-efficient noninvasive magnetic resonance imaging (MRI) techniques. At present, most brain imaging studies of cognition use MRI technology (Deary 2012). Thanks to the increased resolution of MRI scanners, we can now obtain whole-brain images with a spatial resolution of about 300–400 μm (Geyer et al. 2011).

The fast introduction of higher-resolution MRI scanners has been accompanied by a constant improvement of semiautomated statistical methods to quantify and systematically compare morphological and functional differences in brain structures. These methods provide a powerful tool for characterizing individual differences in brain anatomy, as well as in brain activity. Both structural and functional measures of the brain have been associated with cognitive function and dysfunction and have provided more insight into the underlying neural mechanisms of cognitive traits and disease.

I. Chavarría-Siles (✉)
Department of Psychiatry, Mount Sinai School of Medicine,
One Gustave L. Levy Place, Box 1230, New York, NY 10029, USA
e-mail: ivan.chavarria-siles@mountsinai.org

G. Fernández
Donders Institute, for Brainm, Cognition, and Behavior, Radboud University
Nijmegen Medical Center, P.O. Box 9101 6500 HB,
Nijmegen, The Netherlands
e-mail: g.fernandez@donders.ru.nl

D. Posthuma
Center for Neurogenomics and Cognitive Research (CNCR) & Department
of Medical Genomics, Vrij University Amsterdam VUMC W&N building,
room-439 De Boelelaan 1085, 1081 HV Amsterdam, The Netherlands
e-mail: Danielle.posthuma@cncr.vu.nl

D. Finkel, C. A. Reynolds (eds.), *Behavior Genetics of Cognition Across the Lifespan*,
Advances in Behavior Genetics 1, DOI 10.1007/978-1-4614-7447-0_8,

The first part of this chapter provides an overview of the most common imaging techniques used to study the structure and function of the brain. Next, we review how these neuroimaging techniques have been used to elucidate the development of the brain across the lifespan and its relation to cognitive function. Finally, we review some of the most consistent findings on the genetics of neuroimaging measures and the effect genes can have on the brain in relation to normative cognition and some neuropsychiatric disorders.

8.1 MRI-Based Methods to Study Brain Morphology and Function

8.1.1 Structural MRI

In recent years, a number of unbiased, objective techniques have been developed to characterize neuroanatomical differences in vivo using structural MRI. These techniques can be broadly classified into those that deal with macroscopic differences in brain shape and those that examine the local composition of brain tissue after macroscopic differences have been taken into account (Mechelli et al. 2005). The most commonly used MRI measures to study brain morphology in relation with cognition are: *voxel-based volumetry*, *gray matter* (*GM*) *cortical thickness and surface,* and measures of *white matter* (*WM*) *integrity*.

8.1.1.1 Voxel-based brain measures

Voxel-based morphometry (VBM) is one of the most commonly used methods to identify differences in the local composition of brain tissue. This is achieved by spatially *normalizing* all the obtained structural images to a unique stereotactic space; then *segmenting* the normalized images into gray and white matter; followed by *smoothing* the gray and white matter images; and finally performing a statistical analysis to localize significant differences between two or more experimental

Preprocessing steps of brain images for VBM analyses

Spatial Normalization Spatial normalization involves registering the individual MRI images to the same template image. An ideal template consists of the average of a large number of MR images that have been registered in the same stereotactic space.

Segmentation The spatially normalized images are then segmented into GM, WM, cerebrospinal fluid and three nonbrain partitions. This is generally achieved by combining a priori probability maps or "Bayesian priors", which encode the known spatial distribution of different tissues in normal subjects, with a mixture model cluster analysis which identifies voxel intensity distributions of particular tissue types.

Smoothing The segmented gray and white matter images are smoothed by convolving with an isotropic Gaussian kernel. The size of the smoothing kernel should be comparable to the size of the expected regional differences between the groups of brains.

groups (Ashburner and Friston 2000). VBM requires several preprocessing steps, as outlined in Box 8.1.

The VBM analysis output is a statistical parametric map (SPM) showing regions where gray or white matter differs significantly among the experimental groups. These maps can be used to examine differences between high and low cognitive performers, case and controls for a disease state, or between different genotypic groups. VBM has also shown to be useful in characterizing subtle changes in brain structure in a variety of diseases associated with neurological and psychiatric dysfunction (Mechelli et al. 2005).

8.1.1.2 Cortical Thickness and Cortical Surface measures

The human brain GM volume is defined as the amount of GM that lies between the gray–white interface and the pia mater. The total GM volume of the brain is a function of the cortical surface area and cortical thickness; both measurements are globally and regionally independent. Studies of interindividual variation in adult brain size have found that those differences in cortical GM volume are driven almost exclusively by differences in the cortical surface area rather than cortical thickness, such evidence suggests that surface area and thickness are distinct rather than redundant features of cortical structure. In addition, surface area and cortical thickness have been found to be both heritable, but seem to be genetically uncorrelated (Panizzon et al. 2009). It is also important to mention that cortical thickness varies considerably between different cortical areas; these variations across the cortex may reflect differences in cell types or neuron densities (Kanai and Rees 2011).

Several methods have been developed to calculate automatically cortical thickness and surface over the whole-brain based on MR images. Cortical anatomy, which is structured as a corrugated 2D sheet of tissue, can be well represented by surface models, which facilitate the analysis of relationships between cortical regions and provide superior visualization. Intersubject and even interspecies registration can be accomplished using surface-based representations, allowing matching of homologies without relying directly on spatial smoothing as in volume-based methods (Winkler et al. 2010).

Cortical thickness and surface are of great interest to both the study of normal cognitive development as well as a wide variety of neurodegenerative and psychiatric disorders. Changes in the GM that makes up the cortical sheet are manifested in normal aging, Alzheimer's disease (AD) and other dementias, Huntington's disease, corticobasal degeneration, amyotrophic lateral sclerosis, as well as schizophrenia (Fischl and Dale 2000).

8.1.1.3 WM Measures

Diffusion tensor imaging (DTI) is a variation of MRI that measures the diffusion of water in tissues. This method measures and quantifies a tissue's orientation and structure. DTI measures are thought to represent brain tissue microstructure integrity and are particularly useful for examining organized brain regions (Taylor et al. 2004). DTI has become one of the most popular MRI techniques in brain research. DTI enables visualization and characterization of WM tracks in 2D and 3D. Since the introduction of this methodology in 1994, it has been used to study the WM architecture and integrity of the brain (Assaf and Pasternak 2008).

DTI was rapidly accepted by imaging neuroscientists who saw in it a powerful and unique new tool for exploring the structural connectivity of the human brain. However, DTI is a rather approximate technique, and its results have frequently been given implausible interpretations. More recently, diffusion-weighted MRI (DW-MRI), which only measures the dephasing of spins of protons in the presence of a spatially varying magnetic field, has been proposed as the only method capable of mapping the fiber architecture of tissue (e.g., nervous tissue, muscle) in vivo. As DW-MRI has matured, an increasing number of software packages have been developed that allow such data to be analyzed in a push-button manner and then derive a p value, which can be interpreted according to the hypotheses being tested (Jones et al. 2013). In recent years, DW-MRI has been increasingly used to explore the relationship between WM structure and cognitive function. DW-MRI has been extensively employed to investigate how individual differences in behavior are related to variability in WM microstructure on a range of different cognitive tasks and also to examine the effect experiential learning might have on brain structural connectivity. Recent findings suggest that diffusion-weighted imaging (DWI) might even be used to measure functional differences in water diffusion during task performance (Roberts et al. 2013).

8.1.2 Functional MRI (fMRI)

Measuring the blood oxygenation level-dependent (BOLD) signal in humans using fMRI provides a noninvasive and large-scale view of neural activation while subjects perform simple or even complex cognitive tasks (event-related BOLD). fMRI has a primary advantage over other techniques (such as positron emission tomography (PET) or Single-photon emission computed tomography (SPECT) in neuroscience research because of its noninvasiveness, flexibility, and superior temporal as well as spatial resolution (Serences and Saproo 2011). This approach has been used to study a remarkable diversity of topics, from basic processes of perception and memory, to the complex mechanisms of economic decision making and moral cognition (Huettel 2012).

In recent years, the development of high-field MRI methods has resulted in a clearer picture of organization of individual human brains. The dramatic improvement in the quality of in vivo MRI scanning of human brain by increasing the magnetic field to 7 T, and by using a much more sensitive design of radiofrequency receiver coil to detect the MR signal has provided an increase of the signal-to-noise ratio by a factor of 10, allowing whole-brain images with a spatial resolution of only 300–400 μm. In order to meet the goal of *in vivo* mapping of brain's functional areas is necessary to perform systematic high-field MRI studies to provide microscopic anatomical concordance between cortical areas and BOLD (Geyer et al. 2011).

8.1.2.1 Resting-State fMRI (RS-fMRI)

Although the majority of researchers performing functional imaging studies continue to examine changes in brain activity associated while performing a task, some researchers in the field have also studied the spontaneous modulations of brain activity in the absence of an explicit task: RS-fMRI. The strength of this method is that it is paradigm-free, as it more or less ignores the cognitive state of the subject; however, this feature also makes the data analysis considerably more difficult than in standard event-related BOLD, as there is not a task that can be used to model the activation pattern (Norris 2006).

The main difference of this method from regular fMRI is that it looks into differences in connectivity between different parts of the brain and not into brain activity of a particular location. RS-fMRI studies have shown that regional fluctuations of spontaneous brain activity, measured in the absence of an explicit task, are highly organized and correlated across spatially distributed networks in a manner that recapitulates the topography of task-evoked functional coactivation patterns (Fornito and Bullmore 2010).

The majority of approaches to analyzing RS-FMRI data have thus far been spatially model-driven, with strong a priori hypotheses regarding the functional connectivity of a small number of brain regions of interest (ROIs) or individual voxel locations of interest. A characteristic set of coactivating functional systems is found consistently across subjects, stages of cognitive development, degrees of consciousness, and even (to some extent) across species. Interestingly, altered resting functioning of large-scale networks has been found in correlation with individual differences in behavioral performance, as well as in disease and under pharmacological manipulation (Cole et al. 2010). Moreover, individual resting-state networks have been shown to be heritable, thus the interindividual differences found in RS-fMRI studies are expected to be genetically driven (Glahn et al. 2010a).

8.2 Imaging Lifespan Changes of the Human Brain

8.2.1 Development of the Brain

The human brain has a particularly protracted maturation, with different tissue types, brain structures, and neuronal circuits having distinct developmental trajectories undergoing dynamic changes throughout the lifespan. The maturation of specific functional systems underlies the development of increasingly sophisticated cognitive functions from childhood to adulthood, including working memory, attention, and cognitive control (Giedd and Rapoport 2010).

Lenroot et al. (2007) reported the largest longitudinal pediatric neuroimaging study of typically developing children and adolescents (829 scans from 387 subjects, ages 3–27 years); they demonstrated increasing WM volumes and inverted U-shaped trajectories of GM volumes with peak sizes occurring at different times in different regions. Total cerebral volume follows an inverted U-shape trajectory peaking at age 10.5 in girls and 14.5 in boys. In both males and females, the brain is already at 95 % of its peak size by age 6. Across these ages, the group average brain size for males is ~10 % larger than for females. This 10 % difference is consistent with a vast amount of adult neuroimaging and postmortem studies, and often explained as being related to the larger body size of males. However, it has been found in pediatric subjects that the boy's bodies are not larger than girls' until after puberty.

It should be noted that differences in brain size between sexes or other groups should not be interpreted as necessarily imparting any sort of functional advantage or disadvantage. In the case of male/female differences, gross structural measures may not reflect sexually dimorphic differences in functionally relevant factors such as neuronal connectivity and receptor density (Giedd and Rapoport 2010). The shape of the age by size trajectories may be related to functional characteristics even more than the absolute brain size. DTI studies have shown that anisotropy increases and overall diffusion decreases with age (Cascio et al. 2007). WM development is a complex process that continues during childhood and adolescence; whether these changes end in adolescence is not clear. Lebel and Beaulieu (2011) examined longitudinal WM maturation using DTI in 103 healthy subjects aged 5–32 years (each subject was scanned at least twice). They assessed the development of 10 major WM tracts; all tracts showed significant nonlinear development trajectories. Significant within-subject changes occurred in the vast majority of children and early adolescents, and these changes were mostly complete by late adolescence for projection and commissural tracts. Additionally, WM volume increased significantly with age for most tracts, and longitudinal measures also demonstrated postadolescent volume increases in several association tracts.

How structural changes impact functional brain maturation is less well understood; understanding dynamic reconfiguration of brain networks between childhood and adulthood requires identifying changes in structural and functional connectivity during this period (Uddin et al. 2011). Using fMRI approaches in the adult brain, several canonical brain networks have been identified. Three of these can be considered core neurocognitive networks because of their critical roles in high-level cognition: (1) a frontoparietal central executive network (CEN) comprising the dorsolateral prefrontal cortex (DLPFC) and posterior parietal cortex (PPC), related to maintenance and manipulation of information and decision making in the context of goal-directed behavior; (2) a default mode network (DMN), including the ventromedial prefrontal cortex (VMPFC) and PCC, associated with internally oriented and social cognition; and (3) a salience network (SN) with nodes in the right frontoinsular cortex (rFIC) and anterior cingulated cortex (ACC), involved in attention as well as interoceptive and affective processes (Sridharan et al. 2008).

How these systems reconfigure and mature with development is a critical question for cognitive neuroscience, with implications for neurodevelopmental pathologies affecting brain connectivity. Using functional and effective connectivity measures applied to fMRI data, Uddin et al. (2011) examined the interactions within and between the SN, CEN, and DMN. They found that functional coupling between key network nodes is stronger in adults than in children, as are causal links emanating from the rFIC. Specifically, the causal influence of the rFIC on nodes of the SN and CEN was significantly greater in adults compared with children. Developmental changes in functional and effective connectivity were related to structural connectivity along these links. DTI tractography revealed increased structural integrity in adults compared with children along both within- and between-network pathways associated with the rFIC. Their results suggest that structural and functional maturation of rFIC pathways is a critical component of the process by which human brain networks mature during development to support complex, flexible cognitive processes in adulthood.

8.2.2 Brain Aging

Good et al. (2001) described the first optimized method of VBM to examine the effects of age on gray and white matter and CSF in 465 normal adults (age 17–79). They observed accelerated loss of GM volume symmetrically in both parietal lobes and ACC. Additionally, there is accelerated loss of GM concentration in the left middle frontal gyrus, left planum temporale, and transverse temporal gyri bilaterally. There was relative preservation of GM volume symmetrically in the amygdala, hippocampi, entorhinal cortices, and lateral thalami, with relative preservation of GM concentration more diffusely in the thalami. The whole-brain volume and gray and white matter partitions were larger in males compared with females. Furthermore, an interaction of sex with age-related global GM decline was observed, with a steeper age-related decline in males. There was no significant interaction of sex with age for CSF or WM change either globally or regionally. More recently, Peelle et al. (2012) replicated some of these findings by assessing age-related changes in GM volume in a sample of 420 adults evenly distributed between the ages of 18 and 77 years. They found age-related GM decline in nearly all parts of the brain, with particularly rapid decline in inferior regions of frontal cortex (e.g., insula and left inferior frontal gyrus) and the central sulcus.

Postmortem and volumetric imaging data suggest that brain myelination is a dynamic lifelong process that, in vulnerable late-myelinating regions, peaks in middle age. Bartzokis et al. (2012) assessed the adult lifespan trajectory DTI metrics in 171 healthy subjects 14–93 years of age. Their data suggest that the healthy adult brain undergoes continual change driven by development and repair processes devoted to creating and maintaining synchronous function among neural networks on which optimal cognition and behavior depend.

RS-fMRI studies have found that age-related changes in interregional functional connectivity exhibited spatially and temporally specific patterns. During brain development from childhood to senescence, functional connections tended to linearly increase in the emotion system and decrease in the sensorimotor system; while quadratic trajectories were observed in functional connections related to higher-order cognitive functions (Wang et al. 2012)

The aging of the human brain is accompanied not only by changes in cortical and WM structures, but also by functional activity changes and variable degree of cognitive decline. Finkel et al. (2005) used twin data from the Swedish Adoption/Twin Study of Aging (778 individuals tested on four occasions over a 13-year period) to construct four factors from 11 cognitive measures: verbal, spatial, memory, and processing speed. They found that for measures of fluid abilities, the explanatory value of processing speed is paramount for both mean cognitive performance and accelerating decline with age. They concluded that a significant proportion of the genetic influences on cognitive ability arose from genetic factors affecting processing speed. For measures of fluid abilities, it is not the linear age changes but the accelerating age changes in cognition that share genetic variance with processing speed.

Neurocognitive changes in healthy aging have now been reported for almost 2 decades; of these changes, executive functions have received the most attention. fMRI studies of executive control processes report robust differences in brain activity between older and younger subjects, particularly under conditions of high executive control demand. The most commonly reported age-related pattern of brain activity during executive function tasks (e.g., working memory, inhibition, and task-switching) is increased recruitment of lateral aspects of the prefrontal cortex bilaterally (Turner and Spreng 2012).

8.3 Imaging Genetics

8.3.1 Genetic Contributions to Human Brain Morphology

Twin studies have been key to determining the contribution of genetic, common, and unique environmental influences on variation in brain structures (Posthuma et al. 2000). Structural brain measurements are quantitative traits showing considerable variation in human populations; heritability estimates indicate a strong genetic component contributing to these neuroanatomical phenotypes.

Kaymaz and van Os (2009) extensively reviewed the heritability of gross brain structures; they included 24 studies reporting on the heritability of brain structures in healthy subjects. Gross brain structures show higher heritability rates than specific structures. Brain structure volumes have substantial heritability rates ranging from high (70–95 %) for total brain volume, cerebral gray and white matter, and corpus callosum, to moderate (40–70 %) for the hippocampus, the four lobes (frontal, temporal, occipital, and parietal lobe), temporal horn volume, brain parenchyma, WM hyperintensity, and planum temporal asymmetry. Structures formed earlier in development show consistently higher heritability rates than brain structures formed later in development: surface structures seem to be mainly influenced by environmental factors.

Winkler et al. (2010) analyzed surface- and voxel-based representations of brain structure using automated methods, and these measurements were analyzed using a variance-components method to identify the heritability of these traits and their genetic correlations. All neuroanatomical traits were significantly influenced by genetic factors. Cortical thickness and surface area measurements were found to be genetically and phenotypically independent. While both thickness and area influenced volume measurements of cortical GM, volume was more closely related to surface area than cortical thickness.

The surface area of the cerebral cortex is a highly heritable trait, yet little is known about genetic influences on regional cortical differentiation in humans. Chen et al. (2012) created a human brain atlas based solely on genetically informative data using a fuzzy clustering technique with MRI data from 406 twins from the Vietnam Era Twin Study of Aging (110 monozygotic and 93 dizygotic pairs, age range: 51–59). With this method, they described a previously unidentified parcellation system for the human cortex that reflects shared genetic influences on cortical areal expansion. This human brain atlas may provide novel phenotypes that will have greater statistical power for genome-wide genetic association studies in comparison with traditional cortical parcellations. In addition, they found evidence for a hierarchical, modular, and bilaterally symmetric genetic architecture across hemispheres.

8.3.2 Genetic Contributions to Human Brain Function

fMRI is a powerful tool for interrogating the mechanisms of the brain's response to different environmental stimuli. Nonetheless, even with a rigidly standardized stimulus or task, the brain's response is highly variable between people (Blokland et al. 2011). It is, however, challenging to assess the nature of interindividual variation in a spatial process, such as a pattern of neural activity in an fMRI study (Park et al. 2012).

As of today, few studies have addressed the heritability of task-related brain activation. Blokland et al. (2011) reported a voxel-by-voxel genetic model fitting in a large sample of identical and fraternal twins who performed an *n*-back working memory task during fMRI. Patterns of task-related brain response (BOLD signal difference of 2-back minus 0-back) showed moderate heritability, with the highest estimates (40–65 %) in the inferior, middle, and superior frontal gyri, left supplementary motor area, precentral and postcentral gyri, middle cingulate cortex, superior medial gyrus, angular gyrus, superior parietal lobule, including precuneus, and superior occipital gyri. Furthermore, high test–retest reliability for a subsample of 40 twins indicated that nongenetic variance in the fMRI brain response is largely due to unique environmental influences rather than measurement error.

Karlsgodt et al. (2010) assessed the genetic contributions to both working memory performance and structural neuroimaging measures focused on the network of these brain regions associated with working memory. Imaging measures included DTI indices in major WM tracts thought to be associated with working memory and structural MRI measures of frontal and parietal GM density. Their analyses directly addressed whether working memory performance and neural structural integrity were influenced by common genetic factors. While all cognitive measures, GM regions, and WM tracts assessed were heritable, only performance on a spatial delayed response task and integrity of the superior longitudinal fasciculus (a primary frontoparietal connection) shared genetic factors.

The default-mode network is diminished during effortful cognitive tasks and it increases when one's mind wonders. This connectivity pattern may be intrinsic to the primate brain, because it is present in sleeping infants and anesthetized nonhuman primates. Aberrant default-mode connectivity has been reported in individuals with neurological and psychiatric illnesses, suggesting that this intrinsic network is sensitive to pathophysiologic alterations in brain function and structure. Although the exact neurophysiologic mechanisms that regulate default-mode connectivity are unclear and likely differ between illnesses, there is growing evidence that genetic factors play a role (Glahn et al. 2010b).

Establishing the heritability of default-mode functional connectivity would authorize the use of resting-state networks as intermediate phenotypes. Glahn et al. (2010b) estimated the importance of genetic effects on the default-mode network by examining covariation patterns in functional connectivity. The heritability for the default-mode functional connectivity was 42%. Although neuroanatomical variation in this network was also heritable, the genetic factors that influence default-mode functional connectivity and GM density seem to be distinct, suggesting that unique genes influence the structure and function of the network. In contrast, significant genetic correlations between regions within the network provide evidence that the same genetic factors contribute to variation in functional connectivity throughout the default mode.

8.4 Imaging Cognition

8.4.1 Intelligence and the Brain

Individual differences in intelligence are strongly associated with many important life outcomes, including educational and occupational attainments, income, and health (Batty et al. 2007). The relation between intelligence (measured as intelligence quotient (IQ) and the brain has been studied since the end of the nineteenth century (Galton 1888). Structural neuroimaging studies generally report a modest correlation ($r \sim 0.3$) between psychometric measures of intelligence and total brain volume (McDaniel 2005).

The quantity of frontal GM is similar in individuals who are genetically alike; intriguingly, these individual differences in brain structure are tightly linked with individual differences in IQ. The resulting genetic brain maps reveal a strong relationship between genes, brain structure, and behavior, suggesting that highly heritable aspects of brain structure may be fundamental in determining individual differences in cognition (Thompson 2001). Jung and Haier (2007) reviewed 37 neuroimaging studies that focused on the relation between intelligence and neuronal networks. They reported a striking consensus of neuroanatomical and functional data suggesting that variations in a certain distributed network predict individual differences in intelligence and reasoning tasks. They described this network as the Parieto-Frontal Integration Theory (P-FIT); the P-FIT model includes the DLPFC (Brodmann's Areas (BAs) 6,9,10,45,46,47), the inferior (BAs 39, 40) and superior (BA 7) parietal lobule, the anterior cingulated (BA 32), and regions within the temporal (BA 21, 37) and occipital (BAs 18, 19). Colom et al. (2009) tested the P-FIT theory in a sample of 100 young healthy adults. Their findings are consistent with the P-FIT theory, supporting the view that general intelligence involves multiple cortical areas throughout the brain.

Links between intelligence and specific regions of the brain may vary according to developmental stage. In the absence of neurological insult or degenerative conditions, IQ is usually expected to be stable across lifespan, as evidenced by the fact that IQ measurements made at different points in an individual's life tend to correlate well (McCall 1977). Using a longitudinal design, Shaw et al. (2006) found a marked developmental shift from a predominantly negative correlation between intelligence and cortical thickness in early childhood to a positive correlation in late childhood and beyond, suggesting that the neuroanatomical expression of intelligence in children is dynamic. More recently, Ramsden et al. (2011) tested whether variation in a teenager's IQ over time correlated with changes in brain structure; they used longitudinal assessments of 33 healthy and neurologically normal adolescents first tested when they were 12–16-year old (mean 14.1 year) and then retested the same individuals at age 15–20 (mean, 17.7 year); in this way they obviated the many sources of variation in brain structure that confound cross-sectional studies. They found that verbal IQ changed with GM in a region that was activated by speech, whereas nonverbal IQ changed with GM in a region that was activated by finger movements. Surprisingly, their results also suggest the possibility that an individual's intellectual capacity relative to their peers can decrease or increase in the teenage years.

WM integrity has also been associated with differences in IQ. Chiang et al. (2011) reported the first map to demonstrate influences of age, sex, socioeconomic status (SES), and IQ on the heritability of brain fiber architecture. They found moderate but significant modulatory effects of age, sex, intellectual performance (measured by Fluid IQ (FIQ), and SES on the heritability of WM integrity measured by FA. Higher WM heritability was associated with younger age (adolescents), male sex, higher FIQ, and higher SES. They also found that in people with above-average IQ, genetic factors explained over 80 % of the observed FA variability in the thalamus, genu, posterior internal capsule, and superior corona radiata. In those with below-average IQ, however, only around 40 % FA variability in the same regions was attributable to genetic factors.

The use of fMRI to study cognitive abilities has proven more complex than expected; many functional neuroimaging studies have found that a single brain region can be involved in a broad range of tasks. Therefore, it is unlikely that there is always one core region that is crucial for a particular cognitive function. Instead, a region with a structure that correlates with a behavioral measure needs to be interpreted in the context of the known functions of the region and its role in other related behavioral tasks (Kanai and Rees 2011).

8.4.2 Imaging Genetics of Intelligence

Intelligence is known to be highly heritable with estimates ranging from 30 to 40% in childhood and up to 80% in middle adulthood (Posthuma et al. 2009). A handful of candidate genes have been associated at least once with cognitive ability, each explaining only about 1–2% of the variance (Deary et al. 2010). A recent genome-wide association studies (GWAS) for intelligence concluded that intelligence is highly polygenic and thus many genes of small effects underlie the additive genetic influences on intelligence (Davies et al. 2011). The chances of finding these genes may be increased by applying a so-called endophenotype approach[1]. For a measure to be considered an endophenotype, it must be shown to: (1) be highly heritable, (2) be associated with the trait, (3) be independent of clinical state, and (4) the measure must cosegregate with the trait within a family (Glahn 2007).

As a positive correlation between brain size and intelligence has been reported many times, Posthuma et al. (2002) set out a decade ago to test whether this correlation was because of shared genes or shared environmental factors. They found high heritability for total brain GM volume, and a correlation between GM volume and intelligence (0.25; $p<0.05$). They also found a significant correlation between WM volume and intelligence (0.24; $p<0.05$). They concluded that intelligence is related to the volumes of both gray and white matter. Using a twin approach, they decomposed the correlation between brain volumes and intelligence into genetic and environmental components; they showed that the correlation between GM volume and intelligence was completely because of genetic factors and not environmental factors. The same result was obtained for the correlation between WM volume and intelligence.

In a subsequent study, Hulshoff Pol et al. (2006) explored the genetic influence on focal GM and WM densities in magnetic resonance brain images of 54 monozygotic and 58 dizygotic twin pairs and 34 of their siblings. For genetic analyses, they used VBM data to explore the common genetic origin of focal GM and WM areas

[1] It should be noted that the endophenotype approach relies on the assumption that the genetic basis of endophenotypes is easier to analyze than the categorical classification of an end-phenotype, such as a neuropsychiatric disorder. However, a systematic metanalysis of genetic association studies of endophenotypes showed that while endophenotypes measures may afford greater reliability, it should not be assumed that they will also demonstrate simpler genetic architecture (Flint and Munafo 2007). The added value of the endophenotype approach thus remains to be proven.

with intelligence. They found that intelligence shared a common genetic origin with superior occipitofrontal, callosal, and left optical radiation WM and frontal, occipital, and parahippocampal GM (phenotypic correlations up to 0.35). The authors suggested that these findings point to a neural network that shares a common genetic origin with human intelligence.

Joshi et al. (2011) analyzed Brain MRI data from 72 young adult twins of age 21–27 years (194 dizygotic and 178 monozygotic twins) to identify cortical regions in which GM thickness and volume are influenced by genes. They found a strong genetic influence on frontal and parietal regions. In addition, they correlated cortical thickness with full-scale IQ, and several regions where cortical structure was correlated with IQ were under strong genetic control. Genetic variants for brain structures and intelligence thus seem to be largely shared.

Overall, these findings suggest that genes important for brain structure might also be of importance for intelligence, and vice versa, and genes important for intelligence may also be of importance for brain structures. Under this assumption, Ruano et al. (2010) used an innovative functional gene group analysis to identify if synaptic genes were associated with intelligence; they found that a set of functionally related genes coding for G-proteins are associated with intelligence. In order to test if the G-proteins group that was found to be associated with intelligence would also explain differences in brain structure, Chavarria-Siles et al. (2013) tested the effect of this set of genes on local cerebral GM volume using VBM. In this study, the authors found strong associations between four genes encoding heterotrimeric G-proteins (*GNG2, GNAQ, GNA15, GNA14*) with specific, local increase in the medial frontal cortex volume, an area involved in cognitive control (Fig. 11). These findings suggest that individual variation in genes encoding G-proteins may modulate cortical volume and cognitive ability by a common principle probably controlling neocortical development and strengthen the convergent evidence that the medial frontal cortex is an important area for cognitive control (Chavarria-Siles et al. 2013).

8.5 Structural and Functional Brain Imaging of Neuropsychiatric Disorders

The underlying neurobiological pathways of individual differences in human cognitive ability are still poorly understood. Identifying neurobiological pathways for variation in the range of normal cognitive ability could provide important clues to underlying mechanisms of milder but more prevalent forms of altered cognitive functioning. Some of these more prevalent milder cognitive dysfunctions are found in several neurodevelopmental psychiatric disorders such as autism (Mayes and Calhoun 2008), schizophrenia (Ehrlich et al. 2012), bipolar disorder (Glahn et al. 2010a), and attention-deficit hyperactivity disorder (Willcutt et al. 2005). As the life expectancy in the population increases, so does the prevalence of cognitive decline and dementia; up to 50% of adults over 85 years of age are currently suffering from cognitive impairment in the form of AD (Hebert et al. 2003).

Neuroimaging endophenotypes are quantitative indicators of brain structure or function that index genetic liability for an illness. These indices will significantly improve gene discovery and help us to understand the functional consequences of specific genes at the level of systems neuroscience (Glahn et al. 2007). In the last section of this chapter, we provide a non-exhaustive review of the neuroimaging findings for the most common neuropsychiatric disorders that are accompanied by cognitive dysfunction.

8.5.1 Autism

Autism spectrum disorder (ASD) is a heterogeneous disorder characterized by abnormal behavior in the spheres of communication, social relatedness, and stereotyped repetitive behaviors within the first 3 years of life. There are several studies using structural and fMRI trying to identify brain abnormalities in children with ASD. These studies indicate anatomic differences, which although not diagnostic are beginning to elucidate the timing and nature of deviations from typical development (Giedd and Rapoport 2010).

There are five main findings that can be drawn from the literature on structural MRI of ASD (Chen et al. 2010): (1) volumetric studies reveal that young children with ASD have abnormally increased total brain volume. In addition, juveniles and adults with ASD have reduced corpus callosum volume, and children with ASD have increased amygdala volume; (2) VBM studies consistently report increased GM volume in the frontal and temporal lobes in ASD; (3) cortical thickness studies suggest an increased cortical thickness in the parietal lobes in ASD; (4) longitudinal MRI studies of ASD suggest abnormal growth trajectories in the frontal and temporal lobes; and (5) DTI studies of ASD consistently report corpus callosum abnormalities across a wide age range. Differences in prefrontal WM, cingulated gyrus, and internal capsule have also been consistently reported.

Apart from structural studies, fMRI has also been used to understand the neurobiological basis of ASD. Initial studies focused on linear brain–behavior relationships, whereas more recent fMRI studies in ASD have shifted focus toward functional connectivity disturbances. Minshew and Keller (2010) reviewed several fMRI studies of ASD; they consistently found alterations in event-related connectivity in ASD: (1) direct evidence of enhanced activation and connectivity of posterior areas and enhanced reliance on visouspatial abilities for verbal and visual reasoning and reduced frontal systems connectivity; (2) across studies, it was not uncommon for the cortical location of areas to be shifted slightly, perhaps reflecting recruitment of adjacent cortical areas and lack of the usual cortical specialization for task performance; and (3) resting-state connectivity and the DMN also suggested abnormalities in intrinsic mechanism of thinking, feeling, and behaving, and for the regulation of these processes.

8.5.2 *Attention-Deficit/Hyperactive Disorder (ADHD)*

ADHD is the most common neurodevelopmental disorder of childhood, affecting between 5 and 10 % of school-age children and 4.4 % of adults. Cross-sectional anatomical imaging studies of ADHD consistently point to the involvement of frontal lobes, parietal lobes, basal ganglia, corpus callosum, and cerebellum (Giedd and Rapoport 2010).

In a meta-analysis of structural MRI findings for ADHD, Valera et al. (2007) showed that the brain regions most frequently assessed and showing the largest and most significant volume reduction in ADHD patients compared to control subjects include cerebellar areas, in particular the posterior inferior vermis, as well as the splenium of the corpus callosum, total and right cerebral volume and right caudate.

The fMRI studies have reported abnormal activation in prefrontal cortices (including inferior and dorsolateral regions and cingulated gyrus) and striatum (including caudate and ventral stratium) in individuals with ADHD compared with control subjects (Tomasi and Volkow 2011). Some of these changes are normalized by stimulant medications such as methylphenidate and amphetamine, supporting the involvement of Dopamine neurotransmission in these functional changes (Rubia et al. 2007).

Most imaging genetic studies of ADHD have focused on dopamine-related candidate genes; from 14 imaging genetics studies of ADHD, 9 focused on the DAT1 gene and 5 on the DRD4 gene. The combined findings from these studies could explain how these genes may impact the brain at the structural, functional, and biochemical level; however, the effect of neither gene is fully understood yet (Durston 2010).

Several groups have used DTI techniques to study WM integrity in ADHD; fractional anisotropy has been shown to be significantly reduced in right frontostriatal projections and in the right longitudinal fasciculus, among several other areas of cerebral and cerebellar WM (Liston et al. 2011).

Resting-state functional connectivity studies have reported abnormal signal fluctuations in inferior frontal and superior parietal cortices, cingulated cortex, and cerebellum. Higher resting-state connectivity has been observed in anterior cingulum, pons, insula cerebellum, and thalamus; lower resting-state connectivity was observed between putamen and PPC and between superior parietal cortices and cingulum (Tomasi and Volkow 2011).

Finally, there is considerable epidemiological and neuropsychological evidence that ADHD is best considered dimensionally, lying at the extreme of a continuous distribution of symptoms and underlying cognitive impairments. Under this consideration, Giedd and Rapoport (2010) tested whether cortical brain development in typically developing children with symptoms of hyperactivity and impulsivity resembles those found in the ADHD. They found that a slower rate of cortical thinning during late childhood and adolescence, which they previously found in ADHD, was also linked to the severity of symptoms of hyperactivity and impulsivity in typically developing children; this finding suggests neurobiological evidence for the dimensionality of the disorder.

In sum, MRI research in ADHD is a fast developing and very complex field. Every study appears to show differences in brain morphology and in patterns of brain activation between cases and controls; but as of today, the interpretation of such differences is not as straightforward as it may seem.

8.5.3 Schizophrenia

Schizophrenia is a neurodevelopmental disorder that affects 1 % of the population worldwide and is characterized by hallucinations, delusions, and disorganized thinking and speech. Motivation, cognition, memory, executive functioning, affect, and social communication are all altered in schizophrenia.

Before the use of MRI, brain abnormalities were based on crude measurements of the postmortem brains; the major finding of these studies showed enlarged ventricles in patients with schizophrenia. A large proportion of MRI studies of schizophrenia (80 %) also found ventricular enlargement in schizophrenia. Enlargement of the ventricles, however, is not exclusive of schizophrenia, as this is also observed in hydrocephalus, AD, and other neurodegenerative disorders where CSF replaces brain tissue.

Shenton et al. (2010) reviewed several structural MRI studies of schizophrenia; they found a striking consistency of results showing GM abnormalities in chronic schizophrenia including brain regions with the prefrontal, temporal, parietal, and occipital lobe. The list of brain regions reported as abnormal is, in fact, quite long and includes nearly all known brain structures.

Despite their impact on imaging phenotypes, the usefulness of candidate genes for understanding schizophrenia is debated because these a priori hypothesized variants often show an inconsistent effect on the categorical disease phenotype itself. GWAS offer an alternative, hypothesis-free way to identify genetic variants associated with the disease; any genetic variant that survives the threshold for genome-wide significance certainly merits study using intermediate imaging phenotypes (Meyer-Lindenberg 2010).

There has been a rapid growth of fMRI studies in schizophrenia, and abnormal activity has been reported in motor tasks, working memory, attention, word fluency, emotion processing, and decision making. An essential goal of such studies is to demonstrate how failure to activate a neural system leads to behavioral deficits in patients with schizophrenia (Gur and Gur 2010).

Research on brain activity in schizophrenia has shown that changes in the function of any single region cannot explain the range of cognitive and affective impairments in this illness. The RS-fMRI connectivity measures have been used to predict clinical symptoms and cognitive function. Individuals with schizophrenia showed reduced distal and somewhat enhanced local connectivity between the cognitive control networks. Additionally, greater connectivity between the frontal-parietal and cerebellar regions was robustly predictive of better cognitive performance across groups and predictive of fewer disorganization symptoms among patients. These results are consistent with the hypothesis that impairments of executive function and cognitive control result from disruption in the coordination of activity across brain networks and additionally suggest that these might reflect impairments in normal pattern of brain connectivity development (Repovs et al. 2011).

8.5.4 Alzheimer's Disease

AD is the most common cause of dementia in elderly people. Dementia is a disease related to loss of memory and other cognitive abilities of sufficient severity to interfere with activities of daily living (Alzheimer's Association 2011). AD is a complex disease characterized by an accumulation of β-amyloid (Aβ) plaques and neurofibrillary tangles composed of tau amyloid fibrils associated with synapse loss and neurodegeneration (Weiner et al. 2012). AD is not a normal part of aging; however, old age is its single greatest risk factor (Jack 2012; see Chap. 7 for a review of the research on dementia).

As of today one of the best established measurements for the detection and tracking of AD is structural MRI measurements of regional and whole-brain tissue shrinkage. Patients have significantly reduced hippocampal and entorhinal cortex volumes, GM, and cortical thickness, increased ventricular and sulcal volumes, reduced GM or cortical thickness in other cerebral regions, like the precuneus and posterior cingulate, parietal, and temporal cortex (Reiman and Jagust 2012).

Meda et al. (2013) recently summarized the most significant findings on the genetics of AD; the last several decades of research have yielded only one genetic risk factor of large effect for late-onset AD: the apolipoprotein-E, with two copies of the ε4 allele conferring approximately 6- to 30-fold risk for the disease. More recent GWAS have identified and replicated nine additional AD susceptibility genes, including *BIN1*, *CLU*, *ABCA7*, *CR1*, *PICALM*, *MS4A6A*, *CD33*, *MS4A4E*, *and CD2AP*. However, all these have low effect sizes (odds ratios of 0.87–1.23) and cumulatively account for approximately 35 % of population-attributable risk. In order to study alternative methods to understand the imaging genetics of AD, Meda et al. (2013) used quantitative intermediate phenotypes derived from MRI data from the Alzheimer's Disease Neuroimaging Initiative (ADNI) database to test for association with gene–gene interactions within 212 known biological pathways. They tested approximately 151 million SNP–SNP interactions for association with 12-month regional atrophy rates using linear regression, with sex, *APOE* ε4 carrier status, age, education, and clinical status as covariates. They found that 109 SNP–SNP interactions were associated with right hippocampus atrophy, and 125 were associated with right entorhinal cortex atrophy; the SNP–SNP interactions that were overrepresented in those interactions are in the calcium signaling, axon guidance, and the ErbB signaling pathway.

8.6 Conclusions

MRI brain-imaging technology has permitted us to study *in vivo* how the brain develops across lifespan, in addition to compare structural brain differences across individuals. Measuring the BOLD signal in fMRI has also been of great use to understand which brain regions and networks are involved in complex cognitive functions while performing specific tasks or simply at rest. As of today, the use of structural and fMRI technology has become a standard procedure in most neurocognition studies.

The field of imaging genetics provides a unique tool to explore and evaluate the functional impact of brain-relevant genetic variants with the potential to understand their impact on behavior. Some applications of this field include identifying biologic mechanisms and pathways that mediate individual differences in complex behaviors and vulnerability to disease. The fast development of novel statistical methods has allowed the field of imaging genetics to map genes that have been associated with cognitive processes with specific brain structures and functional networks.

Most brain structural measurements have been found to be highly heritable, suggesting a strong genetic component in the interindividual differences found among human brain measures. On the other hand, identifying and replicating genetic influences on the brain has proven very difficult; only a few genes with small effect size have been found to explain some differences seen in the brains of subjects with neuropsychiatric disorders compared to the brains of healthy controls. The lack of consistency of most brain imaging genetics studies could be explained by the fact that most of these studies are underpowered, the high costs of MRI have led to studies with small sample sizes that in combination with small genetic effects have failed to adequately find the genetic variations that explain interindividual brain differences. Some efforts are being taken to increase the sample sizes of imaging studies by creating consortia that combine data from different groups; in addition, some novel statistical models that look at the additive effect of multiple genes involved in common biological pathways are being implemented in brain imaging genetics studies.

Further studies are still needed to elucidate the brain's structural and functional architecture and advance in the diagnosis and treatment of the neuropsychiatric disorders. The increased capability to acquire noninvasive, high-quality brain imaging scans in combination with novel imaging-genetics approaches could lead to customized therapeutic interventions. The future of imaging genetics of cognition holds great promise for brain research and for biological validation of genetic findings in neuropsychiatric disorders (Bigos and Weinberger, 2010).

References

Alzheimer's Association. Thies, W., & Bleiler, L. (2011). 2011 Alzheimer's disease facts and figures. *Alzheimer's & dementia: the journal of the Alzheimer's Association, 7,* 208–244.

Assaf, Y., & Pasternak, O. (2008). Diffusion tensor imaging (DTI)-based white matter mapping in brain research: A review. *Journal of molecular neuroscience: MN, 34,* 51–61.

Ashburner, J., & Friston, K. J. (2000). Voxel-based morphometry—the methods. *NeuroImage, 11,* 805–821.

Batty, G. D., Deary, I. J., Gottfredson, L. S. (2007). Premorbid (early life) IQ and later mortality risk: Systematic review. *Annals of epidemiology, 4,* 278–288.

Bartzokis, G., Lu, P. H., Heydari, P., Couvrette, A., Lee, G. J., Kalashyan, G., Freeman, F., Grinstead, J. W., Villablanca, P., Finn, J. P., Mintz, J., Alger, J. R., Altshuler, L. L. (2012). Multimodal magnetic resonance imaging assessment of white matter aging trajectories over the lifespan of healthy individuals. *Biological psychiatry, 72,* 1026–1034.

Bigos, K. L., Weinberger, D. R. (2010). Imaging genetics—days of future past. *NeuroImage, 53,* 804–809.

Blokland, G. A., McMahon, K. L., Thompson, P. M., Martin, N. G., de Zubicaray, G. I., & Wright, M. J. (2011). Heritability of working memory brain activation. *The Journal of neuroscience: the official journal of the Society for Neuroscience, 31,* 10882–10890.

Cascio, C. J., Gerig, G., & Piven, J. (2007). Diffusion tensor imaging: Application to the study of the developing brain. *Journal of the American Academy of Child and Adolescent Psychiatry, 46,* 213–223.

Chavarría-Siles, I., Rijpkema, M., Lips, E., Arias-Vasquez, A., Verhage, M., Franke, B., Fernández, G., Posthuma, D. (2013). Genes encoding heterotrimeric G-proteins are associated with gray matter volume variations in the medial frontal cortex. *Cerebral Cortex*, 23, 1025–1030.

Chen, C. H., Gutierrez, E. D., Thompson, W., Panizzon, M. S., Jernigan, T. L., Eyler, L. T., Fennema-Notestine, C., Jak, A. J., Neale, M. C., Franz, C. E., Lyons, M. J., Grant, M. D., Fischl, B., Seidman, L. J., Tsuang, M. T., Kremen, W. S., Dale, A. M. (2012). Hierarchical genetic organization of human cortical surface area. *Science, 335,* 1634–1636.

Chen. R., Jiao, Y., & Herskovits, E. H. (2010). Structural MRI in autism spectrum disorder. *Pediatric research, 69,* 63–68.

Chiang, M. C., McMahon, K. L., de Zubicaray, G. I., Martin, N. G., Hickie, I., Toga, A. W., Wright, M. J., & Thompson, P. M. (2011). Genetics of white matter development: a DTI study of 705 twins and their siblings aged 12–29. *NeuroImage, 54,* 2308–2317.

Cole, D. M., Smith, S. M., Beckmann, C. F. (2010). Advances and pitfalls in the analysis and interpretation of resting-state FMRI data. *Frontiers in systems neuroscience, 4,* 8.

Colom, R., Haier, R. J., Head, K., Álvarez-Linera, J., Ángeles Quiroga, M., Chun Shih, P., et al. (2009). Gray matter correlates of fluid, crystallized, and spatial intelligence: Testing the PFIT model. *Intelligence, 37,* 124–135.

Davies, G., Tenesa, A., Payton, A., Yang, J., Harris, S. E., Liewald, D., Ke, X., Le Hellard, S., Christoforou, A., Luciano, M., mcghee, K., Lopez, L., Gow, A. J., Corley, J., Redmond, P., Fox, H. C., Haggarty, P., Whalley, L. J., mcneill, G., Goddard, M. E., Espeseth, T., Lundervold, A. J., Reinvang, I., Pickles, A., Steen, V. M., Ollier, W., Porteous, D. J., Horan, M., Starr, J. M., Pendleton, N., Visscher, P. M., Deary, I. J. (2011). Genome-wide association studies establish that human intelligence is highly heritable and polygenic. *Molecular psychiatry, 16,* 996–1005.

Deary, I. J. (2012). Intelligence. *Annual review of psychology, 63,* 453–482.

Deary, I. J., Penke, L., & Johnson, W. (2010). The neuroscience of human intelligence differences. *Nature reviews. Neuroscience, 11,* 201–211.

Durston, S. (2010). Imaging genetics in ADHD. *NeuroImage, 53,* 832–838.

Ehrlich, S., Brauns, S., Yendiki, A., Ho, B. C., Calhoun, V., Schulz, S. C., Gollub, R. L., Sponheim, S. R. (2012). Associations of cortical thickness and cognition in patients with schizophrenia and healthy controls. *Schizophrenia Bulletin, 38,* 1050–1062.

Finkel, D., Reynolds, C. A., McArdle, J. J., Pedersen, N. L. (2005). The longitudinal relationship between processing speed and cognitive ability: Genetic and environmental influences. *Behavior genetics, 35,* 535–549.

Fischl, B., & Dale, A. M. (2000). Measuring the thickness of the human cerebral cortex from magnetic resonance images. *Proceedings of the National Academy of Sciences of the United States of America, 97,* 11050–11055.

Fischl, B., & Dale, A. M. (2000). Measuring the thickness of the human cerebral cortex from magnetic resonance images. *Proceedings of the National Academy of Sciences of the United States of America, 97,* 11050–11055.

Flint, J., & Munafo, M. R. (2007). The endophenotype concept in psychiatric genetics. *Psychological medicine, 37,* 163–180.

Fornito, A., & Bullmore, E. T. (2010). What can spontaneous fluctuations of the blood oxygenation-level-dependent signal tell us about psychiatric disorders? *Current opinion in psychiatry, 23,* 239–249.

Galton, F. (1888). Head growth in students at the University of Cambridge. *Nature, 38,* 14–15.

Geyer, S., Weiss, M., Reimann, K., Lohmann, G., & Turner, R. (2011). Microstructural Parcellation of the human cerebral cortex—from Brodmann's post-mortem map to in vivo mapping with high-field magnetic resonance imaging. *Frontiers in human neuroscience, 5,* 19.

Giedd, J. N., & Rapoport, J. L. (2010). Structural MRI of pediatric brain development: what have we learned and where are we going? *Neuron, 67,* 728–734.

Glahn, D. C., Thompson, P. M., & Blangero, J. (2007). Neuroimaging endophenotypes: strategies for finding genes influencing brain structure and function. *Human brain mapping, 28,* 488–501.

Glahn, D. C., Almasy, L., Barguil, M., Hare, E., Peralta, J. M., Kent, J. W., Dassori, A., Contreras, J., Pacheco, A., Lanzagorta, N., Nicolini, H., Raventos, H., & Escamilla, M. A. (2010a). Neurocognitive endophenotypes for bipolar disorder identified in multiplex multigenerational families. *Archives of General Psychiatry, 67,* 168–177.

Glahn, D. C., Winkler, A. M., Kochunov, P., Almasy, L., Duggirala, R., Carless, M. A., Curran, J. C., Olvera, R. L., Laird, A. R., Smith, S. M., Beckmann, C. F., Fox, P. T., & Blangero, J. (2010b). Genetic control over the resting brain. *Proceedings of the National Academy of Sciences of the United States of America, 107,* 1223–1228.

Good, C. D., Johnsrude, I. S., Ashburner, J., Henson, R. N., Friston, K. J., Frackowiak, R. S. (2001). A voxel-based morphometric study of ageing in 465 normal adult human brains. *NeuroImage, 14,* 21–36.

Gur, R. E., & Gur, R. C. (2010). Functional magnetic resonance imaging in schizophrenia. *Dialogues in clinical neuroscience, 12,* 333–343.

Hebert, L. E., Scherr, P. A., Bienias, J. L., Bennett, D. A., Evans, D. A. (2003). Alzheimer disease in the US population: prevalence estimates using the 2000 census. *Archives of neurology, 60,* 1119–1122.

Huettel, S. A. (2012). Event-related fMRI in cognition. *NeuroImage, 62,* 1152–1156.

Hulshoff Pol, H. E., Schnack, H. G., Posthuma, D., Mandl, R. C., Baare, W. F., van Oel, C., van Haren, N. E., Collins, D. L., Evans, A. C., Amunts, K., Burgel, U., Zilles, K., de Geus, E., Boomsma, D. I., & Kahn, R. S. (2006). Genetic contributions to human brain morphology and intelligence. *The Journal of neuroscience : the official journal of the Society for Neuroscience, 26,* 10235–10242.

Jack, C. R. Jr. (2012). Alzheimer disease: new concepts on its neurobiology and the clinical role imaging will play. *Radiology, 263,* 344–361.

Jones, D. K., Knösche, T. R., Turner, R. (2013). White matter integrity, fiber count, and other fallacies: the do's and don'ts of diffusion MRI. *NeuroImage, 73,* 239–254.

Joshi, A. A., Lepore, N., Joshi, S. H., Lee, A. D., Barysheva, M., Stein, J. L., McMahon, K. L., Johnson, K., de Zubicaray, G. I., Martin, N. G., Wright, M. J., Toga, A. W., & Thompson, P. M. (2011). The contribution of genes to cortical thickness and volume. *Neuroreport, 22,* 101–105.

Jung, R. E., & Haier, R. J. (2007). The Parieto-Frontal Integration Theory (P-FIT) of intelligence: converging neuroimaging evidence. *The Behavioral and brain sciences, 30,* 135–154.

Kanai, R., & Rees, G. (2011). The structural basis of inter-individual differences in human behavior and cognition. *Nature reviews. Neuroscience, 12,* 231–242.

Karlsgodt, K. H., Kochunov, P., Winkler, A. M., Laird, A. R., Almasy, L., Duggirala, R., Olvera, R. L., Fox, P. T., Blangero, J., & Glahn, D. C. (2010). A multimodal assessment of the genetic control over working memory. *The Journal of neuroscience: the official journal of the Society for Neuroscience, 30,* 8197–8202.

Kaymaz, N., & van Os, J. (2009). Heritability of structural brain traits an endophenotype approach to deconstruct schizophrenia. *International review of neurobiology, 89,* 85–130.

Lebel, C., & Beaulieu, C. (2011). Longitudinal development of human brain wiring continues from childhood into adulthood. *The Journal of neuroscience: the official journal of the Society for Neuroscience, 31,* 10937–47.

Lenroot, R. K., Gogtay, N., Greenstein, D. K., Wells, E. M., Wallace, G. L., Clasen, L. S., Blumenthal, J. D., Lerch, J., Zijdenbos, A. P., Evans, A. C., Thompson, P. M., Giedd, J. N. (2007). Sexual dimorphism of brain developmental trajectories during childhood and adolescence. *NeuroImage, 36,* 1065–1073.

Liston, C., Cohen, M. M., Teslovich, T., Levenson, D., & Casey, B. J. (2011). Atypical prefrontal connectivity in attention-deficit/hyperactivity disorder: pathway to disease or pathological end point? *Biological Psychiatry, 69,* 1168–1177.

Mayes, S. D., & Calhoun, S. L. (2008). WISC-IV and WIAT-II profiles in children with high-functioning autism. *Journal of autism and developmental disorders, 38,* 428–439.

McCall, R. B. (1977). Childhood IQs as predictors of adult educational and occupational status. *Science, 197,* 482–483.

McDaniel, M. (2005). Big-brained people are smarter. *Intelligence, 33,* 337–346.

Mechelli, A., Price, C. J., Friston, K. J., Ashburner, J. (2005). Voxel-Based Morphometry of the Human Brain: Methods and Applications. *Current Medical Imaging Reviews, 1,* 105–113.

Meda, S. A., Koran, M. E., Pryweller, J. R., Vega, J. N., Thornton-Wells, T. A., Alzheimer's Disease Neuroimaging, I. (2013). Genetic interactions associated with 12-month atrophy in hippocampus and entorhinal cortex in Alzheimer's Disease Neuroimaging Initiative. *Neurobiology of aging*, 34, 1518.

Meyer-Lindenberg, A. (2010). Imaging genetics of schizophrenia. *Dialogues in clinical neuroscience, 12,* 449–456.

Minshew, N. J., & Keller, T. A. (2010). The nature of brain dysfunction in autism: functional brain imaging studies. *Current opinion in neurology, 23,* 124–130.

Norris, D. G. (2006). Principles of magnetic resonance assessment of brain function. *Journal of magnetic resonance imaging, 23,* 794–807.

Panizzon, M. S., Fennema-Notestine, C., Eyler, T., Jernigan, T. L., Prom-Wormley, E., Neale, M., Jacobson, K., Lyons, M. J., Grant, M. D., Franz, C. E., Xian, H., Tsuang, M., Fischl, B., Seidman, L., Dale, A., & Kremen, W. S. (2009). Distinct Genetic Influences on Cortical Surface and Cortical Thickness. *Cerebral cortex (New York, N. Y.: 1991), 19*, 2728–2735.

Park, J., Shedden, K., & Polk, T. A. (2012). Correlation and heritability in neuroimaging datasets: a spatial decomposition approach with application to an fMRI study of twins. *NeuroImage, 59,* 1132–1142.

Peelle, J. E., Cusack, R., Henson, R. N. (2012). Adjusting for global effects in voxel-based morphometry: gray matter decline in normal aging. *NeuroImage, 60,* 1503–1516.

Posthuma, D., de Geus, E. J., Neale, M. C., Hulshoff Pol, H. E., Baare, W. E. C., Kahn, R. S., & Boomsma, D. (2000). Multivariate genetic analysis of brain structure in an extended twin design. *Behavior genetics, 30,* 311–319.

Posthuma, D., de Geus, E. J., Baare, W. F., Hulshoff Pol, H. E., Kahn, R. S., & Boomsma, D. I. (2002). The association between brain volume and intelligence is of genetic origin. *Nature neuroscience, 5,* 83–84.

Posthuma, D., de Geus, E. J. C., Deary, I. J. (2009). The genetics of intelligence. In: Terry Goldberg & Daniel Weinberger (Eds.), *The Genetics of Cognitive Neuroscience*. MITT Press.

Ramsden, S., Richardson, F. M., Josse, G., Thomas, M. S., Ellis, C., Shakeshaft, C., Seghier, M. L., Price, C. J. (2011). Verbal and non-verbal intelligence changes in the teenage brain. *Nature, 479,* 113–116.

Reiman, E. M., & Jagust, W. J. (2012). Brain imaging in the study of Alzheimer's disease. *NeuroImage, 61*, 505–516.
Repovs, G., Csernansky, J. G., Barch, D. M. (2011). Brain Network Connectivity in Individuals with Schizophrenia and Their Siblings. *Biological Psychiatry, 69,* 967–973.
Roberts, R. E., Anderson, E. J., Husain, M. (2013). White matter microstructure and cognitive function. *The Neuroscientist: a review journal bringing neurobiology, neurology and psychiatry, 19,* 8–15.
Ruano, D., Abecasis, G. R., Glaser, B., Lips, E. S., Cornelisse, L. N., de Jong, A. P., Evans, D. M., Davey, S. G., Timpson, N. J., Smit, A. B., Heutink, P., Verhage, M., & Posthuma, D. (2010). Functional gene group analysis reveals a role of synaptic heterotrimeric G proteins in cognitive ability. *American journal of human genetics, 86,* 113–125.
Rubia, K., Smith, A. B., Brammer, M. J., & Taylor, E. (2007). Temporal lobe dysfunction in medication-naive boys with attention-deficit/hyperactivity disorder during attention allocation and its relation to response variability. *Biological Psychiatry, 62*, 999–1006.
Serences, J. T., & Saproo, S. (2011). Computational advances towards linking BOLD and behavior. *Neuropsychologia, 50,* 435–446.
Shaw, P., Greenstein, D., Lerch, J., Clasen, L., Lenroot, R., Gogtay, N., et al. (2006). Intellectual ability and cortical development in children and adolescents. *Nature, 440,* 676–679.
Shenton, M. E., Whitford, T. J., & Kubicki, M. (2010). Structural neuroimaging in schizophrenia: from methods to insights to treatments. *Dialogues in clinical neuroscience, 12,* 317–332.
Sridharan, D., Levitin, D. J., & Menon, V. (2008). A critical role for the right fronto-insular cortex in switching between central-executive and default-mode networks. *Proceedings of the National Academy of Sciences of the United States of America, 105*, 12569–12574.
Taylor, W. D., Hsu, E., Krishnan, K. R., macfall, J. R. (2004). Diffusion tensor imaging: background, potential, and utility in psychiatric research. *Biological psychiatry, 55,* 201–207.
Thompson, P. M., Cannon, T. D., Narr, K. L., van Erp, T., Poutanen, V. P., Huttunen, M., Lonnqvist, J., Standertskjold-Nordenstam, C. G., Kaprio, J., Khaledy, M., Dail, R., Zoumalan, C. I., & Toga, A. W. (2001). Genetic influences on brain structure. *Nature Neuroscience, 4,* 1253–1258.
Tomasi, D., & Volkow, N. D. (2011). Abnormal Functional Connectivity in Children with Attention-Deficit/Hyperactivity Disorder. *Biological Psychiatry, 71,* 443–450.
Turner, G. R., Spreng, R. N. (2012). Executive functions and neurocognitive aging: dissociable patterns of brain activity. *Neurobiology of aging, 33*, 826.e1–13. (Epub ahead of print)
Uddin, L. Q., Supekar, K. S., Ryali, S., & Menon, V. (2011). Dynamic reconfiguration of structural and functional connectivity across core neurocognitive brain networks with development. *The Journal of neuroscience: the official journal of the Society for Neuroscience, 31,* 18578–18589.
Valera, E. M., Faraone, S. V., Murray, K. E., & Seidman, L. J. (2007). Meta-analysis of structural imaging findings in attention-deficit/hyperactivity disorder. *Biological Psychiatry, 61,* 1361–1369.
Wang, L., Su, L., Shen, H., & Hu, D. (2012). Decoding lifespan changes of the human brain using resting-state functional connectivity MRI. *PLoS ONE, 7,* 44530.
Weiner, M. W., Veitch, D. P., Aisen, P. S., Beckett, L. A., Cairns, N. J., Green, R. C., Harvey, D., Jack, C. R., Jagust, W., Liu, E., Morris, J. C., Petersen, R. C., Saykin, A. J., Schmidt, M. E., Shaw, L., Siuciak, J. A., Soares, H., Toga, A. W., Trojanowski, J. Q., Alzheimer's Disease Neuroimaging Initiative. (2012). The Alzheimer's disease neuroimaging initiative: a review of papers published since its inception. *Alzheimer's & dementia : the journal of the Alzheimer's Association, 8*, 1–68.
Willcutt, E. G., Doyle, A. E., Nigg, J. T., Faraone, S. V., & Pennington, B. F. (2005). Validity of the executive function theory of attention-deficit/hyperactivity disorder: a meta-analytic review. *Biological Psychiatry, 57,* 1336–1346.
Winkler, A. M., Kochunov, P., Blangero, J., Almasy, L., Zilles, K., Fox, P. T., Duggirala, R., & Glahn, D. C. (2010). Cortical thickness or gray matter volume? The importance of selecting the phenotype for imaging genetics studies. *NeuroImage, 53,* 1135–1146.

Chapter 9
Animal Models of General Cognitive Ability for Genetic Research into Cognitive Functioning

Michael J. Galsworthy, Rosalind Arden and Christopher F. Chabris

There has been ample research into the learning, memory, and cognitive abilities of many animal species. Yet, while there is now considerable documentation of the most impressive cognitive feats performed by one or more representatives of any given species, there has been comparatively little study of the variance in mental abilities within each species. This deficiency is significant because we need to establish whether animals, like humans, show general cognitive factors of genetic origin if we wish to use them as models for human learning, memory, mental disease, and aging. Animals are critical to our research because whereas human genetic studies are limited to natural variation, experimental animal studies also provide opportunity for exquisite genetic and neuroscience manipulation. Identifying the complex web of interacting genes behind even simple cognitive task performance is central to our mission of understanding the underlying "wetware" of human cognition and all the variations on its development, including disease.

Moving from *g* in humans to *g* in animals needs strong clarification upfront due to one very common misunderstanding. The concept of general cognitive ability in humans is often thought of as higher processing distilled from the hardest cognitive tasks. Many authors use *g*, "cognitive ability" and "intelligence" interchangeably. However, comparison of the highest levels of functioning between species such as mice and men can be something of a red herring, distracting us from where the two share a powerful bond; namely, the fundamental gene-driven construction of the brain's learning and memory architecture.

M. J. Galsworthy (✉)
Department of Applied Health Research, University College London, London, UK
e-mail: m.galsworthy@ucl.ac.uk

R. Arden
Institute of Psychiatry, King's College London, London, UK
e-mail: r.arden@kcl.ac.uk

C. F. Chabris
Department of Psychology, Union College, Schenectady, NY USA
e-mail: chabrisc@union.edu

D. Finkel, C. A. Reynolds (eds.), *Behavior Genetics of Cognition Across the Lifespan*, Advances in Behavior Genetics 1, DOI 10.1007/978-1-4614-7447-0_9,

It is this basis upon which we must redefine the essence of *g* in order to open productive channels of communication between animal neuroscientists and human psychometricians. The core essence of *g* in humans is that diverse measures of cognitive functioning share a common etiology and we now know that etiology to be largely genetic (Plomin et al. 2001). This means that there are some genes whose functioning (or lack thereof) will have ramifications throughout the brain, affecting learning, memory, and cognition across a diversity of demands. Single genes whose dysfunctions cause severe mental retardation are the clearest demonstration of this reality. There are also single genes with less aggressive but similarly broad effects that contribute quantitatively to milder intellectual disability (Zoghbi and Bear 2012). Whether those genetic effects are localized to one anatomical region or molecular pathway in the brain or dispersed through the entire central nervous system, their final impact is felt in all cognitive tasks. Accordingly, it is entirely reasonable to suggest that in nonhuman organisms there should also be genes whose variations will have widespread effects on brain functioning. In fact, as we will later see, genetic manipulation studies in mice are now, out of necessity, using batteries of tasks in order to capture the pleiotropic effects so commonly seen with single genes (Crawley 2008).

What we describe here is how the picture has developed. This involves: (1) reviewing what we know about the individual differences structure of cognition in other species, (2) exploring the burgeoning field of genetic investigation of learning and memory in mice, and (3) discussing animals models of cognitive disorders such as mental retardation and Alzheimer's. From all these, we hope to draw a narrative thread about what has been learned so far about the fundamental genetic architecture of cognition. Then, in the conclusion, we hope to extend that thread through future directions.

9.1 From Primates to Invertebrates: Models of g in other animals

9.1.1 Individual Differences in Cognitive Function

"Practically no experimental work has been done upon individual differences and family resemblances in animal behavior. In most cases the behaviorist has been content to study the mass reaction of a group of animals to external stimuli, and in the main, has not attempted to treat the variability of his group because of the relatively small number of animals tested" (Bagg 1920, p. 1).

Sadly, this assessment of the situation nearly 100 years ago remains broadly true today. However, although there is very little research concerning individual differences in other species, yet there is still enough that a picture can be drawn. Quantitative genetic research of cognitive abilities, or "intelligence," in rodents began in the 1920s, when human intelligence research was in its infancy. Edward Tolman attempted to explore the genetics of cognitive ability differences by selectively breeding "bright" and "dull" rats. That work was continued by his student Robert Tryon, and they found that after eight generations of selective breeding for performance on a T-maze, there was no population overlap (Plomin and Galsworthy 2003). Such data evidenced genetic contributions to cognitive task performance, whether via "general" ability or something altogether different. Although there was some evidence for "*g*" in rats from other researchers exploring correlations between tasks (e.g., Thorndike 1935), plus some interesting indications of correlated cognitive tasks in other species (see below), this literature lay effectively dormant until very recently.

During the last decade, a few laboratories have attempted to establish a battery of tasks measuring *g* in mice (Locurto et al. 1998, 2003, 2006; Galsworthy et al. 2002, 2005; Matzel et al. 2003, 2011; Kolata et al. 2005, 2010). These studies were specifically focused on the identification of a general cognitive ability robust across many demands and employed different batteries of cognitive tasks with varying types of motivations and stress levels. For example, the Galsworthy battery included the Hebb-Williams maze (a series of classic "maze" designs motivated by escaping wading-height cold water), spontaneous alternation, the Morris water maze, the puzzle box (requiring animals to dig through sawdust or remove a cardboard plug to access a safe dark area), and novel object exploration. The Matzel (incl. Kolata) lab battery standardly includes fear conditioning, operant avoidance, path integration, odor discrimination, and spatial navigation. The Locurto (1998) battery comprised water-based spatial navigation tasks and the 2003 battery included the Hebb-Williams (swimming), a place learning task conducted in a plus maze, a radial maze, a working memory test, a set of detour problems, and a visual nonmatching to sample task. The 2006 battery comprised olfactory foraging, fear conditioning, radial maze, detour task, and Hebb-Williams maze. All but one of them (Locurto et al. 2003) found evidence for a general factor in mice, with this *g* accounting for approximately 30–60 % of the variance in performance. One study also showed sibling correlations for the general factor (Galsworthy et al. 2005), analogous to findings of family studies of human intelligence. Nevertheless, these studies have also helped demonstrate that standard "cognitive" tasks are strongly influenced by noncognitive factors that overshadow the influence of *g* in most individual tasks. Activity and anxiety are clearly important to individual tasks (and should themselves be validated as cross-task traits; Galsworthy et al. 2012), although they do not correlate with a *g* factor extracted from a battery well-balanced for performance measures and motivational drives (Galsworthy et al. 2002). Exploration (or "curiosity"), however, is consistently a strong correlate of the *g* factor (Matzel et al. 2003; Galsworthy et al. 2005; Kolata et al. 2005). Interestingly, studies of human children have found novelty preferences to be associated with *g* (e.g., Bornstein and Sigman 1986). These studies have also reached beyond the psychometric and into the quantitative genetic exploration of general ability. Using a microRNA analysis to examine 25,000 genes in the frontal cortex of outbred mice, it was seen that some 10 genes were upregulated in animals with high *g* relative to those with a lower *g* scores (Kolata et al. 2010). Table 9.1 provides a compilation of the publications investigating or evidencing *g* across nonhuman species from monkeys to mice.

Table 9.1 Publications on individual differences in at least 3 cognitive tasks

Year	Species	N	Measures and key findings	First author
2012	Apes and chimpanzees	129	Conclusion was "no *g*, but some smart apes." However, only rotated factor analysis used.	Herrmann
2011	Mouse	26	All inter-correlations of 7 tasks were positive, *g* accounted for 37 % variance.	Matzel
2011	Bowerbirds	21	Three factors of which first is predictor of mating success.	Keagy
2010	Mouse	60	*g* factor accounts for 41–42 % of variance across 5 tasks.	Kolata
2009	Cotton-top tamarin	22	11 tasks, Bayesian analysis, low but positive inter-correlations	Banerjee
2006	Mouse	47	*g* factor emerges when control tasks removed from battery	Locurto
2005	Mouse	251	*g* factor accounted for 36–41 % of variance. Significant sibling correlation for *g* factor.	Galsworthy
2005	Mouse	21	*g* factor accounted for 40 % of variance across 7 tasks plus open field.	Kolata
2003	Mouse	56	*g* factor accounted for 36 % of variance across 5 tasks	Matzel
2003	Mouse	60	Modular structure to cognition claimed.	Locurto
2002	Mouse	40	*g* factor accounted for 31 % of variance across 7 tasks	Galsworthy
1998	Mouse	75	5 tasks and activity control. Significant *g* factor, positive inter-correlations	Locurto
1997	Rhesus monkey	83	*g* factor account for 48 % of variance. First PC from 3 of 6 tests accounted for 62 % of variance.	Herndon
1993	Rat	22	2 spatial, 1 novelty task, positive correlations	Anderson
1987	Rat	75	3 spatial, 1 visual task, unsure of results	Thompson
1970	Cat, rat, rabbits	8, 8, 10	Maze, discrimination & activity. Positive inter-correlations between tasks	Livesey
1968	Rat	60	Maze and visual discrimination. Positive correlations.	Rajalakshimi
1961	Cat	21	Discriminations and double alterations. Positive correlations.	Warren
1958	Dog	73	17 measures mostly not cognitive. Low positive correlations.	Anastasi
1935	Rat	64	8 tasks, reliabilities .7–.9. Positive, low inter-correlations in 85 % of matrix.	Thorndike
1934	Rat	132	2 mazes, 1 visual discrimination task. Positive correlation between mazes.	Tomlin

Table 9.1 (continued)

Year	Species	N	Measures and key findings	First author
1933	Chicks	133	4 mazes, low positive correlations.	Dunlap
1932	Rat	152	2 mazes, 1 visual discrimination task. Positive correlation between mazes.	Commins
1932	Rat	28	2 mazes, 1 reasoning task, positive correlation between mazes, negative between maze and reasoning	Campbell
1925	Chicks	90	2 mazes, low positive correlation	Ligget
1920	Mouse	71	2 maze tasks, one multiple choice test, very low correlation between tasks, high correlation between maze tests	Bagg

We note that some studies have not found evidence of *g* in nonhuman animals. A study with 60 outbred mice tested on six tasks showed that the scores fitted a factor structure with four components (Locurto et al. 2003) and a study of $N = 15$ apes showed no *g* factor across eight tasks comprising "color" and "shape" (both learning tasks), "causality" and "exclusion" (both inferential tasks), "quantity," "tools," "shape," and "size" (Hermann and Call 2012). Two intriguing studies in bowerbirds that did not find strong evidence of *g*, nonetheless, found that problem-solving tasks predicted mating success (Keagy 2009; Keagy et al. 2011). Another study relating to the evolutionary importance of cognitive abilities found a significant positive correlation between foraging and learning (red–blue visual discrimination of food wells) scores in sparrows (Katnelson et al. 2012). In addition to the studies in Table 9.1, there are findings of heritable latent inhibition correlated with odor discrimination (reversal) learning in honeybees (Chandra et al. 2000; but see Ferguson et al. 2001 for inconsistent findings), and reliable individual differences at the single-neuron level in mollusks (Matzel and Gandhi 2000).

9.1.2 Brain Structures and Function

Although discussion so far has been exclusively focused on individual differences, any discussion on animal cognition should include cross-species comparisons on cognitive and brain differences. Such comparisons give fascinating insights to the roles of brain properties and where evolution has brought to bear its driving forces. Integrating this with intraspecies findings in humans and model organisms provides ample material for hypotheses on brain development and function.

In human studies, brain volume has a robust association with IQ, accounting for about 10% of the variance, according to a meta-analysis of 37 studies involving 1,530 subjects (McDaniel 2005). Controlling for height typically does not alter the results (e.g., Witelson et al. 2005). Another meta-analysis (Gignac et al. 2003) showed that the IQ–volume association was almost the same for gray matter and white matter. A small study in rats evidenced a general cognitive ability and correlation with brain weight (Anderson 1993), however, this area is very underexplored. Clearly, brain volume cannot exclusively explain intelligence as: (1) it does not explain all the variance in human cognition and (2) some species have much larger brains than humans yet do not appear to have greater cognitive abilities. Whales, bottlenose dolphins, and African elephants have brains that are at least as large as the human brain, but the human brain probably has slightly more cortical neurons (Roth and Dicke 2005). Neuron count is not the only difference between the human and other large brains, however. Generally speaking, white matter volume increases faster than gray matter as total brain size increases (Bush and Allman 2003). Axons are thicker in primates than in cetaceans and elephants, resulting in greater conduction velocity, and distances are shorter (due to the packing of more neurons into a smaller volume), resulting in more efficient communication and synchronization between computational modules (Roth and Dicke 2005). Additionally, it is well documented that in primate evolution the cortex has grown more rapidly than subcortical structures, becoming much more folded, and the frontal cortex has developed more than the rest of the cortex (Bush and Allman 2004; Schoenemann et al. 2005). More specifically, it has been suggested that spindle neurons (found only in great apes and humans) and Brodmann area 10 (located at the frontal pole) may be associated with cognitive capacities in which humans excel, such as self-control (Allman et al. 2002; Allman et al. 2005). Area 10 is at least twice as large in humans (relatively) as in apes and spindle neurons are about 25 times more numerous in humans than in apes.

Although these anatomical differences among species are consistent with several means by which the human brain might be a more efficient information processor, it is difficult to obtain support for causality in a cross-species design. Part of the payoff from developing animal models of within-species *g* will be the ability to plot individual differences in all these anatomical parameters (and more) versus *g*, and to examine even more microlevel information-processing properties of neurons and neural areas with techniques that cannot be used in human studies (e.g., single-cell recording and the statistical analyses of large ensembles of neural firing data). Such studies will also facilitate the identification of genes associated with *g* that can then be compared across species, as was done for *FOXP2* when its effects on speech and language were first characterized in humans, and the gene was found to have changed since the divergence from the common ancestor of humans and chimpanzees (Enard et al. 2002).

9.2 Knockouts and Other Tricks to Study Cognition in Mice

Due to their small size making them easy to house in large numbers, their high reproductive rate and genetically inbred lines, mice have quickly become a major mammalian model in medical research. In fact, as shown in Fig. 9.1, mice overtook rats as the leading rodent model approximately a decade ago—and their use continues to grow rapidly. Within learning and memory research, there has been a slight lag due to this being a classic domain of study in the rat. However, here also mouse research is now overtaking rat research. With Alzheimer's disease, mouse models overtook rat studies over a decade ago and have grown sixfold in annual rate since that time. Overall, the figure indicates that animal models, and particularly mouse models, continue to be fertile paradigms for research on cognitive development and cognitive aging.

Beginning with the use of knockouts, where a gene is rendered inactive, it is important to note that this provides an opportunity to explore the fundamental contribution of every gene, which is relied on for functioning, not just the effects of documented natural polymorphisms. Commenting on natural variation versus the then new fad of knocking out genes in mice, Plomin and Kosslyn (2001) remarked, "… although knocking out a gene can have major effects, such experiments do not imply that the gene has anything to do with the variation responsible for hereditary

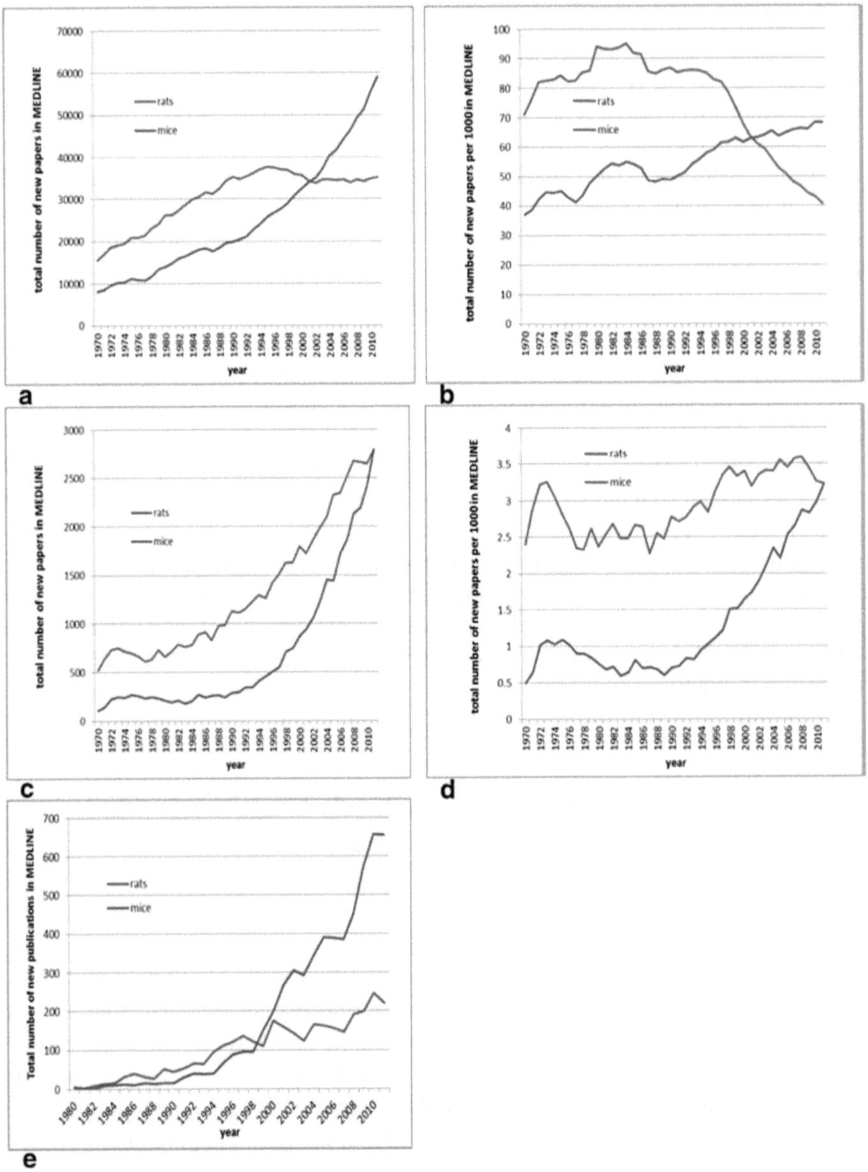

Fig. 9.1. Literature trends in mouse versus rat research. **a** Total papers per annum on mice and rats. **b** Papers as proportion of MEDLINE literature. **c** Learning, memory, and cognition papers. **d** Cognition papers as proportion of literature. **e** Alzheimer's Disease papers. (All graphs were generated by MEDSUM (www.medsum.info)). Panel **a** was generated by running the query "mice" in MEDSUM and clicking the Timeline button, followed by running the query "rats," and clicking Timeline. Panel **b** are the same data divided by the total papers per annum (generated by entering "*" then clicking Timeline) then multiplying by 1,000 to get papers per 1,000. Panel **c** data were generated using Timeline again with the search "<term> AND (learning OR memory OR cognitive OR cognition)" where <term> was either "mice" or "rats." Panel **d** was generated by dividing these numbers by total papers per annum (as before) and multiplying by 1,000. Panel **e** data were generated by the queries "<term>AND 'Alzheimer Disease.'")

transmission of individual differences within a species." However, one decade later, recent findings that many genetic determinants of autism may come from de novo mutations (Neale et al. 2012) suggest that just mapping common variants and their influence will never capture the full genetic etiology of neurological diseases, including intellectual disability (Veltman and Brunner 2012). In addition, there is the issue of a single gene's activity influencing the expression of many others; so even with common variants, explaining the absolute effect of any single gene probably requires understanding of the broader genetic-expression network. Maybe, we simply need to have tried knocking out every gene, then toyed with numerous combinations, in order to build a model system with which we can reasonably predict the quantitative outcome of any given polygenic pattern of insults.

Research with knockouts so far has generally targeted genes associated with systems believed to be crucial in learning and memory processes. Through gene targeting and transgenesis, genes can be disabled, reduced, or increased in expression. Transgenic models have been produced in many species from cattle down to the dwarf surfclam, although mice have by far the broadest array of techniques available (Gama Sosa et al. 2010). Another level of manipulation concerns the turning off or on of genes at chosen moments in time: Turning a gene off later in life is important as complete constitutive knockouts may lead to embryonic lethal phenotypes, which would, of course, preclude study of the gene function in the adult animal (Gama Sosa et al. 2010). Additionally, there is a growing appreciation of the importance of "noncoding" regions in gene expression, leading to increasing popularity of "genomically humanized mice," where large chunks of human DNA can be studied within the mouse system (Devoy et al. 2012).

The number of neurons in the human brain (100 billion) is in the same order of magnitude as the number of stars in our galaxy (300 billion), with a staggering 10^{14} synaptic connections between those neurons (Spires-Jones and Knafo 2012). However, all this complexity rests on another level of complexity that we appear to share with even the humble mouse. In our common genetic heritage, the multitude of genes having an impact on learning and memory processes span a wide range of functional types. They may code for proteins involved in receptor complexes, transporters, signaling cascades, exocytosis, hormone synthesis, or in transcription and translation (Morley and Montgomery 2001).

The first two articles describing a learning and memory deficit in mice after inactivation of a gene appeared over 20 years ago, in 1992. Mice lacking the expression of the fyn-type tyrosine kinase were impaired in long-term potentiation, hippocampal development, and water maze performance (Grant et al. 1992); and mice defective in αCaMKII had similar spatial navigation problems (Silva et al. 1992). Although these studies were very limited in their behavioral phenotyping, subsequent studies with mutant mice confirmed the crucial role of kinases such as the αCaMKII for learning and memory (Elgersma et al. 2004).

Studies of presynaptic proteins involved in neurotransmitter release have implicated a large range of components in learning and memory: αCaMKII, ataxin I, complexin II, GAP-43, PAC1, synapsin I and II, Rab3A, RIM1α, and synaptotagmin (for review, see Powell 2006). Genetic manipulations of growth hormones and their receptors have also appeared to have strong impacts on cognitive abilities, such as manipulation of receptors for mineralcorticoids (Berger et al. 2006) and for glucocorticoids (Oitzl et al. 2000). Deletion of TrkB receptors expressed only during postnatal development in the forebrain was shown to impair spatial learning (Minichiello et al. 1999). Similarly, manipulation of BDNF (Linnarsson et al. 1997) and GDNF (Gerlai et al. 2001) produced spatial learning impairments. The role of the dopaminergic system and BDNF in cognition has been reviewed elsewhere (Savitz et al. 2006).

The role of NMDA receptors in learning and memory is long-established thanks to neuropharmacological tools, so it is not surprising that knocking out the five subunits of the NMDA receptor causes deficits in learning and memory, as has been studied extensively by Joe Tsien's group. Deletion of the NR1 subunit in the CA1 region of the hippocampus impaired spatial learning (Tsien et al. 1996), novel object recognition, context but not cued fear conditioning, and social transmission of food preference (Rampon et al. 2000). In fact, using food additives to selectively turn off the subunit immediately after training impaired a variety of task performances, suggesting a role for the NMDA receptor in maintenance of the memory trace (Cui et al. 2005).

Disruption of NMDA NR2 subunits also adversely affects learning and memory (Miyamoto et al. 2001). However, the more interesting and widely publicized finding was the improved learning and memory performance of mice with NR2B subunits overexpressed (Tang et al. 1999). This result maps nicely with the concept of an increase in general cognitive ability as these mice performed slightly better in a spatial task, contextual and cued fear conditioning acquisition and extinction, and novel object recognition. Researchers have now developed or discovered over 30 additional strains of mice with enhanced learning and memory (Lehrer 2009).

Besides the NMDA receptor, other ionotropic and metabotropic glutamate receptors play a role in learning and memory (Riedel et al. 2003). Mice lacking the AMPA receptor GluR1 (GluRA) subunit were unimpaired in spatial learning (Zamanillo et al. 1999), but a more detailed analysis revealed selective working memory deficits (Reisel et al. 2002; Schmitt et al. 2003). Mice lacking the metabotropic glutamate receptor mGluR1 were slightly impaired in several learning tasks (Conquet et al. 1994). However, part if not all of the deficits might have been due to the simultaneously observed severe motor disturbances. Mice with deleted mGluR5 receptors were impaired in a spatial task and context fear conditioning (Lu et al. 1997), and mice with deletions of the gene for mGluR7 receptors also showed some learning problems (Hölscher et al. 2004; Masugi et al. 1999) but not others (Cryan et al. 2003).

Nonglutamate receptor genes whose manipulation affected learning and memory tasks include the GABAergic receptors (Crestani et al. 2002), nicotinic and muscarinic cholinergic receptors (Drago et al. 2003; Matsui et al. 2004), adrenergic receptors (Kobayashi and Kobayashi 2001), and 5HT1B serotonergic receptors (for review, see Buhot et al. 2003). In summary, whether or not deletion of a receptor gene impaired learning and memory depended on the subunit or receptor type deleted, the type of task, and on the task protocol with deficits showing up more readily when cognitive demands were high.

Learning and memory is based on plasticity in synaptic transmission, which involves pre- and/or postsynaptic enzymes and signaling cascades (Thomas and Huganir 2004). For long-term memory formation, signaling cascades activate transcription factors, which then initiate protein synthesis (Stork and Welzl 1999). Thus, it is not surprising that deletion of genes that code for links in the Ras-MAP kinase signaling cascade differentially affected learning and memory. However, results from different laboratories were inconsistent for mice lacking Ras-GRF1 (Brambilla et al. 1997; Giese et al. 2001) and mice lacking the ERK1 isoform of MAP kinase (Selcher et al. 2001; Mazzucchelli et al. 2002). Interfering with the transcription factor CREB appears to impair long-term memory (Carlezon et al. 2005). CREB comes in several different isoforms, and the type(s) of isoforms deleted might influence results as results have been very mixed with some noncognitive performance abnormalities (Pittenger et al. 2002; Balschun et al. 2003).

Three clear messages emerge from the use of transgenic technologies to study the genetic components of learning and memory in mice. First, the genetic investigation of learning and memory has largely confirmed the neuroscience hypotheses concerning the basic mechanisms of synaptic plasticity that underlie those processes. Second, in line with that complex architecture surrounding synaptic plasticity, there are a great number of single genes whose functioning is critical for performance. Third, a vast number of these genes are very pleiotropic in effect, extending into other areas of behavior and physical development. Manipulating a supposedly "cognitive" gene also may change anxiety, locomotor activity, aggression or other behavioral or physiological properties alongside, or instead of, the learning and memory change. Such effects should have been expected, given what we know about mental retardation in humans, where genetic insults affect not only cognitive features, but also physiology. In fact, turning to mouse models of mental retardation, we can specifically ask whether deletions of the mouse homologues produce comparable insults on both cognitive and physical levels. When this is the case, we know we have a highly useful model with which to dig further and ultimately test therapy, in whatever form that might be.

9.3 Animal Models of Cognitive Dysfunction

9.3.1 Modeling Mental Retardation

Clinical research has identified several dozens of single genes or chromosomal regions whose mutation cause mental retardation (Inlow and Restifo 2004; see Chapter 3 for a review). To model mental retardation, the homologues of human genes known to cause mental retardation have been deleted or replaced by mutated forms in mice (Welzl et al. 2006). The most thoroughly investigated mouse model is that for fragile X syndrome, which in humans is due to a massive triple repeat expansion in the *FMR1* gene on the X chromosome causing hypermethylation and silencing that gene (Jin and Warren 2003). Besides mental retardation, the syndrome includes anatomical features such as elongated faces and large testicles alongside behavioral changes such as hyperkinesia. When a homologous triple repeat expansion was introduced into mice, however, only moderate genomic instability was observed (Bontekoe et al. 2001). However, mice with deleted *FMR1* gene displayed some but not all anatomical features. In more than 90 % of adult mutant mice, testes were enlarged but facial features were normal. Dendritic spines of mutant mice showed anomalies similar to those in human patients, and mutants were only slightly impaired in learning and memory tasks (Kooy 2003).

Mouse models for other forms of hereditary mental retardation similarly yield symptoms weaker than in human patients or even completely absent. Relatively mild impairments in learning and memory tasks accompanied models for Coffin-Lowry syndrome (Poirier et al. 2007), GDI 1 mental retardation (D'Adamo et al. 2002), Rett syndrome (Shahbazian et al. 2002), Agtr2 mutation with X-linked mental retardation (Sakagawa et al. 2000), L1 mutation and CRASH syndrome (Law et al. 2003), and neurofibromatosis type 1 (Costa and Silva 2003).

Generally speaking, syndromes of mouse models only partly replicate the clinical picture. One possible explanation for that observation could be a better compensatory mechanism for deleted genes in mice. Another explanation might be the lack of good tasks to measure *g* in mice compared to the available tasks measuring *g* in humans. Most of the spatial navigation and fear conditioning tasks standardly used in knockout studies in mice are probably too simple, too narrow, and too confounded. We hope that the *g*-battery literature will begin to influence this domain.

9.3.2 Aging and Mental Decline

How well have we done mapping the subtle changes associated with age-related cognitive decline? Although cognitive decline is typically gradual, no form of memory is completely spared (Fratiglioni et al. 2001). Dementias are characterized by progressive and accelerated decline in cognitive function that results from loss of the underlying neuronal architecture. Patients suffering from advanced Alzheimer's disease (AD), frontotemporal dementia, vascular dementia, corticobasal degeneration, or Pick's disease show cognitive impairment that is indicative of widespread neuronal damage. Yet, each disease possesses a unique cognitive phenotype that emerges from a patterned destruction of specific neuronal architectonics (Lee et al. 2001; see chapter 7 for a review).

Alzheimer's Disease is the most prominent of the dementias, and based on cognitive tests accounts for more than 75 % of patients suffering from dementia (Price et al. 1991). AD is characterized by three primary brain pathologies: Senile plaques predominantly containing amyloid-beta (Aβ) protein, neurofibrillary tangles composed of tau protein, and neuronal damage and death mainly in brain regions critical for learning and memory such as the neocortex, hippocampus, amygdala, anterior thalamus, and basal forebrain (Morrison and Hof 1997). Although the first two of these are best known in association with AD, it is the neuron loss that is ultimately directly associated with the cognitive impairments (Whitehouse et al 1982; Li et al. 2012). In addition, the functionality of the monoaminergic and cholinergic systems is reduced (Tong and Hamel 1999).

Very few species spontaneously and naturally develop full-blown Alzheimer's. Nevertheless, since the 1950s, it has been noticed that dogs develop AD-like plaques (Braunmuhl 1956) and since then various species have shown aspects of AD. Amyloid deposits, neurofibrillary tangles, or AD-like symptoms have been seen in aged bears (Cork et al. 1988), dogs (Giaccone et al. 1990), baboons (Schultz et al. 1999), sheep (Nelson et al. 1994), and mouse lemurs (Bons et al. 2006). With selection and genetic manipulation, we now model various aspects of AD in worms, fruit flies, mice, rats, rabbits, dogs, and nonhuman primates (Woodruff-Pak 2008). Such models might not accurately reproduce the anatomical distribution of damage as seen in the human brain, but biochemically they are sufficiently similar to the human condition that they can be combined effectively to understand how the molecular chain of events causes decreasing cognitive performance in humans (Gotz and Ittner 2008; Woodruff-Pak 2008).

The fact that the proteins involved in plaques and tangles (APP and MAPT) have their genes mutated in familial forms of dementia led to the quick generation of mouse models of these genes (Gotz and Ittner 2008). The earliest transgenic models that most convincingly replicated AD-related neuropathology overexpressed various versions of human-mutated APP. These include the Indiana mutation (Games et al. 1995), the Swedish mutation (Hsiao et al. 1996), J20 mice (Mucke et al. 2000), and London mutation (Moechars et al. 1999). Most of the transgenic APP mice rapidly increased amyloid plaque deposition with age, showed behavioral impairment, but no neuronal loss. Combining various APP mutations with tau mutations produced even more variants that better resembled human AD (Gotz and Ittner 2008; Woodruff-Pak 2008). This was augmented by including those gene products that interact with the plaques and tangles, such as secretase and axonal transport mechanisms (Gotz and Ittner 2008). May be, the most successful transgenic model is the triple transgenic AD mice. These consist of APP (Swedish), PS1, and tau mutations (Oddo et al. 2003a; 2003b). These mice develop plaques and tangles in an age-dependent manner, with age-dependent synaptic dysfunction, which preceded plaque and tangle formation, plus an age-progressing memory impairment that correlates with the accumulation of intraneuronal Aβ (Billings et al. 2005).

Independent of transgenics, a mouse model called the senescence-accelerated mouse prone strain 8 (SAMP8) has been presented as a model of AD and mild cognitive impairment (Pallas 2008). The SAMP8 mouse model is based on aging phenotypes not mutation. They show a notable age-related deterioration in learning and memory and so provide an interesting natural model within which to assess triggers and development of mild cognitive impairment and cognitive decline (Woodruff-Pak 2008). Additionally, the cognitive decline observed was correlated with neuronal loss in the hippocampus, as with humans (Li et al. 2012).

Finally, we note the recent work by Lou Matzel and colleagues in which they have linked their research on *g* in mice (and its correlation with attentional processes) to modification of cognitive decline in a genetically diverse CD-1 population of mice. It was found that a group, which was given regular attentional training, had their cognitive decline as measured by various tasks, attenuated by the training (Matzel et al. 2011). It would be of great interest to see such research combined with microRNA analysis as the same team did with high-*g* and low-*g* mice (Kolata et al. 2010), or to test this model of cognitive "rescue" in SAMP8 mice.

In conclusion, although AD has been well described in humans for over a century, it is only in the last two decades that animals have been systematically used to investigate the pathology. Whereas fruit flies and nematode worms have provided powerful models to understand interactions between key molecules of interest (Woodruff-Pak 2008), mouse models provide the best environment in which to manipulate the genetics and explore the behavioral consequences. For this reason, a better understanding of the individual differences structure of cognition in mice will prove important to pinpointing where the neurochemistry of AD begins to affect cognitive and behavioral change.

9.4 Conclusions

One of the main lessons from efforts so far to understand the cognitive consequences of genetic insult is that better quality phenotyping is needed. This is where batteries of cognitive tasks developed from studies of ***g*** in mice (and other species) will come into play. Additionally, while the use of inbred strains was initially believed to be a major experimental advantage of mouse models, we note that genetic background may either exacerbate or compensate for a deleted gene. Mouse strains greatly differ in cognitive task performance (Brooks et al. 2005), and even with standardized conditions and strains, results can vary across laboratories (Crabbe et al. 1999; Richter et al. 2009). Hence the utility of outbred mouse strains, such as those employed in the recent mouse *g* research; the variety in background makes any gene's association more robust and general. Testing inbred control with inbred mutant mice in just one cognitive tasks tells us little about the gene's impact on cognition beyond its effect with that genetic background on some aspect (and not necessarily cognitive) of that task. For a valid conclusion, mice have to undergo a battery that includes several cognitive as well as noncognitive tasks. Only with a full behavioral profile can the gene effect be sensibly characterized. It is critical to have the sensitivity to successfully document the breadth and depth of genetic effects on cognition and behavior. Unfortunately, to date, the neuroscience community has been slow in adopting the quantitative behavior genetic and psychometric methods needed to develop models that are richer in information than binary concepts about abilities derived from crude measures of behavior.

Despite this neglect, we hope that the above studies have shown that animal models contribute much to the understanding of cognition over the lifespan as they have been particularly fruitful at identifying the molecular components of cognition. In addition, a variety of transgenic and natural models from worms to primates have helped us build extremely useful experimental frameworks to explain, on the molecular neuroscience level, how genetic disorders of cognition develop within the human individual.

Finally, we should note some caveats worth addressing. Although *g* is not associated as closely with learning and memory abilities as it is associated with cognitive control, abstract reasoning, and verbal and spatial cognition, there are clear links. First, working memory capacity is highly correlated with *g* (e.g., Gray et al. 2003; Conway et al. 2003), with some researchers even going so far as to suggest that the two concepts are synonymous in humans (Kyllonen and Christal 1990). Second, correlational studies have shown that *g* predicts both explicit and implicit learning (Kaufman et al. 2010; Chabris 2007), with associations ranging from $r=.16$ to .44. It has also been argued that the relationship between the fluid and crystallized facets of intelligence (gF and gC) arises because high-gF individuals are more easily able to acquire and store information and knowledge as they solve problems, and thus they increase their store of gC for use in future problem-solving experiences. Thus, the study of learning and memory mechanisms in animals, where we have experimental control over genes, diets, and long-term environments, has direct relevance for the fundamental understanding of intelligence in humans. We just need better animal experimental psychology to complement the genetics and neuroscience. We hope that rich exploration of all aspects of cognition in a variety of animals will generate a clearer picture of normal development, variation in development (inter- and intra-species), and ultimately allow us to overcome cognitive disorders wherever they occur in a lifetime.

References

Allman, J., Hakeem, A., & Watson, K. (2002). Two phylogenetic specializations in the human brain. *The Neuroscientist, 8,* 335–346.

Allman, J. M., Watson, K. K., Tetreault, N. A., & Hakeem, A. Y. (2005). Intuition and autism: A possible role for Von Economo neurons. *Trends in Cognitive Sciences, 9,* 367–373.

Anastasi, A., Fuller, J. L., Scott, J. P., & Schmitt, J. R. (1955). A factor analysis of the performance of dogs on certain learning tests. *Zoologica, 40*(3), 33–46.

Anderson, B. (1993). Evidence from the rat for a general factor that underlies cognitive performance and that relates to brain size: Intelligence? *Neuroscience Letters, 153,* 98–102.

Bagg, H. J. (1920). Individual differences and family resemblances in animal behavior. *Archives of Psychology, 43,* 1–58.

Balschun, D., Wolfer, D. P., Gass, P., Mantamadiotis, T., Welzl, H., Schutz, G., Frey, J. U., & Lipp, H. P. (2003). Does cAMP response element-binding protein have a pivotal role in hippocampal synaptic plasticity and hippocampus-dependent memory? *The Journal of Neuroscience, 23,* 6304–6314.

Banerjee, K., Chabris, C. F., Johnson, V. E., Lee, J. J., Tsao, F., & Hauser, M. D. (2009). General intelligence in another primate: Individual differences across cognitive task performance in a new world monkey (Saguinus oedipus). (P. F. Ferrari, Ed.) *PLoS ONE, 4*(6): e5883. doi:10.1371/journal.pone.0005883.

Berger, S., Wolfer, D. P., Selbach, O., Alter, H., Erdmann, G., et al. (2006). Loss of the limbic mineralocorticoid receptor impairs behavioral plasticity. *Proceedings of the National Academy of Sciences of the United States of America, 103,* 195–200.

Billings, L. M., Oddo, S., Green, K. N., McGaugh, J. L., & LaFerla, F. M. (2005). Intraneuronal abeta causes the onset of early Alzheimer's disease-related cognitive deficits in transgenic mice. *Neuron, 45,* 675–688.

Bons, N., Rieger, F., Prudhomme D, Fisher, A., & Krause, K.-H. (2006). Microcebus murinus: a useful primate model for human cerebral aging and Alzheimer's disease? *Genes, Brain Behav, 5,* 120–130.

Bontekoe, C. J., Bakker, C. E., Nieuwenhuizen, I. M., van der Linde, H., Lans, H., de Lange, D., Hirst, M. C., & Oostra, B. A. (2001). Instability of a (CGG)98 repeat in the Fmr1 promotor. *Human Molecular Genetics, 10,* 1693–1699.

Bornstein, M., & Sigman, M. (1986). Continuity in mental development from infancy. *Child Development, 57,* 251–274.

Brambilla, R., Gnesutta, N., Minichiello, L., White, G., Roylance, A. J., et al. (1997). A role for the ras signalling pathway in synaptic transmission and long-term memory. *Nature, 390,* 281–286.

Braunmuhl, A. V. (1956). Kongophile angiopathie und senile plaques bei greisen hunden. *Archiv für Psychiatrie und Nervenkrankheiten, 194,* 395–414.

Brooks, S. P., Pask, T., Jones, L., & Dunnett, S. B. (2005). Behavioral profiles of inbred mouse strains used as transgenic backgrounds. II: cognitive tests. *Genes, Brain and Behavior, 4,* 307–17.

Buhot, M.-C., Wolff, M., Benhassine, N., Costet, P., Hen, R., & Segu, L. (2003). Spatial learning in the 5-HT1B receptor knockout mouse: selective facilitation/impairment depending on the cognitive demand. *Learning & Memory, 10,* 466–477.

Bush, E. C., & Allman, J. M. (2003). The scaling of white matter to grey matter in cerebellum and neocortex. *Brain, Behavior, and Evolution, 61,* 1–5.

Bush, E. C., & Allman, J. M. (2004). The scaling of frontal cortex in primates and carnivores. *Proceedings of the National Academy of Sciences, 101,* 3962–3966.

Campbell, A. A. (1935). Community of function in performance of rats on alley mazes and Maier reasoning apparatus. *Journal of Comparative and Physiological Psychology, 31,* 225–235.

Carlezon, W. A. Jr., Duman, R. S., & Nestler, E. J. (2005). The many faces of CREB. *Trends in Neurosciences, 28,* 436–445.

Chabris, C. F. (2007). Cognitive and neurobiological mechanisms of the Law of General Intelligence. In M. J. Roberts (Ed.), *Integrating the mind.* Hove: Psychology Press.

Chandra S. B., Hosler, J. S., & Smith, B. H. (2000). Heritable variation for latent inhibition and its correlation with reversal learning in honeybees (Apis mellifera). *Journal of Comparative Psychology, 114,* 86–97.

Commins, W. D., McNemar, Q., & Stone, C. P. (1932). Intercorrelations of measures of ability in the rat. *Journal of Comparative Psychology, 1*(14), 225–235. doi:10.1037/h0073524

Conquet, F., Bashir, Z. I., Davies, C. H., Daniel, H., Ferraguti, F., et al. (1994). Motor deficit and impairment of synaptic plasticity in mice lacking mGluR1. *Nature, 372,* 237–243.

Conway, A. R., Kane, M. J., & Engle, R. W. (2003). Working memory capacity and its relation to general intelligence. *Trends in Cognitive Sciences, 7,* 547–552.

Cork, L. C., Powers, R. E., Selkoe, D. J., Davies P., Geyer, J. J., & Price, D. L. (1988). Neurofibrillary tangles and senile plaques in aged bears. *Journal of Neuropathology and Experimental Neurology, 47*(1988), 629–641.

Costa, R. M., & Silva, A. J. (2003). Mouse models of neurofibromatosis type I: bridging the GAP. *Trends in Molecular Medicine, 9,* 19–23.

Crabbe, J. C., Wahlsten, D., Dudek, B. C. (1999). Genetics of mouse behavior: interactions with laboratory environment. *Science, 284*(5420), 1670–1672.

Crawley, J. N. (2008). Behavioral phenotyping strategies for mutant mice. *Neuron, 57*(6), 809–818.

Crestani, F., Keist, R., Fritschy, J. M., Benke, D., Vogt, K., Prut, L., Bluthmann, H., Mohler, H., & Rudolph, U. (2002). Trace fear conditioning involves hippocampal alpha5 GABA(A) receptors. *Proceedings of the National Academy of Sciences of the United States of America, 99,* 8980–8985.

Cryan, J. F., Kelly, P. H., Neijt, H. C., Sansig, G., Flor, P. J., & van der Putten, H. (2003). Antidepressant and anxiolytic-like effects in mice lacking the group III metabotropic glutamate receptor mGluR7. *European Journal of Neuroscience, 17,* 2409–2417.

Cui, Z., Lindl, K. A., Mei, B., Zhang, S., & Tsien, J. Z. (2005). Requirement of NMDA receptor reactivation for consolidation and storage of nondeclarative taste memory revealed by inducible NR1 knockout. *European Journal of Neuroscience, 22,* 755–763.

D'Adamo, P., Welzl, H., Papadimitriou, S., Raffaele di Barletta, M., Tiveron, C., et al. (2002). Deletion of the mental retardation gene Gdi1 impairs associative memory and alters social behavior in mice. *Human Molecular Genetics, 11,* 2567–2580.

Devoy, A., Bunton-Stasyshyn, R. K. A., Tybulewicz V. L. J., Smith, A. J. H., & Fisher, E. M. C. (2012). Genomically humanized mice: technologies and promises. *Nature Reviews Genetics, 13,* 14–20.

Drago, J., McColl, C. D., Horne, M. K., Finkelstein, D. I., & Ross, S. A. (2003). Neuronal nicotinic receptors: Insights gained from gene knockout and knockin mutant mice. *Cellular and Molecular Life Sciences, 60,* 1267–1280.

Dunlap, J. W. (1933). The organization of learning and other traits in chickens. Johns Hopkins Press.

Elgersma, Y., Sweatt, J. D., & Giese, K. P. (2004). Mouse genetic approaches to investigating calcium/calmodulin-dependent protein kinase II function in plasticity and cognition. *The Journal of Neurosciences, 24,* 8410–8415.

Enard, W., Przeworski, M., Fisher, S. E., Lai, C. S., Wiebe, V., Kitano, T., Monaco, A. P., Pääbo, S. (2002). Molecular evolution of FOXP2, a gene involved in speech and language. *Nature, 418*(6900), 869–72.

Ferguson, H. J., Cobey, S., & Smith, B. H. (2001). Sensitivity to a change in reward is heritable in the honeybee, apis mellifera. *Animal Behaviour, 61,* 527–534.

Fratiglioni, L., Small, B. J., Winblad, B., & Bäckman, L. (2001). The Transition from Normal Functioning to Dementia in the Aging Population. In K. Iqbal, S. Sisodia, & B. Winblad (Eds.), *Alzheimer's disease: Advances in etiology, pathogenesis and therapeutics* (pp. 3-10). Chichester: Wiley

Galsworthy, M. J., Paya-Cano, J. L., Monleón, S., & Plomin, R. (2002). Evidence for general cognitive ability (g) in heterogeneous stock (HS) mice and an analysis of potential confounds. *Genes, Brainand Behavior, 1,* 88–95.

Galsworthy, M. J., Paya-Cano, J. L., Liu, L., Monleón, S., Gregoryan, G., Fernandes, C., Schalkwyk, L. C., & Plomin, R. (2005). Assessing reliability, heritability and general cognitive ability in a battery of cognitive tasks for laboratory mice. *Behavior Genetics, 35,* 675–692.

Galsworthy, M. J., Madani, R., & Lipp, H.-P. (2012). Identifying reliable traits across laboratory mouse exploration arenas: A meta-analysis. Available from Nature Precedings: <http://dx.doi.org/10.1038/npre.2012.6981.1>.

Gama Sosa, M. A., De Gasperi, R., Elder, G. A. (2010). Animal transgenesis: an overview. *Brain Structure and Function, 214*(2–3), 91–109.

Games, D., Adams, D., Alessandrini, R., Barbour, R., Borthelette, P., Blackwell, C., et al. (1995). Alzheimer-type neuropathology in transgenic mice overexpressing V717F β-amyloid precursor protein. *Nature, 373,* 523–527; doi:10.1038/373523a0

Gerlai, R., McNamara, A., Choi-Lundberg, D. L., Armanini, M., Ross, J., Powell-Braxton, L., & Phillips, H. S. (2001). Impaired water maze learning performance without altered dopaminergic function in mice heterozygous for the GDNF mutation. *European Journal of Neuroscience, 14,* 1153–63.

Giaccone, G., Verga, L., Finazzi, M., Pollo, B., Tagliavini, F., Frangione, B., & Bugiani, O. (1990). Cerebral preamyloid deposits and congophilic angiopathy in aged dogs. *Neuroscience Letters, 114,* 178–183.

Giese, K. P., Friedman, E., Telliez, J. B., Fedorov, N. B., Wines, M., Feig, L. A., & Silva, A. J. (2001). Hippocampus-dependent learning and memory is impaired in mice lacking the Ras-guanine-nucleotide releasing factor 1 (Ras-GRF1). *Neuropharmacology, 41,* 791–800.

Gignac, G., Vernon, P. A., & Wickett, J. C. (2003). Factors influencing the relationship between brain size and intelligence. In H. Nyborg (Ed.), *The scientific study of general intelligence: Tribute to Arthur R. Jensen* (pp. 93–106). Amsterdam: Pergamon.

Gotz J., & Ittner, L. M. (2008). Animal models of Alzheimer's disease and frontotemporal dementia. *Nature Rev Neurosci, 9,* 532–544.

Grant, S. G., O'Dell, T. J., Karl, K. A., Stein, P. L., Soriano, P., & Kandel, E. R. (1992). Impaired long-term potentiation, spatial learning, and hippocampal development in fyn mutant mice. *Science, 258,* 1903–10.

Gray, J. R., Chabris, C. F., & Braver, T. S. (2003). Neural mechanisms of general fluid intelligence. *Nature Neuroscience, 6,* 316–322.

Herndon, J. G., Moss, M. B., Rosene, D. L., & Killiany, R. J. (1997). Patterns of cognitive decline in aged rhesus monkeys. *Behavioural brain research, 87*(1), 25–34.

Herrmann, E., & Call, J. (2012). Are there geniuses among the apes? *Philosophical Transactions of the Royal Society B: Biological Sciences, 367*(1603), 2753–2761.

Hölscher, C., Schmid, S., Pilz, P. K. D., Sansig, G., van der Putten, H., & Plappert, C. F. (2004). Lack of the metabotropic glutamate receptor subtype 7 selectively impairs short-term working memory but not long-term memory. *Behavioural Brain Research, 154,* 473–481.

Hsiao, K., Chapman, P., Nilsen, S., Eckman, C., Harigaya, Y., Younkin, S., Yang, F. S., & Cole, G. (1996). Correlative memory deficits, a-beta elevation, and amyloid plaques in transgenic mice. *Science, 274,* 99–102.

Inlow, J. K., & Restifo, L. L. (2004). Molecular and comparative genetics of mental retardation. *Genetics, 166,* 835–881.

Jin, P., & Warren, S. T. (2003). New insights into fragile X syndrome: from molecules to neurobehaviors. *Trends in Biochemical Sciences, 28,* 152–158.

Katsnelson, E., Motro, U., Feldman, M. W., & Lotem, A. (2011). Individual-learning ability predicts social-foraging strategy in house sparrows. *Proceedings of the Royal Society B, 278,* 582–589.

Kaufman, S.B., DeYoung, C.G., Gray, J.R., Jimenez, L., Brown, J., & Mackintosh, N. (2010). Implicit learning as an ability. *Cognition, 116,* 321–340

Keagy, J., Savard, J. F., & Borgia, G. (2009). Male satin bowerbird problem-solving ability predicts mating success. *Animal Behaviour, 78,* 809–817.

Keagy, J., Savard, J.-F., & Borgia, G. (2011). Complex relationship between multiple measures of cognitive ability and male mating success in satin bowerbirds, Ptilonorhynchus violaceus. *Animal Behaviour, 81*(5), 1063–1070. doi:10.1016/j.anbehav.2011.02.018

Kobayashi, K., & Kobayashi, T. (2001). Genetic evidence for noradrenergic control of long-term memory consolidation. *Brain and Development, 23,* S16–S23.

Kolata, S., Light, K., Townsend, D. A., Hale, G., Grossman, H., Matzel, L. D. (2005). Variations in working memory capacity predict individual differences in general learning abilities among genetically diverse mice. *Neurobio Learn Mem, 84,* 242–246.

Kolata, S., Light, K., Wass, C. D., Colas-Zelin, D., Roy, D., Matzel, L. D. (2010). A dopaminergic gene cluster in the prefrontal cortex predicts performance indicative of general intelligence in genetically heterogeneous mice. *PLoS One, 5,* e14036.

Kooy, R. F. (2003). Of mice and the fragile X syndrome. *Trends in Genetics, 19,* 148–154.

Kyllonen, P. C., & Christal, R. E. (1990). Reasoning ability is (little more than) working-memory capacity? *Intelligence, 14,* 389–433.

Law, J. W. S., Lee, A. Y. W., Sun, M., Nikonenko, A. G., Chung, S. K., Dityatev, A., Schachner, M., & Morellini, F. (2003). Decreased anxiety, altered place learning, and increased CA1 basal excitatory synaptic transmission in mice with conditional ablation of the neural cell adhesion molecule L1. *The Journal of Neuroscience, 23,* 10419–10432.

Lee, V. M., Goedert, M., & Trojanowski, J. Q. (2001). Neurodegenerative tauopathies. *Annual Review of Neuroscience, 24,* 1121–1159.

Lehrer, J. (2009). Small, furry … and smart. *Nature, 461,* 862–864.

Li, G., Cheng, H., Zhang, X., Shang, X., Xie, H., Zhang, X., Yu, J., & Han, J. (2012). Hippocampal neuron loss is correlated with cognitive deficits in SAMP8 mice. *Neurological Sciences*, 1–7 [Epub ahead of print] PubMed PMID: 22872064.

Liggett, J. R. (1925). A note of the reliability of the chick's performance in two simple mazes. *The Pedagogical Seminary and Journal of Genetic Psychology, 32*(3), 470–480.

Linnarsson, S., Bjorklund, A., & Ernfors, P. (1997). Learning deficit in BDNF mutant mice. *European Journal of Neuroscience, 9,* 2581–7.

Livesey, P. J. (1970). A consideration of the neural basis of intelligent behavior: Comparative studies. *Behavioral Science, 15,* 164–170.

Locurto, C., & Scanlon, C., (1998). Individual differences and a spatial learning factor in two strains of mice (Mus musculus). *Journal of Comparative Psychology, 112*(4), 344–352.

Locurto, C., Fortin, E., & Sullivan, R. (2003). The structure of individual differences in Heterogeneous Stock mice across problem types and motivational systems. *Genes, Brainand Behavior, 2,* 40–55.

Locurto, C., Benoit, A., Crowley, C., & Miele, A. (2006). The structure of individual differences in batteries of rapid acquisition tasks in mice. *Journal of Comparative Psychology, 120,* 378–388.

Lu, Y.-M., Jia, Z., Janus, C., Henderson, J. T., Gerlai, R., Wojtowicz, J. M., & Roder, J. C. (1997). Mice lacking metabotropic glutamate receptor 5 show impaired learning and reduced CA1 long-term potentiation (LTP) but normal CA3 LTP. *The Journal of Neuroscience, 17,* 5196–5205.

Masugi, M., Yokoi, M., Shigemoto, R., Muguruma, K., Watanabe, Y., Sansig, G., van der Putten, H., & Nakanishi, S. (1999). Metabotropic glutamate receptor subtype 7 ablation causes deficit in fear response and conditioned taste aversion. *The Journal of Neuroscience, 19,* 955–963.

Matsui, M., Yamada, S., Oki, T., Manabe, T., Taketo, M. M., & Ehlert, F. J. (2004). Functional analysis of muscarinic acetylcholine receptors using knockout mice. *Life Sciences, 75,* 2971–2981.

Matzel, L. D., & Gandhi, C. C. (2000). The tractable contribution of synapses and their component molecules to individual differences in learning. *Behavioural Brain Research, 110,* 53–66.

Matzel, L. D., Han, Y. R., Grossman, H., Karnik, M. S., Patel, D., Scott, N., Specht, S. M., et al. (2003). Individual differences in the expression of a general learning ability in mice. *Journal of Neuroscience, 23*(16), 6423–6433.

Matzel, L. D., Light, K. R., Wass, C., Colas-Zelin, D., Denman-Brice, A., Waddel, A. C., & Kolata, S. (2011). Longitudinal attentional engagement rescues mice from age-related cognitive declines and cognitive inflexibility. *Learning & Memory, 18*(5), 345–356.

Mazzucchelli, C., Vantaggiato, C., Ciamei, A., Fasano, S., Pakhotin, P., et al. (2002). Knockout of ERK1 MAP kinase enhances synaptic plasticity in the striatum and facilitates striatal-mediated learning and memory. *Neuron, 34,* 807–820.

McDaniel, M. A. (2005). Big-brained people are smarter: A meta-analysis of the relationship between in vivo brain volume and intelligence. *Intelligence, 33,* 337–346.

Minichiello, L., Korte, M., Wolfer, D., Kuhn, R., Unsicker, K., et al. (1999). Essential role for TrkB receptors in hippocampus-mediated learning. *Neuron, 24,* 401–14.

Miyamoto, Y., Yamada, K., Noda, Y., Mori, H., Mishina, M., & Nabeshima, T. (2001). Hyperfunction of dopaminergic and serotonergic neuronal systems in mice lacking the NMDA receptor 1 subunit. *The Journal of Neuroscience, 21,* 750–757.

Moechars, D., Dewachter, I., Lorent, K., Reverse, D., Baekelandt, V., Naidu, A., et al. (1999). Early phenotypic changes in transgenic mice that overexpress different mutants of amyloid precursor protein in brain. *Journal of Biological Chemistry, 274,* 6483–6492.

Morley, K. I., & Montgomery, G. W. (2001). The genetics of cognitive processes: candidate genes in humans and animals. *Behavior Genetics, 31,* 511–531.

Morrison, J. H., & Hof, P. R. (1997). Life and death of neurons in the aging brain. *Science, 278,* 412–419.

Mucke, L., Masliah, E., Yu, G. Q., Mallory, M., Rockenstein, E. M., Tatsuno, G., et al. (2000). High-level neuronal expression of abeta 1-42 in wild-type human amyloid protein precursor transgenic mice: synaptotoxicity without plaque formation. *The Journal of Neuroscience, 20,* 4050–4058.

Neale, B. M., Kou, Y., Liu, L., Ma'ayan, A., Samocha, K. E. et al. (2012). Patterns and rates of exonic de novo mutations in autism spectrum disorders. *Nature, 485,* 242–245.

Nelson P.T., Greenberg, S. G., & Saper, C. B. (1994). Neurofibrillary tangles in the cerebral cortex of sheep. *Neuroscience Letters, 170,* 187–190.

Oddo, S., Caccamo, A., Kitazawa, M., Tseng, B. P., & LaFerla, F. M. (2003a). Amyloid deposition precedes tangle formation in a triple transgenic model of Alzheimer's disease. *Neurobiology of Aging, 24,* 1063–1070.

Oddo, S., Caccamo, A., Shepherd, J. D., Murphy, M. P., Golde, T. E., Kayed, R., et al. (2003b). Triple-transgenic model of Alzheimer's disease with plaques and tangles: intracellular abeta and synaptic dysfunction. *Neuron, 39,* 409–421.

Oitzl, M. S., Reichardt, H. M., Joels, M., & de Kloet, E. R. (2000). Point mutation in the mouse glucocorticoid receptor preventing DNA binding impairs spatial memory. *Proceedings of the National Academy of Sciences of the United States of America, 98,* 12790–12795.

Pallas, M., Camins, A., Smith, M. A., Perry, G., Lee, H. G., & Casadesus, G. (2008). From aging to alzheimer's disease: Unveiling "the switch" with the senescence-accelerated mouse model (SAMP8). *J Alzheimers Dis, 15,* 615–624.

Pittenger, C., Huang, Y. Y., Paletzki, R. F., Bourtchouladze, R., Scanlin, H., Vronskaya, S., & Kandel, E. R. (2002). Reversible inhibition of CREB/ATF transcription factors in region CA1 of the dorsal hippocampus disrupts hippocampusdependent spatial memory. *Neuron, 34,* 447–462.

Plomin, R., & Galsworthy, M. J. (2003). Intelligence and cognition. In Cooper DN (ed.), *Nature encyclopedia of the human genome* (Vol. 3, pp. 508–514). London: Nature Publishing Group.

Plomin, R., & Kosslyn, S. M. (2001). Genes, brain and cognition. *Nature Neuroscience, 4,* 1153–1154.

Plomin, R., DeFries JC, McClearn, G. E., & McGuffin, P. (2001). *Behavioral genetics* (4th ed.). New York: Worth Publishers.

Poirier, R., Jacquot, S., Vaillend, C., Soutthiphong, A. A., Libbey, M., Davis, S., et al. (2007). Deletion of the Coffin-Lowry syndrome gene Rsk2 in mice is associated with impaired spatial learning and reduced control of exploratory behavior. *Behavior Genetics, 37*(1), 31–50.

Powell, C. M. (2006). Gene targeting of presynaptic proteins in synaptic plasticity and memory: Across the great divide. *Neurobiology of Learning and Memory, 85,* 2–15.

Price, J. L., Davies, P. B., Morris, J. C., & White, D. L. (1991). The distribution of tangles, plaques and related immunohistochemical markers in healthy aging and Alzheimer's disease. *Neurobiology of Aging, 12,* 295–312.

Rajalakshmi, R., & Jeeves, M. A. (1968). Performance on Hebb-Williams maze as related to discrimination and reversal learning in rats. *Animal Behaviour, 16*(1)

Robbins, T. W., & Murphy, E. R. (2006). Behavioral pharmacology: 40+years of progress, with a focus on glutamate receptors and cognition. *Trends in Pharmacological Sciences, 27,* 141–148.

Rampon, C., Tang, Y. P., Goodhouse, J., Shimizu, E., Kyin, M., & Tsien, J. Z. (2000). Enrichment induces structural changes and recovery from nonspatial memory deficits in CA1 NMDAR1-knockout mice. *Nature Neuroscience, 3,* 238–244.

Reisel, D., Bannerman, D. M., Schmitt, W. B., Deacon, R. M. J., Flint, J., Borchardt, T., Seeburg, P. H., & Rawlins, N. P. (2002). Spatial memory dissociations in mice lacking GluR1. *Nature Neuroscience, 5,* 868–873.

Richter S. H., Garner, J. P., & Würbel, H. (2009). Environmental standardization: cure or cause of poor reproducibility in animal experiments? *Nature Methods, 6,* 257–261.

Riedel, G., Platt, B., & Micheau, J. (2003). Glutamate receptor function in learning and memory. *Behavioural Brain Research, 140,* 1–47.

Rosenthal, N., & Brown, S. (2007). The mouse ascending: perspectives for human-disease models. *Nature Cell Biology, 9,* 993-999. doi:10.1038/ncb437

Roth, G., & Dicke, U. (2005). Evolution of the brain and intelligence. *Trends in Cognitive Sciences, 9,* 250–257.

Sakagawa, T., Okuyama, S., Kawashima, N., Hozumi, S., Nakagawasai, O., Tadano, T., Kisara, K., Ichiki, T., & Inagami, T. (2000). Pain threshold, learning and ormation of brain edema in mice lacking the angiotensin II type 2 receptor. *Life Sciences, 67,* 2577–2585.

Savitz, J., Solms, M., & Ramesar, R. (2006). The molecular genetics of cognition: dopamine, COMT and BDNF. *Genes, Brain and Behavior, 5,* 311–328.

Schmitt, W. B., Deacon, R. M. J., Seeburg, P. H., Rawlins, J. N. P., & Bannerman, D. M. (2003). A within-subjects, within-task demonstration of intact spatial reference memory and impaired spatial working memory in glutamate receptor-a-deficient mice. *The Journal of Neuroscience, 23,* 3953–3959.

Schoenemann, P. T., Sheehan, M. J., & Glotzer, D. (2005). Prefrontal white matter volume is disproportionately larger in humans than in other primates. *Nature Neuroscience, 8,* 242–252.

Schultz, C., Ghebremedhin, E., Sassin I, Braak, E., & Braak, H. (1999). Abnormally phosphorylated tau protein in neurons and glial cells of aged baboons. In K. Iqbal, D. F. Schwaabm, B. Winblad, & H. M. Wisniewski (Eds.) *Alzheimer's disease and related disorders* (pp. 179–185). West Sussex: Wiley.

Selcher, J. C., Nekrasova, T., Paylor, R., Landreth, G. E., & Sweatt, J. D. (2001). Mice lacking the ERK1 isoform of MAP kinase are unimpaired in emotional learning. *Learning & Memory, 8,* 11–19.

Shahbazian, M. D., Young, J. I., Yuva-Paylor, L. A., Spencer, C. M., Antalffy, B. A., Noebels, J. L., Armstrong, D. L., Paylor, R., & Zoghbi, H. Y. (2002). Mice with truncated MeCP2 recapitulate many Rett syndrome features and display hyperacetylation of histone H3. *Neuron, 35,* 243–254.

Silva, A. J., Paylor, R., Wehner, J. M., & Tonegawa, S. (1992). Impaired spatial learning in alpha-calcium-calmodulin kinase II mutant mice. *Science, 257,* 206–11.

Spearman, C. (1904). "General intelligence," objectively determined and measured. *The American Journal of Psychology, 15*(2), 201–292

Spires-Jones T., & Knafo, S. (2012). Spines, plasticity, and cognition in Alzheimer's model mice. *Neural Plast.* 2012:319836. Epub 2011 Nov 28.

Stork, O., & Welzl, H. (1999). Memory formation and the regulation of gene expression. *Cellular and Molecular Life Sciences, 55,* 575–592.

Tang, Y. P., Shimizu, E., Dube, G. R., Rampon, C., Kerchner, G. A., Zhuo, M., Liu, G., & Tsien, J. Z. (1999). Genetic enhancement of learning and memory in mice. *Nature, 401,* 63–69.

Tang, Y. P., Wang, H., Feng, R., Kyin, M., & Tsien, J. Z. (2001). Differential effects of enrichment on learning and memory function in NR2B transgenic mice. *Neuropharmacology, 41,* 779–790.

Thomas, G. M., & Huganir, R. L. (2004). MAPK cascade signalling and synaptic plasticity. *Nature Reviews Neuroscience, 5,* 173–83.

Thompson, R. M., Crinella, F. M., & Yu, J. (1990). *Brain mechanisms in problem solving and intelligence: A lesion survey of the rat brain.* New York: Plenum.

Thorndike, R. L. (1935). Organization of behavior in the albino rat. *Genetic Psychology Monographs, 17,* 1–70.

Tomlin, M. I., & Stone, C. P. (1934). Intercorrelations of measures of learning ability in the albino rat. *Journal of Comparative and Physiological Psychology, 17,* 73–88.

Tong, X. K., & Hamel, E. (1999). Regional cholinergic denervation of cortical microvessels and nitric oxide synthase-containing neurons in Alzheimer's disease. *Neuroscience, 92,* 163–175.

Tsien, J. Z., Huerta, P. T., & Tonegawa, S. (1996). The essential role of hippocampal CA1 NMDA receptor-dependent synaptic plasticity in spatial memory. *Cell, 87,* 1327–1338.

Veltman J. A., & Brunner, H. G. (2012). De novo mutations in human genetic disease. *Nature Reviews Genetics, 13,* 565–575.

Warren, J. M. (1961). Individual differences in discrimination learning by cats. *The Journal of Genetic Psychology, 1*(98), 89–93.

Welzl, H., D'Adamo, P., Wolfer, D. P., & Lipp, H. P. (2006). Mouse models of hereditary mental retardation. In G. S. Fisch & J. Flint (Eds.), R. Lydic & H. A. Baghdoyan (Series Eds.), *Transgenic and knockout models of neuropsychiatric disorders (Contemporary clinical neuroscience)* (pp. 101–125). Totowa: Humana Press.

Whitehouse, P. J., Price, D. L., Struble, R. G., Clark, A. W., Coyle, J. T., Delon, M. R. (1982). Alzheimer's disease and senile dementia: loss of neurons in the basal forebrain. *Science, 215,* 1237–1239.

Witelson, S. F., Beresh, H., & Kigar, D. L. (2005). Intelligence and brain size in 100 postmortem brains: Sex, lateralization and age factors. *Brain, 129,* 386–398.

Woodruff-Pak, D. S. (2008). Animal Models of Alzheimer's Disease: Therapeutic Implications. *Journal of Alzheimer's Disease, 15*, 507–521

Zamanillo, D., Sprengel, R., Hvalby, Ø., Jensen, V., Burnashev, N., Rozov, A., et al. (1999). Importance of AMPA receptors for hippocampal synaptic plasticity but not for spatial learning. *Science, 284,* 1805–1811.

Zoghbi, H. Y., & Bear, M. F. (2012). Synaptic dysfunction in neurodevelopmental disorders associated with autism and intellectual disabilities. *Cold Spring Harbor perspectives in biology, 4*(3).

Part IV
Conclusion

Chapter 10
Future Directions

Deborah Finkel and Chandra A. Reynolds

The twentieth century witnessed the inception and development of behavioral genetics as a valid field of investigation within psychology. As with any new field, that development was marked with plateaus and periods of incredible growth. Even more dramatic were the vicissitudes in the reputation of the field of behavioral genetics during the last century, with the pendulum swinging from genetic determinism to socialization and back again. The response of researchers in behavioral genetics is evident in the titles of books and articles during that period. Even late in the twentieth century, authors and editors were still explaining "heredity, environment, and intelligence" or "nature, nurture, and intelligence" (e.g., Ljungman 1975; McGue 1994; Sternberg and Grigorenko 1997; Vandenberg 1968; Vernon 1979). One of the primary advances we claim for the twenty-first century is development beyond the need to defend or apologize for the findings of behavioral genetic investigations of intelligence. We have moved beyond defenses of heredity of the intelligence quotient (IQ) to more interesting questions of the structure and development of cognitive functioning over the lifespan and the interplay of genetic and environmental influences. To paraphrase Plomin (1997), the current generation of psychologists may wonder what all the nature–nurture fuss was about. Even the title of the current volume reflects this growth: instead of focusing on nature, nurture, and intelligence, we have reached a point when researchers are taking great strides in advancing our understanding of behavior genetics of cognition across the lifespan.

In this chapter, we take the liberty of looking back over the transition from the twentieth century and acknowledging the gains that have been made. Then, we look forward to identify issues that are still in need of attention or resolution and new directions that we feel the field is prepared to explore.

D. Finkel (✉)
Department of Psychology, School of Social Sciences,
Indiana University Southeast, 4201 Grant Line Road, New Albany, IN 47150, USA
e-mail: dfinkel@ius.edu

C. A. Reynolds
Department of Psychology, University of California,
Riverside, CA 92521, USA
e-mail: chandra.reynolds@ucr.edu

D. Finkel, C. A. Reynolds (eds.), *Behavior Genetics of Cognition Across the Lifespan*,
Advances in Behavior Genetics 1, DOI 10.1007/978-1-4614-7447-0_10,

Table 10.1 Summary of recent advances in the study of behavior genetics of cognition across the lifespan

Research issue	Chapters discussing the issue
From Waldman (1997)	
Specific cognitive abilities	1, 4, 5, 7, 8
Extremes of cognitive function	1, 2, 7, 9
Range in environmental background	2, 3, 6, 7
Molecular genetics	1, 3, 4, 5, 6, 7, 8, 9
Gene by environment interplay	2, 3, 5, 6, 7, 9
Developmental behavior genetics	1, 4, 5, 6
Construct validity of intelligence	1, 4, 9
Other Advances	
Sex differences	1, 3, 4, 7
Neuroimaging	3, 6, 7, 8
Intraindividual variability	5
Epigenetics and metabolomics	6, 7

10.1 Looking Backward

At the conclusion of the Sternberg and Grigorenko (1997) volume about intelligence, heredity, and environment, Waldman (1997) reviewed the current state of the field and identified unresolved questions in the behavioral genetics of intelligence at that time. Waldman encouraged researchers to tackle issues including moving beyond general intelligence to focus on specific cognitive abilities; incorporating extremes of cognitive function and environmental background in our studies; searching for specific genetic loci associated with specific cognitive abilities; incorporating measures of environments in genetically informative studies; investigating gene by environment correlation and interaction; using developmental behavioral genetic methods to examine cognitive change; and using behavior genetics as a tool for examining the construct validity of intelligence. It is quite the list of goals, but we claim that in a mere 15 years, the field has made remarkable progress on many of them (see Table 10.1). The fact that each issue is represented in more than one chapter, usually many chapters, highlights the fundamental integration of issues and approaches that characterizes the field of behavior genetics today.

10.1.1 Specific Cognitive Abilities and Construct Validity of Intelligence

As recently as the 1990s, behavior geneticists were still defending the heritability of IQ, but since then the field has begun to investigate genetic and environmental influences on specific cognitive abilities and the related question of how specific abilities contribute to the concept of general intelligence. When we consider specific components of cognitive function, we find differences in both genetic and environmental contributions to functioning across cognitive domains. While established twin studies focus on traditional cognitive domains including verbal ability, spatial ability, memory, processing speed, and academic achievement (Chaps. 1, 4, 5, and 7), recent twin studies have begun to incorporate more nuanced measures of cognitive functioning, including reading ability, math skills, executive function, working memory, and cognitive deficiencies such as mild cognitive impairment and dementia (Chaps. 1, 4, 5, 7, and 8). Recent work in animal models of intelligence focuses on developing cognitive batteries that provide a better approximation of general cognitive ability (Chap. 9). Decomposing cognitive function into its component parts raises the question of the nature of the associations among these components. Evidence from both human and animal studies suggests substantial genetic overlap among specific cognitive abilities (Chaps. 1, 4, and 9). Genetically informed factor analysis confirms the structure of the major cognitive domains and the existence of both shared and independent genetic effects for performance level and change in specific cognitive abilities (Tucker-Drob et al. 2013).

10.1.2 Extremes of Cognitive Functioning

Waldman (1997) bemoaned the fact that behavioral genetic samples typically fail to capture the extremes of functioning or environmental background. Advances have been made in investigations of extremes of functioning both within and beyond the normal range. Thus, Wadsworth et al. report that the etiology of extremely high and low normative cognitive functioning appears to follow the same patterns of heritability as functioning at the center of the normal range (Chap. 1). Advances in studies of cognitive impairment focus on molecular genetic and specific environmental factors in childhood (Chap. 3), late adulthood (Chap. 7), and animal models (Chap. 9). In general, researchers report areas of both overlap and distinction in the etiologies of normative cognitive functioning and cognitive impairment.

10.1.3 Measured Environments and Environmental Extremes

In addition to investigating extremes of cognitive functioning, researchers have also incorporated both fine-grained measures of the environment and focus on extremes of environmental background. Turkheimer and Horn report on the significant impact of extreme ranges of socioeconomic status on heritability of cognitive functioning (Chap. 2). Across studies that incorporate various definitions of intellectual functioning and socioeconomic background, the results consistently demonstrate increasing genetic variance for intelligence with increasing socioeconomic status (SES). In the last 15 years, behavior geneticists have transitioned from anonymous environmental variance to a focus on the role of very specific environmental components on both normative and impaired cognitive functioning (Chaps. 2, 3, 6, and 7). A common thread across the lifespan is the psychological and physiological impact of environmental risk factors and exposures that contribute to reduced cognitive functioning.

10.1.4 Molecular Genetics

Not only have behavior geneticists begun to replace anonymous environmental variance with specific environmental measures, we have also begun to trace anonymous genetic variance to specific genes and gene loci. Molecular genetic approaches to issues of cognitive functioning were in their infancy in the early 1990s (e.g., Fulker et al. 1991) and Plomin's (1997) review of the field speaks more of promise than of confirmed results. It is difficult to overstate the explosion of molecular genetic research into psychological traits, including cognitive functioning, that has occurred since Plomin's review. We chose not to have a chapter devoted to molecular genetics because the methodology has been thoroughly integrated into all areas of behavior genetic research. Consequently, molecular genetic investigations are reported in most of the chapters in the current volume. Numerous genes and gene loci associated with normative and impaired cognitive function have been identified, although as a recent genome-wide association study (GWAS) reports, the complete picture of genetic influences on cognitive function will involve many genes of very small effect (Davies et al. 2011).

10.1.5 Gene by Environment Interplay

In 1997, Waldman discussed possible explanations for the lack of evidence for genotype × environment (G × E) interactions in human research on intelligence. Since then, advances in both measurement of environmental factors and identification of specific genes have resulted in substantial progress in our understanding of gene by environment correlation and interaction in cognitive functioning. These advances allowed researchers to move beyond a reliance primarily on adoption designs to the use of more sophisticated statistical approaches. As a result, two chapters in this volume focus directly on the current state of our understanding of G × E interplay: in childhood (Chap. 2) and in adulthood (Chap. 6). Extensive research reviewed in Chap. 2 demonstrates the association between higher levels of SES and greater genetic variance for intelligence. Nuanced discussions of patterns and types of G × E interplay in Chap. 6 demonstrate that behavioral genetic researchers are beginning to delve deeply into the conceptual issues first described by Waddington (1942), Scarr and McCartney (1983), and Gottlieb (1991), resulting in the development of new theories and approaches to the question. Consequently, the issue of G × E interplay is becoming more fully integrated into behavioral genetic research as a whole, with substantive results also reported in four other chapters in the current volume (Chaps. 3, 5, 7, and 9).

10.1.6 Developmental Behavior Genetics

Although longitudinal twin studies of cognitive functioning in both childhood (Wilson 1978) and adulthood (Jarvik and Bank 1983) were well established in the mid-twentieth century, advanced statistical techniques for tapping the full potential of these databases lagged behind. As more longitudinal twin studies at all points in the lifespan were initiated, statistical methods were developed to support truly developmental behavior genetic analyses (e.g., McArdle et al. 1998; Neale and Cardon 1992). As a result, the field of developmental behavior genetics has expanded dramatically since Waldman's (1997) review. Researchers can identify genetic effects that carry over throughout a developmental period or come online during developmental shifts (Chap. 1). Biometric latent growth curve models allow for the identification of the surprising distinction between genetic variance for level of cognitive performance and genetic variance for changes in cognition with age (Chaps. 4–6).

10.1.7 Other Advances

Two of the most striking advances that Waldman (1997) did not predict are the integration of neuroimaging techniques into behavioral genetic studies and the extension of molecular genetic investigations to incorporate the intricacies of epigenetics and metabolomics. Neuroimaging research became a significant part of investigations of cognitive functioning in the last decade (for a review, see Colom and Thompson 2011), and it is making inroads in genetically informative designs, as well, with the result that sufficient advances have been made to support devoting an entire chapter to neuroimaging research in behavior genetics of intelligence (Chap. 8). Various methods for imaging both structure and function of brain regions have supported investigations into genetic and environmental influences on normative and nonnormative cognitive functioning across the lifespan (Chaps. 3, 6, and 7). Although the concept of epigenetics has been around since Waddington (1942), recent advances in molecular genetic methods and the availability of precise environmental measures have allowed dramatic advances in our understanding of how G × E interplay occurs at the intracellular level to modify the activation of certain genes. The related field of metabolomics focuses on the unique chemical patterns that specific cellular processes produce (Daviss 2005). Several suggestive lines of evidence imply that epigenetic and metabolomic mechanisms play a role in normative and nonnormative cognitive aging (Chaps. 5–7).

Two additional advances in the last 15 years demonstrate how behavior geneticists are adding precision to our understanding of cognitive function: investigations of sex effects and intraindividual variability. Researchers have access to samples and methods that allow for investigation of sex differences in development and decline of normative and impaired cognitive function (Chaps. 1, 3, 4, and 7). Support for stereotypical sex differences in mean cognitive function has been mixed, with sex typically accounting for less than 4 % of variance (see Chap. 1 for a review).

Similarly, results of behavioral genetic investigations of sex differences in etiology of cognitive function report mixed results that vary by domain and phase of development. Generally, modest sex differences in heritability have been reported in some domains, but evidence for sex differences in etiology is rare for normative cognitive functioning. Finally, research is accumulating that indicates intraindividual variability in cognitive performance (e.g., fluctuations across trials or days) can predict subsequent outcomes including cognitive aging (Hultsch et al. 2008). Behavior geneticists are beginning to investigate the etiology of intraindividual variability, per se (Chap. 5), and genetic and environmental influences on the relationships between intraindividual variability and cognitive outcomes (Finkel and Pedersen 2012).

10.2 Looking Forward

Several chapters identified future directions necessary to advance understanding of the etiologies of cognitive performance and change across the lifespan. Common threads in the calls for future work include molecular genetics, environment specificity, cognitive ability phenotypes, a broadening of the concept of interplay between genes and environments, and the continuing need for both quantitative and molecular approaches.

10.2.1 Molecular Genetics

Molecular genetics has not (yet) helped us to resolve an enduring question: apart from chromosomal abnormalities (see Chap. 3), how can we separate typical from atypical cognitive performance at the boundaries between ability and disability and change (e.g., intellectual disabilities in childhood, extremes in functioning within the normal range, dementia in late life)? The lack of consistent findings as to the specificity of gene variants related to extreme performance dimensions within the normal range of childhood cognition reflects the results of studies suggesting that the "Etiology at the extremes of ability appears to follow the same patterns of heritability as ability within the normal range" (Chap. 1). In related domains, however, variants with major effect may be important to reading difficulties, residing in noncoding regions, which may have regulatory consequences on gene expression (see Chap. 1; Smith 2010).

Similarly, molecular genetics has not helped to resolve how to separate typical cognitive aging from pathological cognitive aging. For example, as described in Chaps. 6 and 7, the *APOE* genotype predicts both normative cognitive aging (Davies et al. 2012) and dementia risk, albeit with the strongest findings for Alzheimer's disease and dementia. The lack of specificity on the whole may be a consequence of the inability to establish clear phenotypic separation, although distinct differences in neural structure and function are apparent (Shineman et al. 2010). Moreover, recognition that pathological cognitive change may emerge as a consequence of underlying biological changes occurring decades prior to clinical evidence (see Chap. 7; Sperling et al. 2011) indicates that taking a life course perspective, particularly from a behavioral genetic point of view, may help to elucidate etiological differences and similarities (see also Chaps. 5 and 6; Deary 2012).

Most large-scale molecular (GWAS) studies of intelligence and cognitive abilities have examined common variants. Hence, current work directed to rare variants and variants in noncoding regions (see Chaps. 1 and 5) may provide further details about etiologies of typical and atypical cognitive performance and aging.

10.2.2 Environments

Multiple chapters in this volume called for progress and advancements in the measurement of environments (cf. Chaps. 1, 2, 6, and 7). Putative environmental measures are often complex, showing heritable variation (Plomin et al. 2013), and such findings highlight one reason why behavioral genetic studies are valuable in understanding the underpinnings of cognitive abilities and other behavioral traits. Greater specificity in what is meant by "environments" is needed, not only for its own sake, but with respect to making progress on delineating G × E interplay. To advance an understanding of environmental factors that impact cognitive abilities, both for performance level and change across development, attempts must be made to address known methodological problems systemic in much behavioral research, including biased retrospective recall, single methods of measurement (e.g., self-report) with-out other corroborating sources, and sample attrition in longitudinal studies.

Moreover, cohort differences in etiological contributions to cognitive abilities have not been discussed (mentioned once in Chap. 6, obliquely in Chaps. 2 and 5). The universality of etiological factors can be evaluated with true cohort comparisons. Individuals are born into historical contexts that shape life course pathways and may alter patterns of development and age-related change (Elder 1975, 1998). Hence, etiologies may differ, in part, across cohorts. The changes in health care, environment, and opportunity during the twentieth century may translate to qualitative differences in etiologies contributing to development—differences that are most likely measureable with respect to age-related change for cohorts born in the early and middle parts of the century. Hence, heritability and environmentality may differ as a function of cohort; this has been the case for educational attainment, which has seen marked secular shifts in the last century (e.g., Heath et al. 1985). Consistent patterns of cohort effects on etiological contributions to cognitive ability domains have not been evident from the available work. Indeed, few studies have considered etiological differences between cohorts for cognitive abilities, and such reports have typically relied on age to define the cohorts, as opposed to historical events or eras (Finkel et al. 1995, 1998, 2007; Posthuma et al. 2001; van der Sluis et al. 2008). Fewer studies yet have considered etiological differences for longitudinal change by cohort (Reynolds et al. 2002). With respect to cohort differences in G × E contributions to IQ, a study comparing environmental moderators of IQ in young adulthood (20–35 years) versus middle-aged to older adulthood (36–69 years) from the Netherlands Twin Registry (van der Sluis et al. 2008; see also Chap. 2) suggested that higher parental education and real estate value may be associated with larger common environmental and unique environmental effects, respectively, for older males. Apart from these effects, by and large environmental moderators did not appear to differentially affect genetic and environmental contributions of IQ across cohorts. Future work evaluating the specific contextual factors that may underlie cohort differences is needed beyond age-based definitions. Where considered, cohorts have been identified by birth year or entry age in study, often with splits to achieve nearly equal groups rather than based on other criteria. Age-based conceptualizations of cohort may be responsible for the limited number of significant findings observed. A more refined examination may be warranted. For example, the timing of the introduction of universal health care or shifts in lengths of mandatory schooling are examples of factors that may be important to consider in etiological differences in cognitive performance and change.

10.2.3 Cognitive Ability Phenotypes

While behavioral genetics has contributed to advances in construct validity of cognitive ability traits (e.g., Chaps. 1, 4, and 9), remaining limitations have been referenced (Chaps. 4 and 9). For example, Kremen et al. (Chap. 4) highlight that much behavioral genetic work, particularly that devoted to cognitive aging, has focused on *g* or broad dimensions of cognitive function rather than specific traits. They advocate for the expansion of integrative approaches, including additional behavioral genetic work in the context of cognitive neuroscience methods (see also Chap. 8) and genetic factor modeling approaches to uncover genetic trait dimensions, rather than the more typical decomposition of traits defined by phenotypic factor-analytical approaches. Independent and common pathway modeling approaches may help to some extent to provide hints of common and unique genetic factors contributing to traits but may not be equivalent to performing a factor analysis of genetic variances and covariances (e.g., based on an agnostic full Cholesky decomposition). Examinations of a genetic *g* have been fairly rare (e.g., Alarcon et al. 1999; Luo et al. 1994) and rarer still are examinations of more refined dimensions that are predicted by empirical neuroscience results.

Calls for the refinement of measures and traits are also discussed with respect to animal research reported by Galsworthy et al., suggesting "that better quality phenotyping is needed" (Chap. 9). The extant work to develop cognitive batteries to better approximate general cognitive ability in mouse, rat, primates, and other animals has been limited by low variability and/or low intertest correlations. Work on this topic is critical to advance cross-species comparisons and the development of model systems. Such work may also help illuminate work in humans that considers more refined subdimensions or alternate dimensions more proximal to gene expression (cf. Chap. 4).

In sum, characterizing cognitive phenotypes—in *all* species—is critical to advance an understanding of etiological underpinnings. Moreover, understanding how the dynamics of etiological factors shift across a lifespan is necessary given that both stability and change are ever present.

10.2.4 Broader Notions of Gene–Environment Interplay

Current research suggests that change in one domain influences change in another, e.g., change in emotion or well-being influences physical functioning (Infurna et al. 2011b; Schollgen et al. 2012) and cognitive outcomes (Gerstorf et al. 2007), change in cognitive functioning influences physical functioning (Infurna et al. 2011a), and change in physical functioning influences cognitive aging (Emery et al. 2012). One might claim that such analyses encompass a kind of tautology or unmeasured confounder (cf. Chap. 5) in a sense that one can often find evidence that physical changes predict cognitive change and cognitive changes predict physical changes. However, not all studies find reciprocal effects. Moreover, the larger point is that the nature and timing of changes that occur in particular domains may lead to "environments" for changes in other domains (see Pedersen et al. 2012). For example, increasing depressive symptoms may lead to an environment in which cognitive stimulation is substantially decreased, because of social withdrawal, leading to cognitive decline. In Chap. 6, Reynolds et al. predicted that active G × E correlations such as this would increase with age. Behavioral genetic investigations of cross-domain dynamic processes have been inferred from cross-sectional data (e.g., Chaps. 2, 5, and 6; see also Johnson and Krueger 2005) or from considerations of longitudinal dynamic processes within domains, but thus far only in aging samples (Finkel et al. 2005; McArdle and Hamagami 2003).

The Interplay of Genes and Environment across Multiple Studies (IGEMS) consortia represents an effort to examine G × E interplay across the adult lifespan (Pedersen et al. 2012). To address questions about the impact of proximal and distal social context on late-life outcomes, the IGEMS project is focused on harmonizing early-life adversity and social factor indices as well as physical and psychological functioning, and cognitive functioning traits in nine twin studies originating from Sweden, Denmark, and the USA, comprising a total sample of approximately 17,500 participants aged 25–102 years at study entry, including approximately 2,600 monozygotic (MZ) and 4,300 dizygotic (DZ) twin pairs. Such large-scale efforts are needed to achieve the power necessary to evaluate G × E interplay (Hanscombe et al. 2012).

10.2.5 Gender Differences

Some of the research on cognitive decline and dementia raises yet again questions of differential (quantitative or qualitative) etiologies in cognitive performance and change between men and women. As highlighted in Chap. 6, evidence indicates that there may be a differential risk of mild cognitive impairment (Roberts et al. 2012) and Alzheimer's disease (Genin et al. 2011) in men versus women, suggesting that distinctive age-related risk profiles may be important to consider. Some gender differences in etiology of normative cognitive change have been observed for verbal traits (Finkel et al. 2006). Moreover, risk factors such as serum lipids may differentially predict cognitive decline after age 50 years in men versus women (Reynolds et al. 2010), suggesting that differential phases of change in cholesterol, lipids, and lipoproteins profiles may be important to consider. Altogether, there has not been enough work from a behavioral genetic perspective on gender differences for cognition across the adult lifespan relative to earlier in the lifespan (see Chap. 1). Yet, phenotypic work and the little behavioral genetic evidence suggest that such work may be productive, particularly if it took into account physiological and environmental, indeed cultural, factors that may result in timing, presentation and even etiological differences. Even a brief glance at the emerging literature reveals that studies of both animal cognition (Hebda-Bauer et al. 2007; Mizuno and Giese 2010) and human brain structures (Li et al. 2012) report intriguing sex differences that behavioral geneticists should pursue more fully if we are to develop a complete understanding of the etiology of cognitive function.

10.2.6 The Future of Behavioral Genetic Studies of Cognition

After two decades of advances in the methodologies of gene sequencing and related fields of investigation, it is important to emphasize that quantitative and molecular approaches are both necessary to the continued progress in our understanding of the etiology of cognitive function. One of the evident results from molecular genetic investigations of cognitive performance and change across the lifespan is the relatively limited number of replications, apart from work on cognitive disorders (see Chaps. 3, 7, and 9) that have resulted from large-scale genotyping efforts such as GWAS (see Chaps. 1, 4, 5, and 6). Indeed, one could perhaps say, in a moment of unguarded reflection, that we know little more about the specific genetic etiologies of individual differences in normative cognitive performance using GWAS efforts than we learned from traditional quantitative genetic efforts from twin and adoption studies, i.e., at least half the variance in crystallized and fluid ability traits is due to genetic factors, and some partly unique, emergent, genetic factors may be important to cognitive performance in late adulthood compared with earlier life (Davies et al. 2011, 2012; Deary et al. 2012; Finkel and Reynolds 2009). Yet, as indicated above and in nearly all the chapters in this volume, newer molecular efforts (i.e., gene-sequencing strategies) may be better suited to identify specific genetic factors. Will success from these newer approaches spell doom for traditional quantitative behavioral genetic approaches? The answer is a clear "no," as indicated in a number of chapters in this volume and elsewhere (van Dongen et al. 2012). Indeed, in the realm of human work, twin studies will continue to serve as an important resource for investigations of both genetic and environmental factors (van Dongen et al. 2012).

Beyond heritability, studies using a twin approach can provide unique avenues of investigation to illuminate molecular investigations, particularly evaluating multiple levels of biological and environmental pathways on behavior, through the use, for example, of discordant twin designs as discussed by van Dongen et al. (2012).

Although the chapters in this volume summarize the enormous strides the field has made in the last 15 years, continued progress in our understanding of the behavioral genetics of cognition across the lifespan will require incorporation of all the methods at our disposal: molecular and quantitative genetics; improved measurement of genes, environments, and phenotypes; examination of etiologies across cohorts and genders; and a more comprehensive understanding of gene by environment interplay. Behavior geneticists have come a long way since Galton's first investigation of hereditary genius nearly 150 years ago, but there is still much more work to be done.

Acknowledgments The authors gratefully acknowledge that a portion of this work was supported by the National Institutes of Aging (R01 AG037985). We thank Irwin Waldman for his input on this chapter.

References

Alarcon, M., Plomin, R., Fulker, D. W., Corley, R., & DeFries, J. C. (1999). Molarity not modularity: Multivariate genetic analysis of specific cognitive abilities in parents and their 16-year-old children in the Colorado Adoption Project. *Cognitive Development, 14*(1), 175–193. doi:http://dx.doi.org/10.1016/S0885-2014(99)80023-9.

Colom, R., & Thompson, P. M. (2011). Understanding human intelligence by imaging the brain. In T. Chamorro-Premuzic, S. von Stumm, & A. Furnham (Eds.), *The Wiley-Blackwell handbook of individual differences* (pp. 330–352). New York: Wiley-Blackwell.

Davies, G., Tenesa, A., Payton, A., Yang, J., Harris, S. E., Liewald, D., et al. (2011). Genome-wide association studies establish that human intelligence is highly heritable and polygenic. *Molecular Psychiatry, 16*(10), 996–1005. doi:10.1038/mp.2011.85.

Davies, G., Harris, S. E., Reynolds, C. A., Payton, A., Knight, H. M., Liewald, D. C., et al. (2012). A genome-wide association study implicates the APOE locus in nonpathological cognitive ageing. *Molecular Psychiatry*. doi:10.1038/mp.2012.159.

Daviss, B. (2005). Growing pains for metabolomics. *The Scientist, 19*, 25–28.

Deary, I. J. (2012). Intelligence. *Annual Review of Psychology, 63*, 453–482. doi:10.1146/annurev-psych-120710-100353.

Deary, I. J., Yang, J., Davies, G., Harris, S. E., Tenesa, A., Liewald, D., Luciano, M., Lopez, L. M., Gow, A. J., Corley, J., Redmond, P., Fox, H. C., Rowe, S. J., Haggarty, P., McNeill, G., Goddard, M. E., Porteous, D. J., Whalley, L. J., Starr, J. M., & Visscher, P. M. (2012). Genetic contributions to stability and change in intelligence from childhood to old age. *Nature, 482*(7384), 212–215. doi:10.1038/nature10781

Elder, G. H. (1975). Age differentiation and life course. *Annual Review of Sociology, 1*, 65–190.

Elder, G. H. (1998). The life course as developmental theory. *Child development, 69*, 1–12.

Emery, C. F., Finkel, D., & Pedersen, N. L. (2012). Pulmonary function as a cause of cognitive aging. *Psychological Science, 23*(9), 1024–1032. doi:10.1177/0956797612439422.

Finkel, D., & Pedersen, N. L. (2012). *Intra-individual variability in reaction time shares genetic variance with cognitive functioning in late adulthood.* Paper presented at the annual meeting of the Gerontological Society of America, San Diego, CA.

Finkel, D., & Reynolds, C. A. (2009). Behavioral genetic investigations of cognitive aging. Y.-K. Kim (Ed.), *Handbook of behavior genetics* (pp. 101–112). New York: Springer.

Finkel, D., Pedersen, N. L., McGue, M., & McClearn, G. E. (1995). Heritability of cognitive abilities in adult twins: comparison of Minnesota and Swedish data. *Behavior Genetics, 25*(5), 421–431.

Finkel, D., Pedersen, N. L., Plomin, R., & McClearn, G. E. (1998). Longitudinal and cross-sectional twin data on cognitive abilities in adulthood: The Swedish Adoption/Twin Study of Aging. *Developmental Psychology, 34*(6), 1400–1413.

Finkel, D., Reynolds, C. A., McArdle, J. J., & Pedersen, N. L. (2005). The longitudinal relationship between processing speed and cognitive ability: Genetic and environmental influences. *Behavior Genetics, 35*(5), 535–549. doi:10.1007/s10519-005-3281-5.

Finkel, D., Reynolds, C. A., Berg, S., & Pedersen, N. L. (2006). Surprising lack of sex differences in normal cognitive aging in twins. *International Journal of Aging and Human Development, 62*(4), 335–357.

Finkel, D., Reynolds, C. A., McArdle, J. J., & Pedersen, N. L. (2007). Cohort differences in trajectories of cognitive aging. *The Journals of Gerontology. Series B, Psychological Sciences and Social Sciences, 62*(5), 286–294.

Fulker, D. W., Cardon, L. R., DeFried, J. C., Kimberling, J. W., Pennington, B. F., & Smith, S. D. (1991). Multiple regression analysis of sib-pair data on reading to detect quantitative trait loci. *Reading and Writing: An Interdisciplinary Journal, 3*, 299–313.

Genin, E., Hannequin, D., Wallon, D., Sleegers, K., Hiltunen, M., Combarros, O., et al. (2011). APOE and Alzheimer disease: A major gene with semi-dominant inheritance. *Molecular Psychiatry, 16*(9), 903–907. doi:10.1038/mp.2011.52.

Gerstorf, D., Lovden, M., Rocke, C., Smith, J., & Lindenberger, U. (2007). Well-being affects changes in perceptual speed in advanced old age: Longitudinal evidence for a dynamic link. *Developmental Psychology, 43*(3), 705–718. doi:10.1037/0012-1649.43.3.705.

Gottlieb, G. (1991). Experiential canalization of behavioral development: Theory. *Developmental Psychology, 27*, 4–13.

Hanscombe, K. B., Trzaskowski, M., Haworth, C. M., Davis, O. S., Dale, P. S., & Plomin, R. (2012). Socioeconomic status (SES) and children's intelligence (IQ): In a UK-representative sample SES moderates the environmental, not genetic, effect on IQ. *PLoS ONE, 7*(2), e30320. doi:10.1371/journal.pone.0030320.

Heath, A. C., Berg, K., Eaves, L. J., Solaas, M. H., Corey, L. A., Sundet, J., et al. (1985). Education policy and the heritability of educational attainment. *Nature, 314*(6013), 734–736.

Hebda-Bauer, E. K., Luo, J., Watson, S. J., & Akil, H. (2007). Female CREBalphadelta- deficient mice show earlier age-related cognitive deficits than males. *Neuroscience, 150*(2), 260–272. doi: 10.1016/j.neuroscience.2007.09.019.

Hultsch, D. F., et al. (2008). Intraindividual variability, cognition, and aging. F. I. M. Craik, & T. A. Salthouse (Eds.), *The handbook of aging and cognition* (pp. 491–556). New York: Psychology Press.

Infurna, F. J., Gerstorf, D., Ryan, L. H., & Smith, J. (2011a). Dynamic links between memory and functional limitations in old age: Longitudinal evidence for age-based structural dynamics from the AHEAD study. *Psychology and Aging, 26*(3), 546–558. doi:10.1037/a0023023.

Infurna, F. J., Gerstorf, D., & Zarit, S. H. (2011b). Examining dynamic links between perceived control and health: Longitudinal evidence for differential effects in midlife and old age. *Developmental Psychology, 47*(1), 9–18. doi: 10.1037/a0021022.

Jarvik, L. F., & Bank, L. (1983). Aging twins: Longitudinal psychometric data. In K.W. Schaie (Ed.), *Longitudinal twin studies of adult psychological development* (pp. 40–63). New York: Guilford Press.

Johnson, W., & Krueger, R. F. (2005). Higher perceived life control decreases genetic variance in physical health: Evidence from a national twin study. *Journal of Personality and Social Psychology, 88*(1), 165–173. doi:10.1037/0022-3514.88.1.165.

Li, W., van Tol, M. J., Li, M., Miao, W., Jiao, Y., Heinze, H. J., et al. (2012). Regional specificity of sex effects on subcortical volumes across the lifespan in healthy aging. *Human Brain Mapping*. doi:10.1002/hbm.22168.

Ljungman, C. G. (1975). *What is IQ? Intelligence, heredity, and environment*. London: Gordon Cremonesi.

Luo, D., Petrill, S. A., & Thompson, L. A. (1994). An exploration of genetic g: Hierarchical factor analysis of cognitive data from the Western Reserve Twin Project. *Intelligence, 18*(3), 335–347. doi:http://dx.doi.org/10.1016/0160-2896(94)90033-7.

McArdle, J. J., & Hamagami, F. (2003). Structural equation models for evaluating dynamic concepts within longitudinal twin analyses. *Behavior Genetics, 33*(2), 137–159.

McArdle, J. J., Prescott, C. A., Hamagami, F., & Horn, J. L. (1998). A contemporary method for developmental-genetic analyses of age changes in intellectual abilities. *Developmental Neuropsychology, 14,* 69–114.

McGue, M. (1994).Why developmental psychology should find room for behavioral genetics. C. E. Nelson (Ed.), *Threats to optimal development: Integrating biological, psychological, and social risk factors* (pp. 105–149). Hillsdale, NJ: Lawrence Erlbaum.

Mizuno, K., & Giese, K. P. (2010). Towards a molecular understanding of sex differences in memory formation. *Trends in Neurosciences, 33*(6), 285–291. doi:10.1016/j.tins. 2010. 03.001.

Neale, M. C., & Cardon, L. (1992). *Methodology for genetic studies of twin and families*. Boston: Kluwer.

Pedersen, N. L., Christensen, K., Dahl, A. K., Finkel, D., Franz, C. E., Gatz, M., et al. (2012). IGEMS: The Consortium on Interplay of Genes and Environment Across Multiple Studies. *Twin Research and Human Genetics: The Official Journal of the International Society for Twin Studies, 16*,1–9. doi:10.1017/thg.2012.110.

Plomin, R. (1997). Identifying genes for cognitive abilities and disabilities. In R. J. Sternberg & E. Grigorenko (Eds.), *Intelligence, heredity, and environment* (pp. 89–104). Cambridge: Cambridge University Press.

Plomin, R., DeFries, J. C., Knopik, V. S., & Neiderhiser, J. M. (2013). *Behavioral genetics* (6th ed.). New York: Worth Publishers.

Posthuma, D., de Geus, E. J., & Boomsma, D. I. (2001). Perceptual speed and IQ are associated through common genetic factors. *Behavior Genetics, 31*, 593–602.

Reynolds, C. A., Finkel, D., Gatz, M., & Pedersen, N. L. (2002). Sources of influence on rate of cognitive change over time in Swedish twins: An application of latent growth models. *Experimental Aging Research, 28*(4), 407–433.

Reynolds, C. A., Gatz, M., Prince, J. A., Berg, S., & Pedersen, N. L. (2010). Serum lipid levels and cognitive change in late life. *Journal of the American Geriatrics Society, 58*(3), 501–509. doi:10.1111/j.1532-5415.2010.02739.x.

Roberts, R. O., Geda, Y. E., Knopman, D. S., Cha, R. H., Pankratz, V. S., Boeve, B. F., et al. (2012). The incidence of MCI differs by subtype and is higher in men: The Mayo Clinic Study of Aging. *Neurology, 78*(5), 342–351. doi: 10.1212/WNL.0b013e3182452862.

Scarr, S., & McCartney, K. (1983). How people make their own environments: A theory of genotype-environment effects. *Child Development, 54,* 424–435.

Schollgen, I., Huxhold, O., & Schmiedek, F. (2012). Emotions and physical health in the second half of life: Interindividual differences in age-related trajectories and dynamic associations according to socioeconomic status. *Psychology and Aging, 27*(2), 338–352. doi:10.1037/a0026115.

Shineman, D. W., Salthouse, T. A., Launer, L. J., Hof, P. R., Bartzokis, G., Kleiman, R., et al. (2010). Therapeutics for cognitive aging. *Annals of the New York Academy of Sciences, 1191* (Suppl 1), E1–15. doi:10.1111/j.1749-6632.2010.05532.x.

Smith, S. D. (2010). Learning disabilities. In J. Nurnberger & W. Berrettini (Eds.), *Psychiatric Genetics.* Cambridge: Cambridge University Press.

Sperling, R. A., Aisen, P. S., Beckett, L. A., Bennett, D. A., Craft, S., Fagan, A. M., et al. (2011). Toward defining the preclinical stages of Alzheimer's disease: Recommendations from the

National Institute on Aging-Alzheimer's Association workgroups on diagnostic guidelines for Alzheimer's disease. *Alzheimer's & Dementia: The Journal of the Alzheimer's Association, 7*(3), 280–292. doi:10.1016/j.jalz. 2011. 03.003.
Sternberg, R. J., & Grigorenko, E. (Eds.). (1997). *Intelligence, heredity, and environment.* Cambridge: Cambridge University Press.
Tucker-Drob, E., Reynolds, C. A., Finkel, D., & Pedersen, N. L. (2013). Shared and unique genetic and environmental influences on changes in multiple cognitive abilities over 16 years of late adulthood. *Developmental Psychology* (in press).
van der Sluis, S., Willemsen, G., de Geus, E. J., Boomsma, D. I., & Posthuma, D. (2008). Gene-environment interaction in adults' IQ scores: Measures of past and present environment. *Behavior Genetics, 38*(4), 348–360. doi:10.1007/s10519-008-9212-5.
van Dongen, J., Slagboom, P. E., Draisma, H. H., Martin, N. G., & Boomsma, D. I. (2012). The continuing value of twin studies in the omics era. *Nature Reviews. Genetics, 13*(9), 640–653. doi:10.1038/nrg3243.
Vandenberg, S. G. (1968). The nature and nurture of intelligence. In D. C. Glass (Ed.), *Biology and behavior: Genetics* (pp. 3–58). New York: Russell Sage Foundation.
Vernon, P. E. (1979). *Intelligence: Heredity and environment.* New York: W. H. Freeman.
Waddington, C. H. (1942). Canalization of development and the inheritance of acquired characters. *Nature, 150*, 563–565.
Waldman, I. D. (1997). Unresolved questions and future directions in behavior-genetic studies of intelligence. In R. J. Sternberg & E. Grigorenko (Eds.), *Intelligence, heredity, and environment* (pp. 552–570). Cambridge: Cambridge University Press.
Wilson, R. S. (1978). Synchronies in mental development: An epigenetic perspective. *Science, 202*, 939–948.

Index

D. Finkel, C. A. Reynolds (eds.), *Behavior Genetics of Cognition Across the Lifespan*, Advances in Behavior Genetics 1, DOI 10.1007/978-1-4614-7447-0,

MIX
Papier aus verantwortungsvollen Quellen
Paper from responsible sources
FSC® C105338

FSC
www.fsc.org

If you have any concerns about our products,
you can contact us on
ProductSafety@springernature.com

In case Publisher is established outside the EU,
the EU authorized representative is:
Springer Nature Customer Service Center GmbH
Europaplatz 3, 69115 Heidelberg, Germany

Printed by Libri Plureos GmbH
in Hamburg, Germany